WITHDRAWN

DISEASE MAPS

DISEASE
MAPS

EPIDEMICS ON THE GROUND

TOM KOCH

THE UNIVERSITY OF CHICAGO PRESS / CHICAGO AND LONDON

Tom Koch (http://kochworks.com) is a writer, researcher,
and public speaker in bioethics, disability studies, geron-
tology, medical geography, and public health. He holds
an interdisciplinary doctorate (geography, medicine, and
philosophy) and is adjunct professor of geography at the
University of British Columbia. He is the author of more
than three hundred articles and thirteen books, including
Cartographies of Disease (2005).

The University of Chicago Press, Chicago 60637
The University of Chicago Press, Ltd., London
© 2011 by Tom Koch
All rights reserved. Published 2011
Printed in the United States of America

18 17 16 15 14 13 12 11   1 2 3 4 5

ISBN-13: 978-0-226-44935-7 (cloth)
ISBN-10: 0-226-44935-1 (cloth)

Library of Congress Cataloging-in-Publication Data

Koch, Tom, 1949–
Disease maps : epidemics on the ground / Tom Koch.
     p. cm.
Includes bibliographical references and index.
ISBN-13: 978-0-226-44935-7 (cloth : alk. paper)
ISBN-10: 0-226-44935-1 (cloth : alk. paper)
1. Medical mapping—History. 2. Medical geography—
Maps. 3. Medical geography—Methodology.
4. Epidemics—History. 5. Cholera—History. I. Title.
RA792.5.K633 2010
614.4'2—dc22

                              2010008809

♾ The paper used in this publication meets the minimum
requirements of the American National Standard for
Information Sciences—Permanence of Paper for Printed
Library Materials, ANSI Z39.48–1992.

CONTENTS

List of Illustrations vii

PART I. THE IDEA THAT IS DISEASE

1 Moving Forward: Cartographies of Disease 1

2 Mapping Symptoms, Making Disease 8

3 Body and World: The Sixteenth Century 30

4 Disease in Cities: The Neighborhoods of Plague 48

5 The Yellow Fever Thing 72

PART II. CHOLERA: THE EXEMPLAR

6 "Asiatic Cholera": India and Then the World 95

7 Bureaucratic Cholera 118

8 John Snow's Cholera 142

9 South London Choleras: William Farr, John Snow, and John Simon 164

10 Choleric Broad Street: The Neighborhood Disease 192

11 Cholera, the Exemplar 216

PART III. THE LEGACY AND ITS FUTURE

12 Cancer as Cholera 245

Afterword 275

Acknowledgments 281

Notes on the Illustrations 283

Illustration Credits 293

Notes 295

Works Cited and Consulted 307

Index 323

ILLUSTRATIONS

1.1 Death sketch of a cholera victim / 3

1.2 Death sketch of a male cholera victim / 3

1.3 Death sketch of a cholera victim / 3

1.4 London Board of Health physicians "hunting" for cholera / 6

2.1 Gould's maps of AIDS in the United States / 9

2.2 Map of West Nile virus in the United States in 2000 / 11

2.3 Map of the projected effects of a late-stage influenza pandemic / 11

2.4 Map of the spread of H1N1 virus via travelers / 15

2.5A-E Maps of West Nile virus in humans in the United States / 17

2.6A-D Maps of a diarrheic outbreak in Greater Vancouver / 20–21

2.7A-C Analysis of a diarrhea outbreak in Greater Vancouver / 23

2.8 Where is it? What is it? / 27

2.9 What do we do about it? / 28

3.1A-B Vesalius, *De Humani Corporis Fabrica* / 32–33

3.2 Skeleton at desk from *De Humani Corporis Fabrica* / 35

3.3 Detail of torso from *De Humani Corporis Fabrica* / 35

3.4 Skeleton in landscape from *De Humani Corporis Fabrica* / 36

3.5 Double cordiform world map, 1536 / 38

3.6A-B Ortelius's *Theatrum Orbis Terrarum* / 40, 42

3.7A-B London, *Civitates Orbis Terrarum*, 1584 / 43–44

4.1 Merchant, Holbein's *Death of Death*, 1538 / 50

4.2 Ox cart, Holbein's *Death of Death*, 1538 / 50

4.3 Map of Italy from Ortelius's atlas / 52

4.4 Arrieta's first map of plague in Bari, 1690 / 53

4.5A-B Arrieta's second map of plague in Bari, 1690 / 55–56

4.6 Arrieta's plague: diagram of thinking / 57

4.7 A woodblock print of Thomas Sydenham / 59

4.8 The 1667 Bill of Mortality / 65

4.9A-C The Faithorne-Newcourt map of London, 1658 / 66, 68, 71

5.1 Title page of William Hillary's *Observations*, 1759 / 75

5.2 Mathew Carey, temperature records, 1793 / 76

5.3 Playfair's table of British–West Indies trade, 1786 / 78

5.4 Title page of Mathew Carey's *A Short Account of the Malignant Fever*, 1793 / 79

5.5   Burial records from Philadelphia / 81

5.6   Valentine Seaman's map of yellow fever, New York / 83

5.7   Valentine Seaman's map of odiferous sites, New York, 1798 / 85

5.8   Pascalis's map of yellow fever, New York, 1819 / 88

6.1   Jameson's index map of cholera in India, 1819 / 97

6.2A-B   Brierre de Boismont's map of cholera in Poland, 1831 / 99–100

6.3A-B   Hamett's map of cholera in Dantzick, ca. 1831 / 102

6.4A-B   *Lancet* map of cholera diffusion, 1831 / 105, 107

6.5   Schnurrer's world map of cholera, 1831 / 109

6.6A-B   Christie's map of cholera, 1817–1830 / 111

6.7A-B   Corbyn's map of cholera in India, 1832 / 113

6.8A-B   Brigham's world map of cholera, 1832 / 114–115

7.1   British postal reform map, 1838 / 121

7.2   Postal reform map, London area / 122

7.3   London boroughs, 1832 / 124

7.4   London registration districts, 1851 / 125

7.5   Reese's map of cholera in New York, 1832 / 126

7.6   Sources of cholera in New York / 128

7.7A-B   Cholera in Rouen, France, 1833 / 129–130

7.8   Grainger's map of cholera in Hamburg, 1832 / 131

7.9A-B   Chadwick's map of poverty and disease in Leads, 1844 / 132–133

7.10   Grainger's map of cholera in metropolitan London, 1849 / 134

7.11A   Petermann's cholera maps, England, 1848 / 137

7.11B-C   Petermann's cholera map of London / 138–139

7.12   John Lea's map of cholera on a street in Cincinnati, 1848 / 140

8.1   London Sewer Commission grid map, 1850 / 145

8.2A-B   Wyld's London Sewer Commission map, 1850 / 146

8.3   Portrait of John Snow / 148

8.4A-B   Schematic of cholera at Albion Terrace, 1849 / 150

8.5   Shapter's epidemic curve of cholera in Exeter / 156

8.6A-B   Shapter's map of 1830s cholera in Exeter, 1849 / 157, 159

8.7   Portrait of William Farr / 162

9.1   Portrait of John Sutherland / 165

9.2   Title page of William Farr's cholera study, 1852 / 166

9.3   Farr's chart of cholera, diarrhea, and climate / 168

9.4   Farr's map of cholera in England, 1849 / 170

9.5   Farr's coxcomb of mortality and temperature / 171

9.6A-B   Metropolitan London "table-map" of cholera variables / 173, 175

9.7A-B   Farr's "map-diagram" of cholera and altitude / 177–178

9.8   The inverse relationship between altitude and mortality / 179

9.9A-C   Snow's map of South London water company catchments, 1855 / 183–184

9.10   Snow's summary calculations / 186

9.11   Portrait of John Simon / 187

9.12   Simon's coarse findings / 189

9.13   Graph of Simon's fine-grained findings / 189

9.14   Snow's analysis of Simon's data, 1856 / 191

10.1A-B   St. James Parish, 1720 / 194

10.2   Whitehead's first cholera map, 1854 / 196

10.3   Map of St. Luke Parish boundaries and cholera activity, 1854 / 198

10.4   John Snow's map of cholera in St. James and vicinity, 1855 / 201

10.5   Snow's polygon centered on the Broad St. Pump / 202

10.6   Cooper's cholera map / 205

10.7A-B   Whitehead's second map of cholera, 1855 / 206, 208

10.8   Cholera in Columbia, Pennsylvania / 212

10.9   Acland's map of cholera in Oxford / 213

10.10   Portrait of Reverend Henry Whitehead / 215

11.1   Sedgwick's map of Broad Street cholera, 1901 / 217

11.2   Frost's graph of typhoid fever incidence, 1910 / 219

11.3   Frost's map of typhoid fever in Williamson, West Virginia / 220

11.4   Hamilton's map of typhoid fever in Chicago, 1903 / 222

11.5   Hamilton's map of typhoid fever in the Nineteenth Ward, Chicago / 223

11.6   Frost's map of polio in Mason City, Iowa, 1910 / 225

11.7   Frost's map of polio in school district 8, Mason City, Iowa / 226

11.8   Frost's map of the spread of polio in Iowa, 1910 / 227

11.9   A GIS map with original map spatially adjusted beneath it / 235

11.10   A GIS map of pest-field and sewer analysis / 236

11.11A   Map of cholera mortality in Broad Street's central service areas / 238

11.11B   Map of relation of old plague burial site and sewer lines, 1850s / 238

12.1   Haviland's geographical distribution of cancer / 247

12.2   Haviland's geographical distribution of diseases / 248

12.3A-B   Power's map of cancer in a British Village / 249–250

12.4   Arnaudet's map of cancer in the French village of Cormeilles / 251

12.5   Green's map of cancer in the region of Leyburn / 253

12.6   Green's map of cancer in a neighborhood / 254

12.7   Green's map of a proposed correlation between cancer and air quality / 255

12.8   Cancer intensity dot map / 257

12.9   Stock's cancer tables / 259

12.10A   Stock's map of all cancers reported in British registration districts / 260

12.10B   Stock's map of lung cancer in England, 1920–1930 / 262

12.11   Howe's map of tracheal, lung, bronchial cancer / 264

12.12   U.S. oral and esophageal cancer in women, 1950–1969 / 266

12.13   Sunlight and cancer, 1980–2006 / 269

12.14   Cancer and Vitamin D / 270

13.1   Cholera tramples all, from *McLean's* magazine, 1832 / 278

PART I

THE IDEA
THAT IS
DISEASE

CHAPTER 1

MOVING FORWARD: CARTOGRAPHIES OF DISEASE

Nasal congestion, a drippy nose, and a bit of a cough: you're sick but not seriously ill. It's allergy season and you've had this before; it's cold season and almost expected. Either way, there is nothing to worry about. Put a few tissues in your pocket and off you go to school or work, confident that health will return. Add aching joints and a fever to the mix and it's something serious enough to keep you at home drinking chicken soup and watching ridiculous television shows; you have no energy for anything else. Now add to this a hacking cough, raspy and perhaps painful breath, and difficulty walking and breathing at the same time. It's time for the doctor—or the ambulance.

This is disease in its descent from wellness, a progressive severity of symptoms that help us define the illness and propose a treatment. If you're the only one who is sick, it is for you an unfortunate reality but for the community at large not particularly worrisome. If others you know share your symptoms, what you are suffering is part of an outbreak of "that thing" that is going around. Let a larger number of people in your city, province, or state have the same congress of complaints, and the thing that is causing you distress is an epidemic and a matter of broad concern. If your symptoms are shared simultaneously by folk in a number of countries we call it a pandemic and the world's resources are marshaled to combat it.

This is how we think about the curious thing called disease. We know it by the severity of its symptoms and their effect on our body's systems. But we categorize it by the extent to which its symptoms are shared: outbreak, epidemic, or pandemic. With the effect of the symptoms and the extent of their presence we name the thing (cold, flu, pneumonia) and plan a course of treatment.

There are today, by conservative estimate, more than fifty thousand different diseases that can affect us (Sadegh-Zadeh 2008, 208–9). Some are "token ailments" that affect us one by one, eccentrically. Others are simultaneously shared by thousands, and in some cases, millions, of people. These are the public health challenges that cannot be combated by local physicians and nurses working alone but in their general occurrence and severity demand a unified and public response. Whether it is one person's misfortune or a community's health crisis the critical questions—"what is it?" "How do we treat it"? and "How do we prevent a recurrence?"—can only be answered by a methodology generally accepted as credible and impartial, a "science" of the day.

As an object of treatment, this or that disease is a collection of physical symptoms (congestion, nausea, fever, diarrhea, jaundice, and so forth) with which we construct a condition (cold, yellow fever, tuberculosis, cancer, cholera, plague, AIDS, and so forth) for which one or more theories of origin and diffusion are first proposed and then tested in various ways. To name it is to propose a conclusion based on the applicability of a theory of disease and its method of study: "it" is a distinct collection of symptoms that through this or that theory is specifically understood.

For many centuries, Western medicine grouped symptoms under a system of humours—blood, phlegm, black bile, and choler (yellow bile)—by which disease itself might be understood. In China and Japan the same symptoms were seen within a completely different system based upon a theory of *Chi*, body energy shared with the universe. Today we put the symptoms together and categorize them by the microscopic agents that are their cause—bacterial or viral—invisible animalcules made visible through technology.

In the passage from not knowing toward knowing, from supposition toward testable assumption and then confirmed diagnosis, the importance of visualizing this or that disease at one or another scale cannot be overstated. The naming and then treatment of disease is about *seeing* and then thinking about what we see. From Vesalius's anatomy of the normal to Morgagni's pathology of the abnormal, it was in *seeing* the body and its parts that modern medicine took root (Nuland 1988). It was and is the symptoms of illness writ large on the patient that has always signaled the type and nature of the affliction to be treated (figs. 1.1–1.3). In the late nineteenth century, previously invisible "animalcule," proposed as disease agents causing these symptoms, this or that disease, were revealed in Gram-stained slides viewed under the microscope as bacteria resident at once in the environment and in the patient. In the twentieth century, a progression of imaging technologies added the virus and later the prion to imaged things, informing disease definitions and patient prognoses.[1]

At another scale, it was in maps of disease as a public reality that the individual pathology was transformed into a public health event affecting communities and nations. Epidemics and pandemics are spatial phenomena; mapping them is how the public nature of a disease threat is seen and studied. As Gunnar Olsson puts it, in the mapping we "lay bare the familiar of the unknown" (Olsson 2007, 4). For centuries the map has been a mechanism by which the rolls of the dead and the dying became shared realities whose relation to local environmental conditions could be assessed.

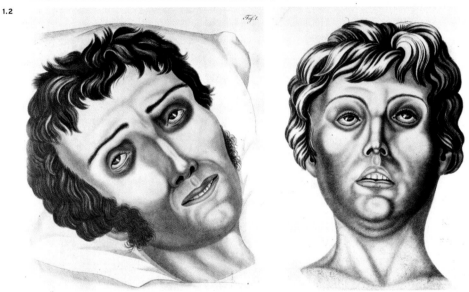

FIGURE 1.1 This nineteenth-century image of the cholera patient was used as a clinical illustration to assist physicians in recognizing the effects of the disease.

FIGURE 1.2 In this 1832 image, the sunken eyes, thin face, and darkened coloration of a deceased cholera patient were used to describe the effects of the disease.

FIGURE 1.3 This diagnostic sketch served as a definition of the effects of cholera on those whom it killed.

Mapping this or that shared congress of symptoms invited comparisons between patterns of disease incidence and local characteristics (congested housing, a fetid local waste site, a marshy swamp, and so on) that, according to this or that disease theory, might promote health or encourage a specific illness. *Here* was plague transposed to a map in the seventeenth century, there the *cordon sanitaire* that sought its containment. Embedded in the map was a radical idea: plague was a portable disease carried by humans. A century later, it was in the mapping of yellow fever in port cities in the United States that the proximity of mortal cases to sites of noxious waste was used to "prove" a miasmatic theory of the disease (Pascalis 1819, Seaman 1798). It was in nineteenth-century maps of cholera that its waterborne nature was first argued (Snow 1855a). In other maps of the era cholera as an airborne thing was similarly advanced.

The central argument of this work is that to understand disease and its history we need to think about *seeing* at every scale. It is in the *seeing*—of the animalcule, the parts of the infected body, and the shared set of symptoms evidenced across maps of the city, nation, and world that the unknown is made real, its public nature asserted. In mapping, boundaries are set and the content within those boundaries organized to create the context in which disease theories are proposed and tested. In this way science and the reality it seeks to consider are created.[2]

From this perspective, disease is not simply a fact experienced but a conclusion reached on the basis of investigations carried out within the framework of a theory. Each new condition goes through a series of stages in which its symptoms are identified, listed, and then somehow projected in the search for clues to its origin and possible prevention. In the seventeenth century it was plague that was most investigated, in the eighteenth century it was yellow fever. In the nineteenth century it was cholera whose nature and causes were debated from the first European report of 1819 (Jameson 1819) until Robert Koch identified definitively the bacterial agent in 1883. In recent years we have seen this process played out with AIDS, H1N1 influenza, severe acute respiratory syndrome (SARS), and West Nile virus (Koch and Denike 2007a). The process of identification in the body and the population is as old as cancer and plague, as new as tomorrow's epidemic.

MAPS AND MAPPING

The medium of the map ties together the broad collecting of knowledge seekers and the subjects of their investigations in a way that makes a disease visible in the public domain. It is *in* the map that a consistent collection of symptoms is located by a dot, bar, or "x" to mark the location of the person who suffers. In the map those cases are joined in a manner that asserts their commonality in a way all can see. The map presents the specific condition not as an isolated thing—a community of unfortunates—but more importantly in relation to geographic and social elements that may (or may not) promote or inhibit this or that congress of deathly symptoms. In the map, knowledge is assembled to permit ideas about disease to be brought forth, in relation to the environment, through the study of this or that specific experience.

It was *in* the map that British anesthesiologist Dr. John Snow argued his theory of cholera as a waterborne disease by laying out the pattern of local deaths in the Broad Street area of St. James, Westminster, in relation to local water sources (Snow 1855a, 1855b). It was in *his* map that Reverend Henry Whitehead concluded an independent study of cholera deaths in the same neighborhood (Whitehead 1855). It was *through* mapping that London Sewer Commission engineer Edmund Cooper disputed popular charges that fetid air emanating from sewer lines laid in the early 1850s caused the cholera outbreak in the Broad Street area that Snow and Whitehead separately described (Cooper 1854). It was in another set of maps that apothecary, bureaucrat, and statistician William Farr earlier had argued a principally airborne cholera in a study of its patterns of occurrence in the registration districts and subdistricts of England, and especially South London (Farr 1852a).

It is the evidentiary nature of the disease map that is of importance here, the manner in which maps present the discrete elements of an epidemic or pandemic occurrence—so many dead, so many made ill—as a unified event. In the transformation from one person's disease to the population's epidemic a set of ideas is also lodged in the map, a rationale for the collection of specific sets of data (What is important? What is not?) in a manner encouraging the development of different theories whose strengths are tested in attempts to correlate events in the map. In a real sense we do not map data. Instead, we map theories using data to present them in place.

Mapping is not unique in this. It shares with other tools of knowledge assemblage (for example, charts and graphs) a common underpinning: records of mortality and morbidity. Like charts and graphs, mapping is inherently numerical; the sum of individual occurrences tallied in the table is posted in the map. And, again like them, mapping is often statistical, offering across its surface not simply the discrete, isolated location of this or that occurrence but its general pattern of incidence in a spatially embedded population (five or fifteen or twenty-five deaths per one hundred persons). And, finally, all such modes of presentation encapsulate ideas of incidence and their explanation (numerical, statistical, and so forth).

Mapping, however, uniquely encourages comparisons between the variety of disease experiences and the environmental realities that may promote or inhibit its incidence and spread. Where diagrams, charts, and tables distill and summarize a set of data, mapping as a medium of knowledge assemblage and exploration places that distillation in a context that is at once communal, ecological, and multifactorial. Mapping is *always* about things together in place at a time and never out of space and removed from time. As important, mapping transforms the rows of a database from discrete, singular cases into equal, shared events whose elements are located in an environment whose relation to this or that disease event demands consideration. It is in these two transformations, from the individual into the aggregate and the aggregate into an environmentally grounded reality, that maps serve and have served practically and theoretically.

Mapping disease makes science when it distills the particular and eccentric into the shared on the testable basis of theories of disease and health. That science becomes medicine when it transforms tentative into accepted theories of health and disease that apply ideas about contagion and control to specific events. In maps of

FIGURE 1.4 Every age hunts for the cause of a disease, wondering how its symptoms work together. This 1832 magazine illustration by Robert Seaman shows members of the then-new public Board of Health "hunting" for the causes of the mysterious diarrheic condition called Asiatic cholera.

disease and health, subject and object are ineluctably joined. They create a subject, this or that disease state, which is then transformed into an object set in relation to elements of economic, geographic, social, and political landscapes. The result is a way of thinking about disease "on the ground," not a theoretical construct but a reality experienced, within the science of the day and the politics that defines it in society.

THE PAST IN THE PRESENT

All this is a history that resonates in the present, a past that continues into our future. Today a host of emerging infectious diseases is presenting new challenges. At the same time, older diseases (such as cholera, influenza, tuberculosis, and syphilis) have mutated into new and often more virulent, difficult-to-treat forms. With each new health crisis, with each new disease, the old history is relieved and reformed as resources are marshaled to confront yet another new health threat.

Simultaneously, new technologies enable new methods of knowledge assembly that we hope will be applicable to this or that new congress of symptoms. For example, over the last half-century the increasing availability and computational power of computers have changed almost every avenue of our knowing (Koch 2005, chap. 10). They have transformed mapping from a print-based medium into one that is increasingly digital, encouraging, with ever more powerful statistical programs, increasingly rigorous methods of assessing progressively greater amounts of data.

Important as this is, it is only technology. The critical questions remain what we see and what we think about what we see. The limits of knowing remain, our theories of disease and its causes are still incomplete. We remain half-ignorant, desperately seeking tentative solutions to problems that are at best only partially understood. In the struggle to know, each generation laughs at the assumptions of its predecessors, prideful in its own state of half-knowledge, half-ignorance. The means by which learning is carried out is the real legacy each age passes on. Here the methodologies of that thinking are found embedded in the maps, traces of the continual reformation of our knowing. The resulting study is messy because history is messy; proceeding not in a simple line but in fits and starts. The need for the exercise is manifest, however, not simply to better understand the disease challenges we have faced, century by century, but also to prepare for those we face today and will confront tomorrow.

CHAPTER 2

MAPPING SYMPTOMS, MAKING DISEASE

Nobody knows how many disease maps are produced each year. There are neighborhood, city, state, national, and international maps of epidemic and endemic conditions produced by agencies ranging from neighborhood activist groups and local public health departments to state and federal health agencies. The UN, Pan-American Health Organization, and the World Health Organization each produce annually hundreds of disease maps. We know this because many are published in academic journals, daily newspapers (from the Akron, Ohio, *Beacon Journal* to the *Times* of London), general periodicals, and of course, across the Internet. Typically these maps are based upon the work of medical researchers, but others bring forward conditions at the fine scale of very local health concerns, mapped realities advanced by citizens that officials have overlooked or rejected.[1] Maps of disease and health are collected in atlases and produced in books whose job it is to unravel the skeins of often conflicting research conclusions about this or that health threat. Together, all these maps emphasize this conclusion: what we know is dwarfed by what we have yet to learn. In that learning, maps are, and for centuries have been, a critical medium of our studies.

There are maps of bovine spongiform encephalopathy, for example. Bad as it is for Bossy and her bovine friends, it is worse for us in its human variant, Creutzfeldt-Jakob disease. Its agent is a proteinaceous infectious particle, the prion, unknown until recently (Pattison 1998), and we have little idea about how these nonnucleated, self-reproducing, crystal-like proteins propagate or work (Rhodes 1997). We do know that in 2003, a cow from Washington State discovered to have BSE was likely infected in Canada (CDC 2004). We know this because commercial herd movements have been mapped as possible disease vectors.[2]

There have been hundreds, perhaps thousands of maps of HIV/AIDS since its first reported case in North America. These range from global maps of disease diffusion

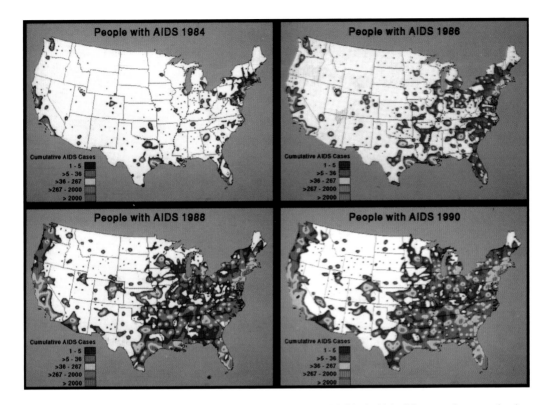

FIGURE 2.1 Peter Gould's famous map series describes the spread of AIDS in the United States at the county level as part of the work toward a model that could accurately predict future diffusion in U.S. communities.

to maps of patient infection in local and regional communities (Verghese, Berk, and Sarubbi 1989). Famously, in the 1980s geographer Peter Gould and coworkers not only mapped AIDS at the county level in the United States (fig. 2.1) but did so in a manner generating the first accurate predictive model (mapped) of its spread, identifying contiguous and hierarchical elements to its pattern of diffusion (Gould 1993). HIV is a retrovirus, mutating rapidly to maximize its circumstances. We didn't know retroviruses existed prior to the 1970s.

By contrast, West Nile virus (WNV) is almost a relief. In at least a general way we think we understand this emergent virus and its principal vector, infected mosquitoes noshing on birds and mammals. Once confined to a small area of Uganda, WNV migrated from Africa to North America via Romania and Israel (CDC 2001). From the first U.S. case reported in New York State in 1999 (Nasci et al. 1999) through the end of 2007, federal Web sites posted myriad maps of WNV by state and county both for each year and in each year for five major species groups (birds, humans, mosquitoes, veterinary mammals, and sentinel chickens) (USGS 2006). Also available are national maps with county data of cognate viral conditions, including St. Louis encephalitis, eastern and western equine encephalitis variants, La Crosse encephalitis and Powassan virus (USGS 2007).

We see the virus in its host populations in the map. This *is* West Nile virus, the map says. In figure 2.2 we see its progression from the 1999 index cases in New York State. There is a slow southwest spread of the disease in mosquitoes and both a northerly and easterly spread in infected birds. Human infections remained centered in the area around New York but among veterinary mammals the spread is more diffuse. Over time, map-to-map, we can trace the progress of the virus in its hosts and in that tracing seek patterns of explanation. Is WNV promoted by certain climates favoring specific species of mosquitoes? Do specific avian migration patters explain the virus's spread? What is the relation of the infection in humans and veterinary mammals to that of the other species (Koch and Denike 2007a)? Each question generates more work, more maps of the spatial foundations of viral interactions and spread.

For centuries maps have served in this manner, collecting and then projecting data in a manner that permits this or that disease event to be seen and then considered carefully. With each new disease we look more closely, map more carefully the data that are also reviewed in other ways (genetically at one level, statistically at another). Emerging diseases like WNV are not all newfound threats, however. Since 1983 more than twenty-three variants of long-established conditions have emerged, mutating old familiars into new and real threats (Callahan 2006, 135). The history of their mapping is also a history of how we have learned to think about these conditions, about how we see disease.

Cholera, for example, has been consistently mapped since 1819, a history of disease construction and investigation detailed in the middle chapters of this book. The seventh international pandemic, whose bacterial agent was a new variant, the El Tor serotype, was first identified in 1961 and spread into this century, traveling the world in cargo ships and with airline travelers (Salim et al. 2005).

Influenza in its many, ever-mutating forms is mapped annually. There are maps of past epidemics and maps of current disease excursions. Practically, some map imbalances in flu vaccine supplies—the distance between where they are stored and where they will be most needed—to assure the maximum level of protection for a population (Davenhill 2005). The 1918 influenza pandemic with its devastating death toll remains for many the very definition of a potentially devastating epidemic occurrence. For this reason, some have used it as a model on which to build predictive maps of the effect of another, similarly severe epidemic. Figure 2.3 is one of a series of maps that modeled the effect today of a 1918-style, killer outbreak in the Canadian province of New Brunswick and the U.S. state of Maine. By day eighty of a pandemic the surface of the map is largely uniform. The virus has diffused, and while there are some spots of activity, the greatest mortality has occurred. Across the series the early hot spots signaling the viruses incursion spread across the region, slowly fading until, at the end of the epidemic, the pattern of continuing mortality is a series of small hillocks against the greater geography of the health of the state.

Similarly, new maps proliferate of tuberculosis in all its variants: classic presenta-

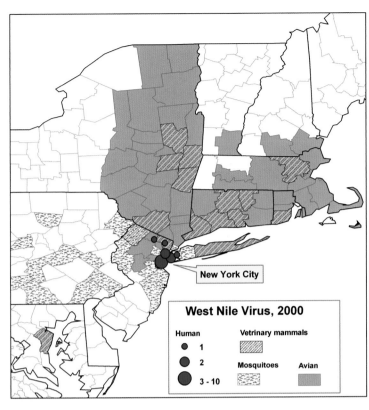

FIGURE 2.2 West Nile Virus infects a range of different species. Introduced into the United States in 1999, it spread continuously in birds and mosquitoes in the year 2000. Human incidence seems to be simple outcome of these infections.

West Nile Virus, 2000

Human
● 1
● 2
● 3 - 10

Vetrinary mammals

Mosquitoes

Avian

New York City

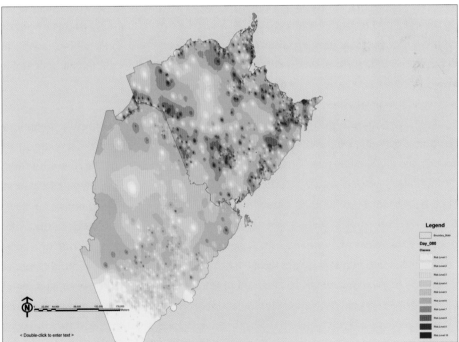

FIGURE 2.3 Working with historical data from the 1918 influenza pandemic, New Brunswick researchers modeled the possible effects, day by day, of a similarly severe occurrence. On day 80, shown in the map, the outbreak has subsided and only minor eruptions can be seen.

tion, multidrug-resistant tuberculosis and extremely drug resistant tuberculosis. What was a single pulmonary condition is now a family of related diseases whose members are separately considered and distinctly named.[3] The variants share similar symptoms, but each is caused by a different if related bacterial agent and each requires a distinct treatment. Current estimates are that perhaps a third of the planet's population is infected with at least a latent form of one or another tuberculosis and that in the United States alone ten to fifteen million people are affected (Markel 2004, 16).

The list goes on but the message is clear: if it is a disease we map it at every stage of our knowing, from the first collection of symptoms across decades of investigation. What are we doing when we map these problems and what do we learn from the exercise? More importantly, perhaps, what is it that really we map?

MAPS AND MAPPING

Maps make arguments about disease, their pattern of incidence, and their method of diffusion. They are workbenches on which we craft our theories about the things that cause health to fail, imaging data collected in this or that disease outbreak. They are not, as some argue, either mere representations of the world or simple illustrations of work completed in other media (Miller and Wentz 2003). It is true but insufficient to say that "disease maps provide a rapid visual summary of complex geographic information and may identify subtle patterns in the data that are missed in tabular presentations" (Elliott and Wartenberg 2004, 999). Certainly, mapping permits a visual statement of spatially grounded data accumulated on the basis of theories whose exposition—and mode of analysis—is lodged in the map itself. More fundamentally, however, maps are cognitive instruments whose mode of argument and not the techniques of imaging is their *raison d'être*.

As Franco Moretti put it more generally: "Mapping it—is not the conclusion of geographical work; it's the *beginning*. After which begins in fact the most challenging part of the whole enterprise: one looks at the map, and *thinks*. You look at a specific configuration—those roads that run towards Toledo and Sevilla; those mountains, such a long way from London; those men and women that live on opposite banks of the Seine—you look at these patterns and try to understand." (Moretti 1998, 7–8; original emphasis).

It is this visual thinking, grounded and spatial, that the map promotes. To map the incidence of a thing is to create a shared range of experience in a manner that is not merely descriptive but more importantly "constative," bringing forth a meaning accessible to analytic address (Austin 1975). Descriptive statements report and reflect a "simple" reality, while "constative" statements indicate "the circumstances in which the statement is made or reservations to which it is subject or the way in which it is to be taken and the like" (p. 3). This constative essence enables the map's performative ability to argue how a subject, in this case a recurring class of disease symptoms, is to be understood. Put another way, mapping is a method of assemblage within which ideas are constituted and then argued about specific experiences.

In its method of assemblage the map changes the individual and particular, transforms discrete rows in tables of occurrence into a common experience, an event *class* whose members are assumed to be related somehow, one to another, through the evidence of disease symptoms. In this way the individual rows of morbidity and mortality tables are transformed in the two-dimensional plane of the map into a unitary thing, a single reality composed of multiple occurrences of similar attributes. By making a map—*any* map—relationships are asserted between sets of phenomena in a manner that is inherently both analytic and experimental. "The former is about specifying the content of the 'known,' the latter about putting together elements and controlling them to create new phenomena (or old phenomena in new ways)" (Pickstone 2000, 12).

The map's intellectual service lies in this conjunction of analytic presentation and experimental argumentation in a visual exposition. The elements assumed to define this or that disease state, and this or that place of occurrence, are spatially arranged in an assemblage proposing that a congress of symptoms (cramps, dehydration, diarrhea) is one thing (amoebic dysentery) or another (cholera) whose reality we seek to document as a single condition. We do this in the hope that we can limit, or better, eliminate, its presence in our communities. By including in the map environmental elements that may be related causally (air temperature, food outlets, sewer lines, water sources, and the like) we create a medium in which their likely relation to this or that illness is asserted in a manner inviting systematic assessment.

In this the map is both the subject, an evidentiary statement in its own right, *and* an object on which ideas are played out in the two-dimensional plane of the map. In this way mapping produces a type of knowledge, one rooted in a relational space, which has been critical to disease studies for centuries. This is not a characteristic unique to disease mapping but an attribute of mapping generally. As David Turnbull argues, "Maps are the paradigmatic examples of the kind of spatial knowledge that is produced in the knowledge space we inhabit. Not only do we create spaces by linking people, practices and places, thus enabling knowledge to be produced, we also assemble the diverse elements of knowledge by spatial means" (Turnbull 2000, 89).

All this is to say mapping is an experimental system that is at once graphic, numerical, spatial, and theoretical. "Experimental systems are the units within which the signifiers of science are generated. They display their meanings within spaces of representation, in which graphemes, that is, material traces such as fraction patterns or arrays of counts are produced, articulated, and placed, displaced, replaced . . . scientists create spaces of representation through graphemic concatenations that represent their epistemic traces as engravings, that is, generalized forms of 'writing'" (Rheinberger 1997, 2).

This particular experimental system employs the map surface as a type of inscription device, a technology that embeds in the map page a set of arrangements for labeling, naming, and counting things. In that process relations between elements of a set are proposed, creating a range set in relation to other, similarly constructed ranges. Based on the constative backcloth of its form, the argument advanced is performative, concrete, and visible. "It is a set of arrangements for converting relations from non-trace-like into trace-like forms," things we can see as specific types of relations between elements in the map are assessed (Law 2004, 23–29).

Mapping does this by fusing signs upon the map surface under the control of a series of codes that structure the graphic surface and thus order the data presented. These codes have a logic that permits a coherent argument to be articulated across the map (Wood and Fels 1986). Map symbols organized by these codes create at a concrete level a "geographical matrix" in which a set of events is considered on the basis of relevant, subject-specific data accumulated to advance an argument or test a hypothesis based upon it (Chrisman 1997, 24–26). It is this matrix that is constantive, asserting potential linkages among mapped attributes located on the page in relation to each other. Maps thus carve specific territories from the blank space of an empty page, creating ranges of related elements at a specific scale (international, national, regional, and local). On the map page the rows of data, the elements of the matrix, take visual form as a single reality (this is . . . cholera, cancer, influenza, WNV, and so forth) in a manner that permits questions to be asked and theories first to be generated and then tested.

H1N1 INFLUENZA: THE PANDEMIC

The 2009 H1N1 influenza pandemic is an example of the service of mapping in the address of public diseases. In an article on the diffusion of the disease from its Mexican index case, a group of Canadian researchers published a map in the *New England Journal of Medicine* describing its relation to air travel volumes (fig. 2.4). Using International Air Transport Association (IATA) data, the goal was to identify the threshold of travelers (the threshold of travel) required to spread the disease from one city to another (Khan et al. 2009). In the map major cities around the world are identified by name, and with bars calculated to present the number of travelers arriving from Mexico, the origin of the new outbreak. Implicit in each bar is the assumption that cities receiving Mexican air travelers also were cities where H1N1 influenza could or would soon be found.

Embedded in the map are a series of ideas. First, the map asserts a single space, a shared world. We take this for granted but its construction was an enormous achievement of the sixteenth and seventeenth centuries (as chapter 3 will argue). Within this space the map creates an urban sphere, a collection of cities among which travel occurs. Again, we take for granted this fact, another boon of those centuries. Third,

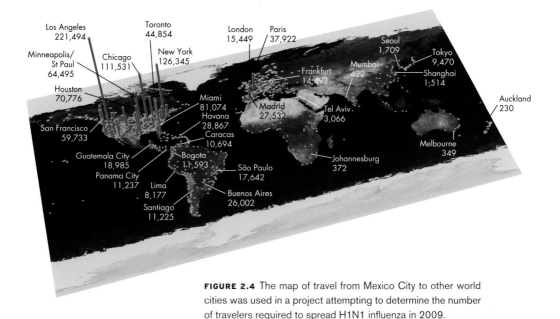

FIGURE 2.4 The map of travel from Mexico City to other world cities was used in a project attempting to determine the number of travelers required to spread H1N1 influenza in 2009.

the map proposes that this world city space is linked by air travel carrying people from one city (Mexico City) to many others.

Fourth, the numbers of travelers included at each point in the map insist we can precisely calculate the astonishing degree of traveler between one city, Mexico, City, and other cities around the globe. Totaled in the map are 2.3 million passengers who flew in early 2009 from Mexico City to 1,918 cities in 164 cities as far flung as Santiago, Chile, Shanghai, China, and Johannesburg, South Africa. We trust the numbers in the map to the extent that we trust its source, IATA. We trust the researchers in part on the basis of their vehicle of publication, the prestigious *New England Journal of Medicine*. The creation of a bureaucracy capable of this type of record keeping, and the method by which bureaucratic data are considered and reported, is the subject of the middle chapters of this book.

The point of the map is not IATA's record-keeping ability or even the vast human migrations it chronicles. They are the backcloth that supports the mapped image and the research it presents. The map is about a new version of influenza, a disease we have known for millennia, and its diffusion from Mexico City to . . . everywhere. In reading the map we reflexively translate air travel into a map of influenza diffusion. We do this because the map is in an article on this new variant of a well-known virus and because the article, the "paratext" (Wood and Fels 2008, 4) in which the map is embedded, tells us to. We see the map this way because we think of influenza as a disease transported by humans. Finally, with all this behind it, the map makes this proposition: if this virus travels along human pathways then it should be possible to calculate the amount of travelers required to transmit this virus from city to city.

Each virus has its own level of intensity. Some require more exposure than others before its introduction takes hold in a community. In this case, in 2009, this virus required 1,400 air passengers to transplant the virus from one city to another (Khan et al. 2009). The map of travel volumes is the basis for work that can be done modeling diffusion over time (add dates of first cases in each city) and of the relation between diffusion first to larger rather than smaller cities (the "gravity" of large bodies, it's called). The H1N1 map raises the potential for studies using these ideas, showing us the data in a way that invites further exploration.

WEST NILE VIRUS: THE U.S. EXPERIENCE

What is mapped is not this or that thing, the discrete datum, but things *together*. These things-together, lodged in data sets, create factual statements that are propositional in nature. There is nothing surprising in this. Propositions are statements affirmed or denied by their predicates.[4] All evidence is propositional, its validity accepted or rejected on the basis of prior assumptions that are the backcloth out of which the fabric of an argument is constructed (Neta 2008, Williamson 2000). This means the data we use in our deliberations are conditional, "IF this is true and accurate then. . . ." Consider figure 2.5, a set of maps of humans infected with West Nile virus in the United States from 2000 to 2004. Each map separately asserts the presence of a viral condition in citizens resident in a specific geography during a specific time period. Residence is defined not by home, street, or city, but by county. The rationale for this jurisdictional frame is simple: the data are derived from reports submitted by county health officials reporting WNV incidence to Centers for Disease Control and Prevention agents collecting data on this virus (Hayes et al. 2005). The maps are not of WNV but of reports of WNV that were collected by local and state officials and reported to national officials who brought the data together.

First, the maps each propose a geographic thesis: that is, (a) there is a distinct region, the continental United States, in which (b) the presence of this virus can be appropriately studied. This area is divided into counties that are the reservoirs of data to be considered. Were this untrue the maps would have a different shape. A second class of propositions asserts a class of events, viral infections, that can be definitively categorized at the resolution of counties and which occurred during a specific time period. Each map then insists that all events within a class of occurrence (viral infection) located in the map can be considered a single entity, creating an event *class* out of tens, hundreds, or thousands of things. The argument that results is this: *if* this thing (WNV) in this county is the same as that thing (WNV) in that county, then both are the same thing, together. Through identical symbolization and consistent lettering the map asserts the consistent presence of this thing, WNV as a single event occurring across continental U.S. counties.

2000

2001

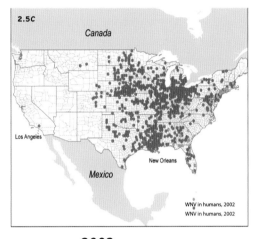

2002

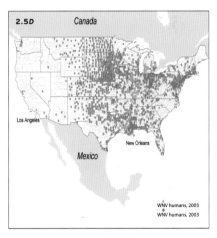

2003

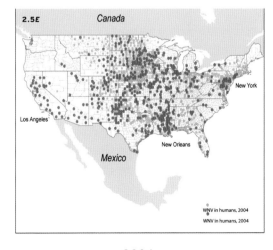

FIGURE 2.5A-E These maps show the yearly progress of West Nile Virus in humans after its 1999 introduction in the United States. All are based on data collected and by the Centers for Disease Control and published in table and map form on federal websites.

2004

The repetition of these propositions, map by map, gives to each (and to the map set) added weight. We see in the maps the progress of the virus. In the seeing, the maps propose a way of thinking about events (in this case, WNV in humans) occurring in jurisdictional boundaries assumed to be the appropriate scale at which disease data can be collected, posted in the map (and its database), and trusted. In these maps, that scale is that of the United States, its resolution that of U.S. counties in which cases were reported. Individually and together the maps and their data assert the broad geography of the continental United States and its local political entities as both existential realities and as appropriate vehicles for WNV study. We accept this as real and relevant geography even though northern and southern boundaries are wholly porous political fictions that exist without a physical geographic referent.

These borders are not relevant landforms (not like the oceans to west and east, for example) and not necessarily relevant to any study of bacterial or viral incidence. Insects, birds, humans, and trade goods all constantly travel across them, carrying the bacteria and viruses that become disease. And yet the map insists its geography is sufficient; on the basis of that posting we assume the appropriateness of U.S. county, state, and national borders—bureaucratic artifacts with little natural rationale. The map says the borders are as real as mountains and . . . we accept that. The map says "we'll see WNV in this frame at this resolution" and we accept that, too, without reflection. Were it otherwise, it would be mapped differently. We might then include Canadian and Mexican WNV to transform the American epidemic into a North American pandemic.

Each dot in the map posts not a case of WNV but a U.S. county in which West Nile virus was reported to local health officials. It does not say how many cases occurred or the percentage of the population that were infected, only that one or more persons were diagnosed with this condition in a certain time period. We could use graduated symbols to show the number of cases, grouping them in classes (1–50, 51–150, 151–300, more than 300, for example). And we could calculate the incidence per, say, one million persons. These statistical refinements, which have become standard, were part of a revolution in medical mapping and statistics in the nineteenth century (and will be described in later chapters).

The map of WNV activity in 2000—the virus was first introduced to the United States in late summer of 1999—shows a very limited number of cases in the New York City area. In 2001 the virus progressed contiguously, spreading out down the eastern seaboard to Washington, D.C., and then jumped—there is no better word—to Florida. There is little activity in the Carolinas or Georgia where one on the simple basis of proximity one might have expected it. By 2002 WNV was rampant across the Midwestern states as well as the eastern seaboard. In 2003 the virus spread more slowly and contiguously, from the Midwest to the Southwest and into California. Its travel north, to either Washington State or Maine, seems to have been inhibited by something. By 2004 incidence was declining and the level of activity seemed diminished.

We trust these maps not because we trust the data collectors in this or that U.S. county (we know nothing about them), or because we trust the anonymous mapmakers. Assumptions of validity rest upon our trust in the CDC and its system of data collection. We believe in the maps to the degree we believe in the CDC and its competence. We also believe in the integrity of the continental United States, a political reality defined by borders drawn on the map but not the land. Similarly, we believe in the integrity of U.S. counties as catchments for bureaucratic data. Simply, the geography posted in the map is appropriate because the CDC (and scores of other federal agency and maps) says it is so. For the purpose of American disease studies it appears irrelevant that WNV also attacked thousands of people in Canadian provinces and Mexican *estados*. After all, if it were important it would be in the map.

There is nothing magical in this. It is simply to say that maps aggregate facts in a manner that creates evidentiary classes in service of an idea. The evidentiary value of any one datum, any single row in a table of incidence by U.S. county is assumed to be the same as any other. Data from Cerro Gordo County, Idaho, Sonoma County, Washington, Worcester County, Massachusetts, and Yonkers, New York, are assumed to hold an equal truth value; contributing to the whole. If this were not true, if points of mapped fact are disputed or plain fiction, the mapped class dissipates into conjecture. It is not that "all maps are wrong," as the saying goes, but that all maps are as right or wrong as the data collected on the basis of the assumptions made about its importance and validity.

The utility of all data is bounded by our confidence in it and the assumptions we seek to argue on its basis. What makes mapping unique, and a uniquely valuable tool, is that it anchors the isolated, evidentiary fact—an entry in a table of incidence—in an event class at a location. The map then builds from its fields of common incidence to assert their relation. *If* this was WNV in continental U.S. counties in 2000 and *this* was WNV in U.S. counties in 2004, *then* the epicenter of the viral attack expanded westward from New York in those years. The central argument of the map space asserts both the validity of data classes and the importance of their juxtaposition.

A LOCAL EXAMPLE

West Nile virus was a national concern and one heavily studied (and continuously mapped) by entomologists, epidemiologists, public health experts, and veterinarians. It may be easier to see how disease mapping works—intellectually and practically— in a larger scale, more common example of a simple, local disease outbreak. Consider an outbreak of severe diarrhea that occurred in metropolitan Vancouver, British Columbia, in 2000. As an example it serves admirably because there was nothing unusual about it. It was one of scores—probably hundreds—of similar outbreaks in that year in North America. If it is special it is only because data about this outbreak were made accessible.[5]

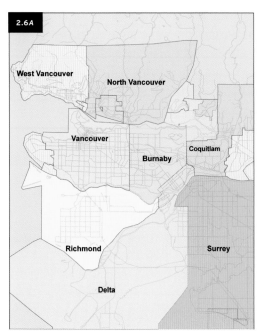
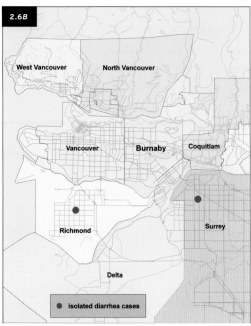

FIGURE 2.6A-D Maps of Greater Vancouver and the elements of a local diarrhea outbreak in 2000. *A*, Greater Vancouver; *B*, isolated cases; *C*, hospital-reported cases in the Greater Vancouver area; *D*, suspected sources. Severe diarrhea, often but not necessarily from food poisoning, is one of the most frequent type of outbreaks faced by local public health departments.

Figure 2.6*A* is a map of Greater Vancouver, an area containing about 2.3 million persons in Canada's westernmost province. The map is an arrangement of lines symbolizing streets within the political boundaries of a region bounded by the sea and its intrusions upon the land, the creeks, inlets and rivers that are also mapped. It was created using a computerized mapping program, ArcGIS 9.2. With the program came prepared files of geographical data we accept as real (the boundaries of nations and local jurisdictions, of mountains, and rivers, and the like).

We projected elements of this data set onto the map page as Greater Vancouver. Different mapping would create a different city, perhaps one of sewer and power lines but without political boundaries. While the buildings and homes we know exist on the streets are absent in this map they could be added if we thought them important to this project. We could populate the map with economic data, elevation contours, or population housing density . . . whatever seemed relevant to the problem at hand. So *this* Vancouver, created for the problem of understanding a set of cases (persons reporting severe diarrhea to hospital officials) is different from, say, the Vancouver produced in a Google map created for a tourist seeking local restaurants or museums.

During the outbreak health officials registered sixty-one cases of hospital-treated diarrhea for which home addresses were reported. In figure 2.6*C*, each of these cases is located in the map near the home address the person gave to hospital staff

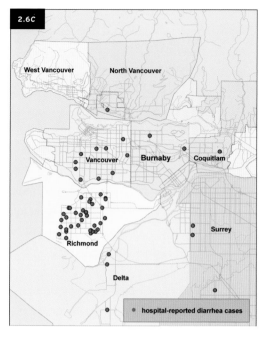

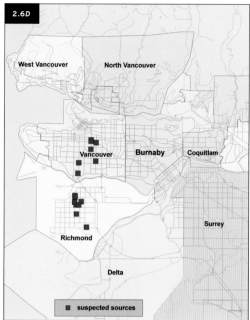

prior to treatment. Each dot symbolizes an individual case, asserting, "a person with these symptoms reported living here." Their shared mapping transforms the individual cases into a single disease event constructed of a range of cases whose commonality is assumed to outweigh any differences that might exist between members of the class of mapped patient homes.

The map implies accuracy and a completeness it cannot deliver. To protect patient confidentiality, all addresses in the official database were randomized to within two blocks of their actual location, creating a set of cases each of whose members is inscribed near to but not at the precise coordinates of a patient's home. Nor is the completeness of the map guaranteed. Each dot is a case reported to local health officials but, almost certainly, there were additional but unreported cases. Some people experiencing only moderately severe diarrhea may not have sought hospital treatment but self-medicated in their homes. Others may have consulted a family physician whose treatment program was not reported to officials. Even in a small database like this at least one or two cases were probably misreported, entered incorrectly into the database with a keystroke error. And even if all were reported without error, there were likely patients treated at a hospital whose location was not reported because they were visitors without a local address or homeless persons with no fixed address.

A single case, or two, or even three in dispersed areas of the region would create a set of cases that we would not care about. In figure 2.6B, for example, two cases of persons hospitalized for severe diarrhea are imagined. One is in the suburb of Richmond, the other in Surrey. They are widely separated in the space of the map and, we assume, therefore in the Greater Vancouver regional district that the map presents.

Hospitals frequently see patients presenting with severe diarrhea, typically from food poisoning, and no epidemiologist would worry about two cases so dispersed within a single time frame.

But in figure 2.6C we have a *lot* of cases and there seems to be a clustering of them. In fact, there are several clusters. Nothing in this figure says anything about what caused the patients' distress, only that it was reported at hospital. A range of agents can cause severe diarrhea, some of them inhaled (classes of airborne poisons, for example) but many more of them ingested. Since Robert Koch identified the cholera bacillus *Vibrio cholerae* in 1883, and Theobold Smith named the salmonella bacillus *Salmonella choleraesuis* in 1885,[6] we have come to assume diarrheic outbreaks most commonly result from ingesting tainted food or water contaminated with one or another family of bacteria. On the basis of this assumption, two theories of this particular set of cases were discussed in the summer of 2000. First, it was feared that local sewer lines under construction in the Vancouver suburb of Richmond might somehow have contaminated local drinking water. Second, it was thought that contaminated food products sold by one or more local food outlets might be the source of the agent causing the class of diarrhea symptoms. Both these ideas have histories grounded in nineteenth-century disease studies and are discussed in later chapters of this book.

Sewer line construction was localized in Richmond while in the map the diarrhea outbreak extended north into Vancouver, south into Delta, and east into Burnaby and Surrey (fig. 2.6A). But were sewers the sole source of the problem, experience suggests more people in Richmond likely would have been affected and fewer people elsewhere. We know this because we've seen it before. The mapped argument, based on a simple visual examination, did not support the idea of a sewer contamination as likely and is not included here. To consider tainted food as a source meant considering products from all the food outlets where the sixty-one hospital-treated patients had eaten in the eight to twelve hours prior to the onset of their symptoms. Local health officials gathered that data, critical if these were, indeed, cases of food poisoning. As a group, diarrhea suffers reported eating or ordering food from one of twenty-one food outlets, restaurants, and stores offering a variety of cuisines and sale products. These are mapped separately in fig. 2.6D. In the map, their common symbolization—a blue square—made of all a single class: suspect food outlets patronized by one or more of the diarrhea patients during the outbreak.

By merging the maps of diarrhea patients and of where at least one member of the patient class had eaten or purchased food (figure 2.7A) we advanced a set of assumptions. First, that one can describe an area called Greater Vancouver within which this outbreak occurred (*if* it is accurate and complete, *then* it is the area within which, and only within which, these related events occurred). Second, that a class of diarrhea patients reporting to hospital could be considered members of a single disease class (*if* these cases are the same thing, *then* we can consider them together).

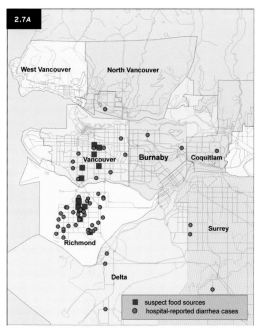

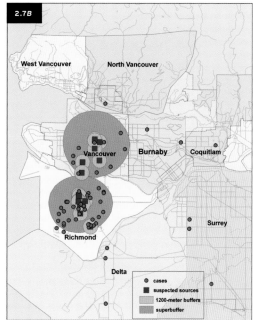

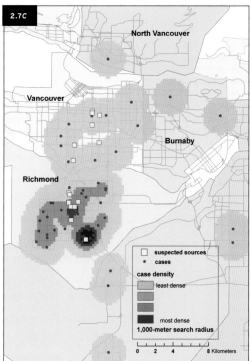

FIGURE 2.7A-C Combining maps of both reported cases of severe diarrhea and food outlets where patients had eaten (*A*) permitted the idea of a food source as the origin of the outbreak to be tested using buffers (*B*) and surface analytics (*C*).

Third, the map proposed that food outlets patronized by these persons could be identified and similarly located (*if* a diarrhea patient ate or purchased food at a place, *then* it was a place in this set).

Accepting these propositions for the purpose of testing, we then made a meta-argument based on them (Wood and Fels 2008): *if* these reported diarrhea cases (red dots) have a similar source in a contaminated food product (blue squares), *then* proximity between members of the two sets will reveal a causal relation between them. The idea that geographic proximity might imply causality is itself a proposition, a "thing" whose validity is based on prior assumptions and whose history in medicine is told later in this book.

Sometimes (although rarely) mappings like these reveal one and only one possible source. You can look at the map and say, "This is it!" More commonly the result is a map like figure 2.7A in which no single, suspected source (or collection of sources) is clearly indicted. A coworker, Ken Denike, and I later carried out two types of analysis to demonstrate how a relationship that was not immediately evident in the map could be uncovered (Koch and Denike 2007b). First, we drew overlapping buffers (circles) around individual food outlets in a way that allowed their combination, building successive circles until we had two "superbuffers," each with its own epicenter (figure 2.7C). We counted the cases in each and found the southern buffer was the one with by far the most cases and that a small number of food outlets were at its center. In this way we identified a subset of suspected sources that, on the basis of proximity, became "prime suspects" as the origin of the outbreak.

In another, more sophisticated test, we used an analytic combining surface density (kerneling) and a measure of proximity between cases and nearby food outlets (nearest neighbor analysis). In the first case the goal was to identify areas with the highest density of cases. The assumption was that the source of the outbreak would likely be centered in the area where disease incidence is greatest. Using the "nearest neighbor analysis," we calculated the distance between each reported diarrhea case and the nearest food source on the assumption that the source would be near the most cases that were closest to a food outlet. Combining these sets created a new space in which the density of cases was simultaneously mapped in a metric that relates food outlets to nearby diarrhea cases. In this manner the incidence of disease was generalized across the surface of the map and the dots of individual incidence became irrelevant except as data permitting a precise argument about intensity and density to be inscribed in the map. The second approach (figure 2.7C, which shows the density gradient) identified a very small set of suspect food outlets in the southern buffer.

In this map we noticed a north-south bias to the data (it's pretty obvious) and upon testing, there was indeed a narrowly oval, north-south, elliptical pattern to both data sets, but especially to the location of suspected food sources. This "deviational ellipse" suggested—because we had seen it before—not a single stable origin point but one that traveled, something or someone moving across the range of food outlets. Because Vancouver is the most densely populated municipality in the region we were not surprised to see a spike in the central part of the city where the northern buffer's epicenter had been. Inserting into the map the highway and road system for metropolitan Vancouver showed the food outlets were almost all clustered along a very few north-south arterial roads. One of the suspected sources was a bakery whose product was sold by local suppliers to a number of area restaurants. It was later discovered that the bakery used raw eggs in its custard products. Richmond officials tentatively identified that bakery as the origin of the outbreak, and its products the source of the hospital-reported diarrhea cases.[7]

A provisional report on the outbreak by British Columbia Centre for Disease Control researchers was published in the *Canada Communicable Disease Report* in October 2000; a final, definitive report was published several years later (Strauss and Fyfe, 2005). Those reports took so long to produce because they required bacteriological testing of patient stool samples and of foods from suspected sources. The result identified the bacterium *Salmonella enteritidis* as the agent whose source was the pastry product sold by truck across the lower mainland. Just as the map promised completeness it did not deliver; so did the bacteriology. Samples for all sixty-one persons were not tested, only enough were to assert with statistical confidence that the majority of the cases were caused by this single agent. Nor was every possible food product tested, only those that seemed most likely. That one or another of the sixty-one diarrhea suffers may have eaten another, contaminated food product elsewhere was possible but, in light of the general report, inconsequential. Like the map, the bacteriological work proceeded on assumptions of commonality and completeness that was necessarily limited.

This type of mapping goes on all the time. Often, there is no bacteriology or virology, no handy laboratory to identify the specific agent and its source. In those cases the mapped study is the only way an outbreak can be quickly investigated. A recent survey of CDC reports of suspected food-related diarrhea outbreaks in the United States found that "in 64 percent of all food-related outbreaks in the U.S. state and local health departments failed to isolate the specific bacterium or virus responsible; the cause was officially listed as unknown" (Hargrove 2007, 30). Without such testing, one only has mapping that may point to the origin of an outbreak but cannot clearly identify its agent. Bacteriology, where available, may identify the agent of an outbreak and thus confirm (or disprove) a theory of the disease's likely source identified in the map.

These maps of food poisoning and those of West Nile virus seem to be of a similar nature. They use points and the mapped geography of the space in which disease has occurred. In this case the purpose was to apply a well-accepted theory of a disease, one whose conclusions could be tested through bacteriology. For this we did not need population denominators or more sophisticated statistics. The maps of West Nile virus began in a similar manner and revealed what appeared to be a more complex pattern of multispecies incidence that proposed a new idea about the spread of that virus. While the diarrhea outbreak maps were hypothesis confirming, testing an idea based on experience, the West Nile virus maps were hypothesis generating, using patterns of spatial incidence to propose a human vector based on travel and trade. For West Nile virus and its finer resolution data, the mapping was theory applying, and thus required different types of proof in its application.

THREE QUESTIONS

These are commonplace, mundane examples. But things that are mundane and common are not necessarily simple. A series of complex things were required before these cases of severe diarrhea could be transformed into a bacterial disease whose source was traced to a local bakery. In the map and in the laboratory the study of these cases rested upon the application of a scientific ideal about the nature and construction of truth that came into being in the seventeenth century in the contested worlds of Robert Boyle and Thomas Hobbes (Shapin and Schaffer 1985). Boyle's construction of truth as a testable proposition whose results are replicable and judged by an informed jury of the knowledgeable underpins the assumptions of commonality, completeness, range, and testability that were critical to the diarrhea study (Shapin 1994).

In our mundane case the enterprise rested upon an idea of disease as a testable thing (1) for which accepted protocols (2) of data accumulation (3) and argumentation, (4) have been established and are accepted. The analysis required a method of visualization with which the disease outbreak might be imaged in relation to suspected food sources. In the process of "seeing" part and whole, the diarrhea cases became a single range of spatially located incidence set in relation to the food sources that class of persons patronized prior to their severe diarrhea. In the end, the map becomes a thing of its own, at once a statement of occurrence, a theory of what caused it, and a testing field for that theory.

So many *things* had to be in place before three basic questions could be asked about the diarrheic symptoms of patients in Greater Vancouver in the summer of 2000. These are the questions always asked of symptoms that affect the bodies we inhabit, the means by which those symptoms become things that can be named and treated: "Where is it?" "What is it?" and "Who is responsible for it?" All three are related, one to the other, part of the assemblage with which we structure our thinking about illness and its investigation. In a very real way, these questions become the cognitive landscape, the "mental space" within which disease is produced from the symptoms of sufferers. At the least, they provide a convenient structure within which the often-messy realities of disease study can be made clearer.

What and Where

"To name a state is indeed to put it in legible order, to interpret it according to visible symptoms—and often, naming a condition is a way of establishing a diagnosis (Arikha 2007, 277). Naming requires both a "what" and a "where." To think about location is to consider three related elements: origin, source, and the mode of transmission from here to there.

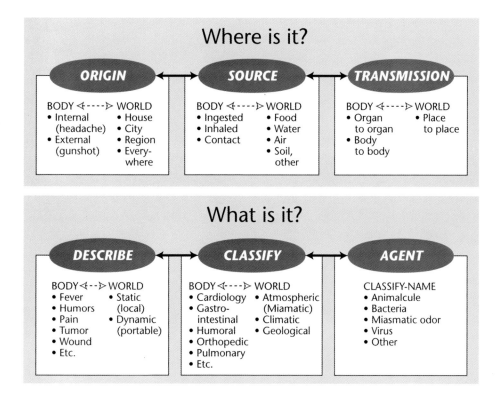

Where is it?

ORIGIN	SOURCE	TRANSMISSION

BODY ◁----▷ WORLD
• Internal   • House
   (headache)   • City
• External   • Region
   (gunshot)   • Every-
            where
	BODY ◁----▷ WORLD
	• Ingested • Food
	• Inhaled • Water
	• Contact • Air
	           • Soil,
	            other
		BODY ◁----▷ WORLD
		• Organ • Place
		   to organ to place
		• Body
		   to body

What is it?

DESCRIBE	CLASSIFY	AGENT
BODY ◁--▷ WORLD • Fever • Static • Humors (local) • Pain • Dynamic • Tumor (portable) • Wound • Etc.	BODY ◁----▷ WORLD • Cardiology • Atmospheric • Gastro- (Miamatic) intestinal • Climatic • Humoral • Geological • Orthopedic • Pulmonary • Etc.	CLASSIFY-NAME • Animalcule • Bacteria • Miasmatic odor • Virus • Other

FIGURE 2.8 Asking where a thing is and what it might be are intimately related attempts to define the nature of a condition and its means of address.

We often name disease after the location where it first manifested. We have, for example, the nineteenth-century "Asiatic cholera" that had to be distinguished from "English cholera," the former originating in India while the latter was common in the summer months in Great Britain. There is West Nile virus, born in Uganda but found as well in Romania and Israel; and the Hong Kong flu variant, first identified in, of course, Hong Kong. Sometimes the origin is a kind of flag of convenience in which politics overrides geography. Spanish flu, for example, was born in a U.S. military camp at Fort Riley, Kansas, before being exported to the European war theater in World War I.[8] If a condition is epidemic or pandemic a critical element of its knowing is its origin as well as the specific local source. Separating origin and source has taken us several hundred years of thinking.

To understand this or that disease the question becomes where it might be in the body. Is it an internal complaint (such as breathing problems, headache, or stomach pains) or the result of external injury (like a gunshot, knife wound, or penetrating arrow)? If it is an internal problem what is the organ most effected? Did it begin in the home, the neighborhood, the city, the region, or is it somehow everywhere at once? Thinking about geographic and internal origins influences our thinking about sources of the symptoms that affect us: was it inhaled (pneumonia), ingested (diarrhea), introduced through contact (poison ivy rash) or perhaps always there, latent in the person's genome, or in classical medicine, his or her humoural makeup? If inhaled, what in the air made us ill and if it is ingested what in the food or water caused this complaint? And perhaps most importantly, is it a static and unchanging agent unique to one place (or one family) or is it mobile and dynamic, moving from place to place, infecting populations as it moves?

"What" and "where" are collaborating elements of the same question. "What" describes the symptoms that result from its influence on the body, wherever it originates. Is it a fever, a humour (the pus of yellow bile, the darkened stool of black bile), the pain of a ruptured appendix, or the hard nodule of a cancerous lymph node? Is its source local or imported, unique to a few or a complaint common among many? Its description determines how we categorize it, seeking an agent that may be addressed experimentally and therapeutically. If it is in the heart it is cardiology; the gut gastrointestinal, a broken bone is orthopedic, and onwards. All this ultimately permits the symptoms to be named and a disease constructed in a manner that permits the practical question: what do we do about it?

The "what" and "where" influence our thinking toward the condition and responsibility for it. If the origin lies in individual lifestyle (gluttony, for example), the answer to

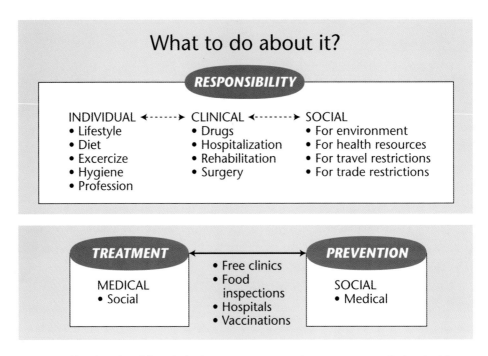

FIGURE 2.9 Knowing where it is and what its symptoms are permits a response—medical and public—to a disease event.

it may lie with the patient and not the physician or society. But if the disease is pervasive and many are affected, then individual responsibilities rarely serve as a sufficient explanation. Treatment and ultimate prevention become a communal affair, a task for society at large. In the Vancouver example the patients were blameless and responsibility assignable in part to the local bakery whose cream custards were the apparent source of the outbreak. But the local baker bought supplies from wholesalers and they carried a predicate responsibility. Ultimately, the final responsibility rested with the hospitals that treated the patients and the health agencies that in theory but not always in practice assure restaurants and food producer practices are safe.

THAT CURIOUS THING

These questions are an integral part of a process of knowing slowly constructed over centuries to permit the study of the conditions that affect our populations . . . a way of thinking that was neither inevitable nor intuitive. One may construct this history in a variety of ways, taking any of several points in time as a beginning. The question I began with is: when did we first begin to see disease in a certain way, to visualize its reality in populations as well as individuals? The answer to that question is the subject of the next chapter's brief review of the sixteenth-century marriage of intellectual and technological advances. It was in this time that Western societies first presented body and world as spatial, visual realities amenable to a type of critical study. Seeing changed everything, contributing to the backcloth of medicine and science in a way that was fundamental.

CHAPTER 3

BODY AND WORLD:
THE SIXTEENTH CENTURY

Nobody knows who made the first map of an illness—who first used pen, ruler, and compass to create on paper a visual statement about sickness in a population. Maybe a monk noted the location of plague deaths in London or Paris or Seville in the fourteenth century. Maybe another located the houses in which lepers had lived in Barcelona or Bristol or Lucerne. It is unlikely, however, because of the many things that had to come together to create the idea of disease as a thing that could be visualized in space, one whose location mattered. If such maps were made they are now lost, rare attempts at knowledge that perished over the centuries.

We do know the conditions that finally enabled the printed disease map to be made, however. It required a technology of paper production and a means of print reproduction permitting ideas, texts, and images to be reproduced efficiently and in relative quantity. Without paper those texts would be unlikely to be produced except as curiosities hoarded by the wealthy few. Parchment and vellum were too expensive and time consuming to make for their texts to be anything but rarities. The 1455 Gutenberg Bible, for example, required three hundred sheep carcasses to produce the parchment surface for a single volume (Hunter 1978, 17–18).

Paper created from rag cloth was a far less expensive medium with which increasingly efficient production technologies permitted the growth of what historian Raleigh A. Skelton called a "paper world" where works were produced in multiple copies. This permitted first a type of thinking to come into being and second for that thinking to be made permanent in published texts. "The paper world, therefore, did not simply provide a means for accumulating and storing what everyone knew. Rather it was a matter of inventing the conceptual means for coordinating the bits of geographical, biological, mechanical, and other forms of knowledge required from many sources into an adequate and common frame of reference. This common frame of reference became the theoretical model into which local knowledge was inserted and reorganized" (Skelton 1958).

THE BODY

With printing and paper came the means to produce a new kind of knowledge. At one scale it brought forth the mapped world; the print presentations of cities, regions, nations, and international communities; the spatial field in which disease was to be considered. At another scale, the sixteenth-century printing revolution brought forth the corporeal body, sick or healthy. That body, uncovered in autopsies whose findings were published, would be where injury and illness would find their place to be seen in the individual. Both scales indulged a self-consciously spatial perspective that focused on the relationship between individual parts, separately observed, and a whole in which they could be conjoined. This was a new way of seeing and of verifying as real those epistemic things, the body and the environment, in which health and disease would be located.

One sees this in Andres Vesalius's great anatomy textbook, *De Humani Corporis Fabrica*, which presented the body simultaneously as a landscape within whose components disease prospered or health resided, and as a landscape *within* the landscape of the world shared by all. Anatomical studies and teaching texts were certainly not new. Mondino published the first classroom manual for dissection in 1316, a guide that served Leonardo da Vinci in his later studies and continued to be used by some into the 1600s (Mayor 1984, 37). And in the sixteenth century there were others, for example, Jacopo Berengario da Carpi, whose anatomy texts included woodblock illustrations (Rifkin and Ackerman 2006, 13–14).

What distinguished Vesalius and his work was not simply the artistry of his illustrations but the systematic method of his technique and his organization of the results. Andrew Pickstone, historian of science, technology, and medicine, has argued the fundamental change Vesalius's work presented: "In medieval anatomy texts, all the illustrations are diagrammatic; in Vesalius, all is naturalistic—bodies stripped of skin, or variously splayed out, or skeletons posed in landscapes" (Pickstone 2000, 63). In taking as his subject the dissected body, Vesalius created an object of visual study that was the locus of disease. By orienting those cadavers as figures within a landscape Vesalius argued for a distinct process of knowing and a different temporality in which knowledge could be created.

Before Vesalius, anatomy was taught through the recitation of the canon of Galen of Pergamon, the second-century Greek physician who, with Hippocrates, was medicine's unquestioned historical authority. Lecturers read passages aloud as underlings, barber-surgeons, simultaneously carried out inexpert dissections. The dissection was secondary to the text, the corpse to the corpus of Galen's writings that in the sixteenth century had been newly translated into Latin by, among others, Vesalius's own anatomy instructors. Before Vesalius, dissection served principally to illustrate the text whose authority, while extrinsic—from the historical canon rather than the body itself (Hacking 2006, 33)—was assumed to be absolute. When the reality of the corpse appeared to

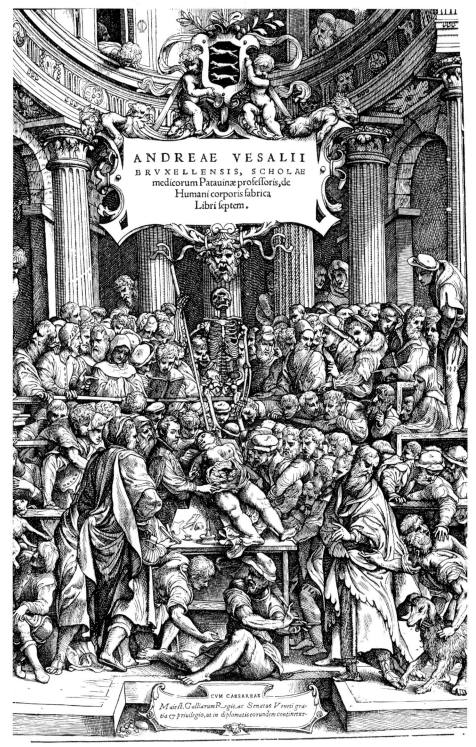

FIGURE 3.1A The frontispiece of Andreas Vesalius's anatomy book shows an autopsy in progress. He and some of his contemporaries make cameos in the engraving.

CVM CAESAREAE

FIGURE 3.1B This detail from the frontis-piece of Vesalius's *De Humani Corporis Fabrica* shows him dissecting a corpse before an audience of physicians and townspeople. The robed figure to the right, looking away, is thought to represent the older tradition of medicine, whose anatomy he sought to dispute.

contradict the textual description, the fault was assumed to lie not in Galen's authoritative description but the eccentricity of the splayed body on the table.

Radically, Vesalius insisted that knowledge was not an unchanging historical truth to be transmitted without question but instead something to be investigated in the world. Galen, who Vesalius revered but whose work he criticized, also insisted on evidence as something to be perceived rather than received. But Galen was no friend of the Skeptics, walking out of an anatomy demonstration because of a question about evidentiary issues (Athanasios, Pavlos, and Theodors 2007).

Galen's anatomy was based more on vivisection than dissection; the importance of that fact in the texts he prepared had been lost in the succeeding centuries (Mayor 1984, 108). Vesalius argued in a fashion radical in his day that where dissection revealed different anatomical structures than expected the anatomist should trust what is seen rather than what is read. "When Vesalius one morning showed his Bolognese audience the correct insertion of an abdominal muscle, the indignant Corti, piqued by the younger man's presumption, rose to invoke the irreproachable Galenic authority to disprove him. Vesalius did not hesitate. Boldly and without equivocation he stated that whenever he disagreed with the text, he—Andreas Vesalius—could prove that he was right and Galen was wrong" (Nuland 1988, 80).

To prove that he was right that, for example, the heart lay not central in the body but to the left, Vesalius offered as evidence surgically opened corpses his audience could see. In this way Vesalius made of dissection a way of testing historical, extrinsic knowledge through the medium of immediate visual evidence. As importantly, Vesalius promoted a replicable, standardized process of dissection that anyone could engage. And it was this process of systematic dissection, as much as the findings, which was Vesalius's fundamental contribution. To transform his lectures into a teaching text that promoted method, Vesalius required a system of organization in which parts were related to the whole, a way to make sense of the bones, muscles, organs and tendons that are the physical components of the human body. To make sense of the complexity of the whole of the body demanded a general classificatory system whose logic would fit the parts together.

Vesalius's anatomy also required a method of visual reportage, of graphic presentation more detailed than any that had gone before. Other sixteenth-century anatomists had published illustrated texts. Gregor Reisch published his *Margarita Philosophica* with anatomical drawings in 1503; in 1522 Berengario da Carpi published another (Rifkin and Akerman 2006). The illustrations in those texts were stylized and less accurate, however; the anatomy that resulted argued without the same sense of organization or realism that made Vesalius's *De Humani Corporis Fabrica* revolutionary. They lacked as well Vesalius's radically experimental insistence that to choose between Galen's *De usu partium* (or Reisch's *Margarita Philosophica*) and *De Humani Corporis Fabrica*, one had only to *look* at the dissected cadaver, and the disputed texts, to see the truth of the matter, experiencing it for oneself (Saunders and O'Malley 1973).

Vesalius was not particularly interested in taking Galen's place as an unquestioned sage for future generations. Instead he wanted others to adopt with him a type of knowledge creation that was experiential and observable rather than extrinsic and historical. *De Humani Corpus Fabrica* thus challenged others to trace for themselves Vesalius's intellectual journey through the process of dissection he taught: *here* were the bones in relation, one to another and to the muscles and tendons that both held them together and permitted their use. *There* were the internal organs displayed in all their overlapping complexity. *This* was the process of reportable discovery, experiential and experimental in its process, that if done properly would reveal the physical realities that were men and women.

In this way Vesalius brought forth the internal sites where disease and injury might reside. In the autopsy one could not help but see the tendrils of cancer that grew on the pancreas or the tumor in the brain, pleurisy's inflammation in the lung, scrophula in the enlarged lymph nodes of the neck. As Foucault put it in another context, "The human body defines, by natural right, the space of origin and of distribution of disease: a space whose lines, volumes, surfaces, and routes are laid down, in accordance with a familiar geometry, by the anatomical atlas. But this order of the solid, visible body is only one way, in all likelihood neither the first, nor the most fundamental—in which one spatializes disease" (Foucault 1973, 3).

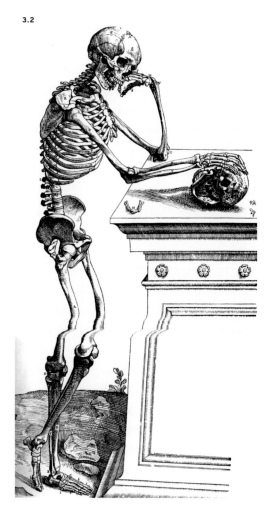

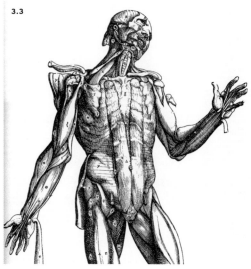

FIGURE 3.2 Vesalius dissected the body and organized its components but at the same time was aware, as a physician, of the place of the deceased in society. In this case, he imagined the remains of an anatomy subject at his desk.

FIGURE 3.3 Vesalius dissected the body to catalog its parts and their anatomical connections. But in this work he emphasized the human whose body was the object of his work, asserting their humanity in the process.

What permitted this space to be shared widely was the printing revolution then underway. Production capacity increased and costs decreased sufficiently so that young physicians could purchase *De Humani Corporis Fabrica* and keep it as a reference text. Nor was the anatomy Vesalius promoted limited to a single trade or class. "Furtive" sheets of multilayered anatomical drawings from Vesalius's and similar texts, engravings pirated and sold independently, "delivered the latest available information about the body's makeup even to those who could not read" (Arikha 2007, 144). Privileged knowledge became public knowledge; the lessons of the anatomy lecture the provenance of all. Just as Vesalius put his skeletons and muscled corpses in the landscape, commercial printing put his work in the greater world.

The symptoms that together defined a disease might be in the body, but the person whose body the cadaver had been had lived in the world. The dead had been scholars who worked at their desks or farmers who labored in the fields that surrounded the towns and cities of Renaissance Italy. In a series of extraordinary drawings that placed his cadavers' skeletons back in the world, Vesalius emphasized

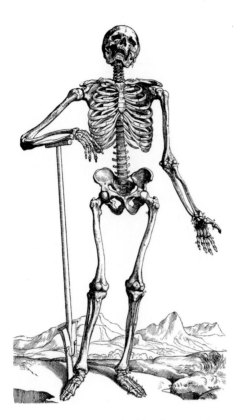

FIGURE 3.4 An anatomist, Vesalius was also a physician who understood the Hippocratic relation between environment and patient. For example, much like the laborers working in the hills outside Vesalius's Padua dissecting theater, this skeleton stands with a shovel.

the relationship of the body of the person who had been with the world he or she had inhabited. Vesalius was a physician who had not forgotten the essential Hippocratic insistence on the relationship between health (or disease) and place. The illustrations in his text at once dissected the body into its components and simultaneously reintegrated them in the landscape of the world.

None of this happened over night. It took time for Vesalius's ideas to be accepted. There were critics who insisted on the provenance of Galen and a knowledge whose validity rested upon an historical authority. The "scientific revolution" was still a century in the future. In the sixteenth century Giovanni Morgagni would develop a systematic pathology that defined the "space and origin" of disease through the correlation of patient symptoms (shortness of breath) and the evidence of physical changes (enlarged heart and diseased lung) observed upon autopsy (Nuland 1988, 150–70). But for that space to be created, medicine first needed an exemplar whose work created a norm, a reference space within which physical health and disease might be distinguished. That exemplar was Vesalius and his textbook promoting direct observation whose results could be seen by any anatomist, whose textbook brought forth the human body through a close, detailed, organized description of its related parts.

MAPPING THE WORLD

In 1564, Bartolomeo Eustachi (the Eustachian tubes in the ear bears his name), a critic of Vesalius, published his *Tabulae anatomicae*, in which he placed images of his dissected bodies in maplike frames whose coordinates were rulers giving the size of the body depicted, or at a different scale, an association of its individual parts. In this way a student might locate this or that bone, muscle, or organ through reference to the graphic coordinates just as they would come to locate cities in then evolving world maps (Rifkin and Ackerman 2006,103). The conjunction of map frame and anatomy was neither surprising nor accidental. As Vesalius and his fellow anatomists were creating the reference space in which future generations would investigate disease close in, others were mapping the world in which the anatomists, their subjects, and readers resided. The same elements—observation, spatialization, and visualization of the whole—that made *De Humani Corporis Fabrica* so critical similarly informed the cartographic exploration of the world at large.

Just as the new anatomy advanced knowledge of the body beyond the confines of the old texts and their thinking, a new cartography was advancing a way of knowing and seeing the world and its parts in a manner that Ptolemy, the authority of the ancients, could not have imagined. Here, too, experiential knowledge of the world, in this case gained by traders and travelers, was used to reform a knowledge base that was classical and incomplete. And as with anatomy, so with the world, this new view would be embedded in book pages that visualized new knowledge for all.

In 1536, for example, the medically trained mathematician Oronce Finé published a cordiform projection of the world,[1] an equal-area projection shaped like a human heart. The evolving anatomy of Vesalius showed the skeletal body posed in the landscape of the world. The cordiform map was a mathematically constructed projection of the world as an anatomically shaped organ in which the known continents became the chambers of a heart that was the earth. Kish (1965) catalogs sixteen maps of the era that took this form, including one by Abraham Ortelius.

Finé's map was titled *Recens et integra orbis descripto* and promised a new and "integral description of the world" through a system of computation that resulted in accurately sized landmasses in the cordiform projection. No less than Vesalius's anatomy, this and other similar maps were about seeing with unprecedented accuracy the parts and whole together. The heart was "the primary receptor of sensation, the seat of memory, the portal to wisdom, and above all, the governor of the microcosmic human body" (Cosgrove 2007, 107). Why not use mathematics to project a heartfelt world within which the microcosms of nations and cities were situated?

In this projection was cosmology as well. "In the same way that the heart is the seat of divine impulse and the spirit of charity, so the earth, in the centre of the universe, is the place where humanity has to demonstrate ethical choice and witness faith in a constant exchange between spirit and matter" (Mangani 1998, 66). Almost certainly, the projection was as much allegorical as it was evidentiary. But in the cordiform map was a triumph of mathematics whose argument was available to any who chose to read Finé's description of the map's construction. Here, too, was Vesalius's sense of transparency, of "here is what I did and you can, too," that made *De Humani Corporis Fabrica* so remarkable a piece of not simply anatomy but also pedagogy and science.

Empowered by new techniques of cartographic rendering and improved technologies of printing, maps of cities and countries and the world in which they existed were transformed in another sixteenth-century innovation, the atlas. British historian Paul Binding described it this way: "The most successful book of the entire sixteenth century—the first century to open with the printed book as a fact of western life—was *Theatrum Orbis Terrarum* (Theater of the countries of the world). It did something no

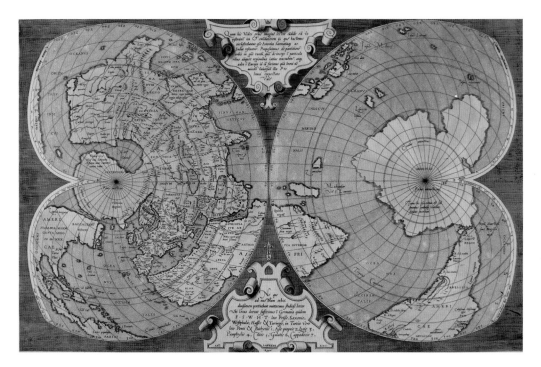

FIGURE 3.5 The sixteenth-century cordiform map joined anatomic imagery and cartographic methodologies in its imaging of the world. This double-cordiform map by Anton Lafréry and Gerhard Mercator was produced in 1560.

previous book had done. Here was the world with its many component parts, and it was shown to be both a place of extraordinary varieties and a singular whole. The *Theatrum*, published in Antwerp on 20 May 1570, was the world's first-ever atlas" (Binding 2003, 3). A uniform collection of map sheets bound in book form, Abraham Ortelius's atlas attempted a systematic organization of the world through the careful interrogation of its parts. It can be distinguished from its predecessors in part through its focus on the *tabulæ modernæ*, one known by experience rather than the historical geographic perspective of Ptolemy's *Geographica* (Wood 2004). The world it called forth, one grounded in politics and trade, was presented in a manner that made its interlocking kingdoms and nations natural features of the landscape.

Just as Vesalius had gone into the body to see its parts, so too explorers and merchants were traveling the globe, returning to describe the places they had been. Those descriptions enlarged the status of the states whose outposts were the end points of those explorations, the newfound sources of trading wealth. Nor was this process of exploration and survey limited to the far away. "Beginning in the sixteenth century, vast swaths of territory were subjected to systematic surveys by newly self-conscious states" (Wood and Krygier 2009). Abraham Ortelius, the *Theatrum*'s creator, collected, collated, and standardized the maps that were the products of that work, joining kingdoms and principalities into geographic regions whose relationships, plate by plate, brought forth the world that the *Theatrum* presented its readers.

This was not the second-century world of Ptolemy's geography, composed of a catalog of eight thousand points of longitude and latitude. Sixteenth-century cartographers began to fashion maps that transformed Ptolemy's old world into the new seafaring expanse progressively experienced by explorers and other travelers. In so doing they literally recreated the world. The cordiform was useful in promoting an idea of the world as organic and whole, less so a description of its varying mercantile parts in relation to one another. In 1564 Ortelius produced a cordiform map but it did not serve the world he was imagining in the *Theatrum* (Mangani 1998). In that atlas the world took on a different form, one shorn of the allegorical and imbued with a sense of commonplace construction of part and whole that Vesalius would have understood.

The *Theatrum* was first available in 1570 in three different versions and at three different prices. A small black-and-white version cost less than 7 guilders, a larger version, still uncolored, about a guilder more, while a colored *Theatrum* sold for 16 guilders (Skelton 1964, 7). The brilliance of the *Theatrum* resided in its scale and its resulting global argument made through individual plates of continents and embedded countries, some of which were individually investigated at a finer scale in separate plates. Part and whole, together they made the appearance of commonality, and unity, inevitable.

Consider a map of Europe from a color edition of the *Theatrum* preserved today at the Newberry Library in Chicago (fig. 3.6*A*). Its countries were those best known by Ortelius and his readers, an area in which trade was widespread and travel commonplace. Here was a civilized mercantile community coming into being as a powerful political reality. The power of the map rested in the weight not simply of its subject, or of detailed plates of individual countries, but in the authority of the entire Atlas collection. This was one map among many, *all* of which together asserted a reality seen and experienced piecemeal by individuals who had described this or that country and visited this or that city.

The atlas created a field not principally for doubting experimenters but rather for satisfied observers. One benefited from the exploration of others to the extent one assumed the veracity of their reports. As the editor and compiler of a later, atlas of world cities, the *Civitates Orbis Terrarum*, Georg Braun, wrote in a 1581 edition of his work (Braun and Hogenberg 1966): "What could be more pleasant than in one's own home far from danger, to gaze in these books at the universal form of the earth . . . adorned in the splendor of cities and fortresses and, by looking at the pictures and reading the texts accompanying them, to acquire knowledge which could scarcely be had by long and difficult journeys?" (repr. in Skelton 1964, vii).

The knowledge that appeared so constantive in the plates of the atlas was, from our more modern view, at best tentative and uncertain. The maps were made from the reports of travelers who sailed so many days in such a direction at a speed estimated by inexact methods. Distance was roughly estimated on this basis. When there was sufficient sun navigators would take a noon sighting for latitude, but it would be the

FIGURE 3.6A Ortelius's *Theatrum Orbis Terrarum* mapped the mercantile world, within which the distinct geographies of individual nations were joined by the trading ships that sailed between them.

eighteenth century before exact longitude could be as easily seen (Brown 1956). In the lands visited there was no way to correctly estimate the area of huge landmasses known principally only from costal visits by traders. The cordiform in some ways was a better representative of the state of the knowledge base from which the maps of the world were constructed. Its allegorical shape announced practical limits hidden in the world the *Theatrum* brought forth.

Ortelius's text invested his map of Europe with bountiful resources and Christian governments. In a 1606 English-language version Ortelius described his Europe as an area of good trade with like-minded Christians: "This Europe of us, besides the Roman Empire which is held in reverence all over the World, has all in all twenty-eight Christian Kingdoms (including those fourteen which Damianus from Goes already counts in Spain alone). This will allow you to estimate the worthiness of this region." That worthiness was to be measured not in miles but in the mercantile measure of trading goods produced across the Europe he mapped, and in the climate and conditions that made it possible to grow the grapes from which wines were made. "It is a place extraordinary fruitful, and the natural disposition of the weather is very temperate. In all sorts of Grain, Wines, and in its abundance of Wood, it is inferior to none, but comparable to the best of others" (Ortelius 1606).[2]

There is no map of produce in the *Theatrum*, however, no legend detailing the tonnes of grain or barrels of wine transported from this mapped place to another. Commercial mapping with legends that were tables of goods shipped would not evolve for centuries, just as the circulatory system of the body's arteries and veins described by Vesalius would require advances in science for a complete understanding.

The power of the atlas resided not only in its more familiar parts, the presentation of European countries more or less known to its readers (like England, Spain, France, and Germany). The real attraction was in its maps of lesser-known places. There were the still exotic Americas, whose promise was only becoming known and whose parts were gathered together on a single plate. Turn the page and there was China, in all its mystery, and again on another page Africa sketched in outline but mostly empty because it was unexplored and thus unseen. This was Ortelius's great genius. By gathering together individual maps from different sources and pasting them together in a manner that argued a coordinated whole, each map gained authority from its neighboring plate. Each became an evidentiary piece of the greater world the atlas brought forth. The *Theatrum* became the world explored, a geographical anatomy that offered the reality of experience, defined plate by printed plate, without the inconvenience of personally traversing the globe. And if experience argued a different place, Ortelius encouraged readers to send in new materials to be incorporated into future editions. Like Vesalius's *De Humani Corporis Fabrica*, the atlas of Ortelius encouraged new knowledge over simple, historical assumptions.

The conventions employed in producing Ortelius's Europe appear surprisingly familiar, the map something we can easily read despite the four hundred–some years that have passed since its first printing. It is not simply that the Europe of today is more or less recognizable in the geography of Ortelius but that the codes of its presentation are similar to those we continue to employ. "Title, legend box, map image, text, illustrations, inset map images, scale, instructions, charts, apologies, diagrams, photos, explanations, arrows, decorations, color scheme, type faces are all chosen, layered, structured to achieve speech: coherent, articulate discourse" (Wood 1992, 112). The presentational code demanded a colorful compass rose in the upper right of the map, and a scale bar at the lower left promising directional and aerial accuracy in this rendering of the world. Tectonic codes assured that rivers, coastlines, and mountains were presented, their accurate location assured by the compass rose and the scale bar in the map. Unlike the cordiform map projection, this one is familiar, we've seen it so often we do not wonder at its use in work over 450 years old.

All this disguises the fundamentally unnatural (perhaps artificial is a better word) nature of the world the *Theatrum*'s plates brought forth, one in which political realities were the map's most critical assertion. This is best seen in the, for us, fanciful "Europe," stationed off the coast of England and France. Here the mythological floats uneasily with the geographic, animal and human symbols a way station on the way to the geographic. Prosaically, kingdoms are distinguished typographically:

FIGURE 3.6*B* In the precise cartography of the *Theatrum*, classical mythologies were continued in, for example, this image of Europa floating with a bull in the Atlantic Ocean off the coast of England and France.

Font size and style are different for England, for Wales and for Scotland. And then there is the hand coloring that gives the whole its modern feel: Scotland is a dark green above the light yellow space that is England, to the left of which Ireland is lighter and tinged with red. At the center of the map beige France sits among a light yellow Spain, a dark yellow Italy, and a dark green Prussia-Germany. How better to present countries whose borders were not physical realities but instead political constructions? Vesalius constructed the body from its various parts, setting them in the mundane landscapes of Italy. Ortelius and his contemporaries created countries as landscape forms from nothing but fonts differentially sized, and for the wealthy purchaser with 16 guilders, an arbitrary color palette.

The map promised a precision evidenced in its scale bar. There are three distinct scales imposed upon the map, German (*Germanica*), Italian (*Italica*), and English (*Gallica*). Each represented a different national system of distance calculation. All three were necessary for the atlas to be useful to an expanding, international audience of book buyers. And because the individual plates were based on maps and measurements from different sources, the three together promised a consistency, some type of commonality among the diversity of different national survey systems. The separate measures seem somehow out of place, cumbersome in the unity the rest of the map proposes. Over the years the variations would disappear as the separate nations came to agree on a single measure of distance.

CIVITATES: CITIES IN THE WORLD

In retrospect it seems inevitable that other mapped compendiums would be created to compete with Ortelius's *Theatrum* just as it now seems inevitable that individual map pages, either torn from an atlas or printed for sale as individual sheets, would become ever more popular knowledge artifacts of the age. Nor is it surprising that once the idea of accumulated knowledge in books of maps was actualized, other atlases with other foci were introduced. In 1572 the first edition of *Civitates Orbis Terrarum* was published. The sixth and final volume of this first atlas of world cities was published in 1617, a commercial and intellectual triumph. In many ways, it was a companion to the *Theatrum*, sharing with it a host of contemporary references and a range of stylistic similarities. Produced by Georg Braun and largely engraved by

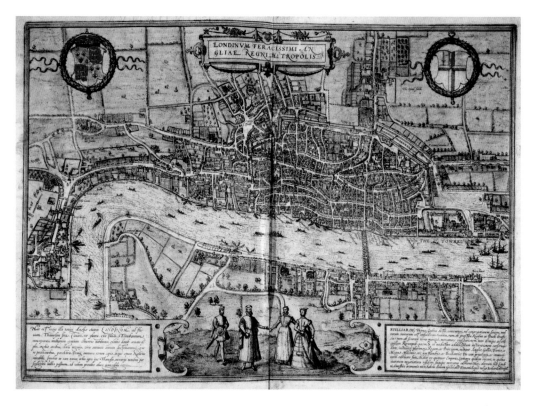

FIGURE 3.7A The London of *Civitates Orbis Terrarum* in 1584 is dominated by the Thames River, along which urban traffic rowed. The map shows broad streets and streets on which churches, each representing a local parish, dominated a dense aggregation of buildings presented in a bird's eye view.

Franz Hogenberg, the first edition contained 546 bird's-eye and map views of cities from all over the world.

A feel for the whole can be gained from the map of London and especially the old, walled city of history north of the river (fig. 3.7A). This London is dominated by largely local river traffic on the Thames, each pier on the river carefully named. Broad streets dissected a city in which buildings were packed around churches distinguished by crosses on their spheres, each church symbolizing the parish for which it stood as administrative center. To the south of the river were the bull and bear baiting rings, places of entertainment for the well-to-do. At the top of the page, laid across the agricultural fields are the heraldic emblem of the city to the west and a heraldic cross in a leafy wreath to the east. *Londinum Feraccisimi Angliae Regini Metropolis*, the map's title proclaims.

This atlas was less a report of exploration than an artistic collaboration whose production was a commercial rather than an overtly scientific enterprise. It observed and reported the world it produced, naturalizing the cultural elements of its description. Like Ortelius's *Theatrum* it was also an intellectual and technical tour de force. The work of over a hundred artists and cartographers was combined to produce the engravings that became the published work.[3] In its early editions Braun and Hogenberg actively solicited city maps from readers for incorporation in future editions, building an ever-more complete collection of urban images based on a growing experiential database embedded in city maps produced across the successive volumes of their work. Ortelius's *Theatrum Orbis Terrarum* became the landscape in which the cities of *Civitates* were produced for general consumption, a shift in scale that transformed the dots on Ortelius's maps of Europe, the Americas, and Asia into the urban centers within which the citizens of nations dwelled.

Braun added to many of his maps representative human figures in what he took to be native dress. In his map of London, for example, he placed these figures in the rural, cultivated lands south of the Thames. Those he imaged were not peasants or tradesmen but the landed and well-to-do, persons perhaps not out of place at the royal courts of the day. Vesalius placed his skeletons in a landscape to emphasize the world in which their lives had been lived; Braun placed figures in his maps to assert the lived space his maps produced. The insertion of figures in the maps of world cities simultaneously gave to each an additional sense of the specificity of the place and of the lives pursued in the landscapes the map created.

In the disease maps that would follow the geography of the map would be the receptacle in which diseases affecting citizens, like those idealized in the map, could be located and then studied. Just as Vesalius investigated the anatomy in its vary-

FIGURE 3.7B To the map of London, as he did in maps of other cities, Georg Braun added the figures of local gentry. Later, maps would be filled with other populations, sick and well, as the map took up a role in disease studies within broad urban populations.

ing parts, detailing the connection of muscles and ligaments to the skeletal whole, Braun and Hogenberg focused upon the parts that contributed to Ortelius's whole. Because this is neither a history of the map nor of the atlas compendium, a detailed consideration of atlas production and the specifics of their images can be overlooked here. What cannot is that the scale of the *Civitates* maps argued the importance of the evolving city in the changing world. And it would be at the scale of the city as much as the scale of the nation and world in which disease studies would become rooted in the seventeenth and eighteenth centuries. It was *in* the cities and regions of the world that the "where" of disease would be investigated in the maps of the world's individual places. It was at the scales of city, nation, and world that various theories of disease origin and causation, the "what," would be investigated.

KNOWLEDGE AS LABOR

It is common to talk of Braun and Hogenberg's *Civitates*, Ortelius's *Theatrum*, and Vesalius's *De Humani Corporis Fabrica* as if they were the works of individuals, the creations of geniuses uniquely responsible for the embodied knowledge. This reflects less the reality of the work than our desire for heroes. As Ferdinand Braudel says disapprovingly, "Generations of historians have accustomed us to the simply dramatic, truncated narrative," about individuals whose superiority can change the world (Braudel 1980, 27). Lost in this focus on the dramatic, heroic story is the complex reality of the objects of our interest, what Michel Foucault described, in a phrase repeated approvingly by historian of science Steven Shapin (1994, 36–37), as a truth-producing "system of ordered procedures for the production, regulation, distribution, circulation, and operation of statements" (Foucault 1980, 131–32).

Consider the extensive system of labor required to bring forth the epistemic map or anatomic drawing, the intense communities of production, regulation, distribution, and circulation. The paper that was at the heart of the sixteenth-century print revolution was manufactured from cloth collected by rag pickers for recycling by papermakers who had first to await its fermentation and then, after the rags' maceration, refashion it into paper on frames (there was a guild for their formation) made of wood and paper. The linen (one business) was made into clothes (a second) that wore out and were collected (a third) and processed (a fourth) on frames (a fifth) to produce the paper that once sold in specialty shops was used to make the map or drawing included in a published document (Hunter 1978).

For Vesalius's work to be transformed into a book there had to be the trades that created the printing technology, the producers of the press and the type on which the pages were made. For the drawings to be reproduced first there had to be the knowledge of pen and ink rendering at a level capable of creating the carefully witnessed images of dissection as well as a community of artists capable of applying that knowledge to the construction of the precise images printed in *De Humani Corporis Fabrica*.

Here were two trades, that of the graphic artist and that of the woodblock artist who could transform the image into something that could be printed in the book.[4] All this employment, all these trades required an audience with sufficient means to pay for the final book. Economies of scale were required to make this a textbook and not simply a collector's item. Vesalius's genius rested not simply on his unparalleled skills as an anatomist but also in his ability to obtain the services of skilled artists and printers who would produce the anatomical illustrations in a text he believed would be of sufficient interest to warrant profitable publication.[5]

Vesalius was no solitary genius, no lonely visionary, but a physician and scholar in a community that valued both. He was a professor at the University of Padua and the student of professors who had translated Hippocrates and Galen, making them accessible in printed form to students like Vesalius. He was the successor of those like Johannes de Ketham, whose 1494 *Faciulo de medicina* was a crude but necessary predecessor to the work that Vesalius would create (Rifkin and Ackerman 2006, 1011). And not least, Vesalius was the inheritor of a movement in art that gained immeasurably from the precise anatomies constructed by those, like Leonardo da Vinci, who similarly sought to understand the body, part and whole.

Like other professors before and after, Vesalius required an audience of students and professionals to transform the dissection from a private desecration of the body into a publicly sanctioned teaching exercise. It was as a textbook that *De Humani Corporis Fabrica* was published, not simply as a rarified text on the limits of Galen's anatomy. The anatomy theater in which Vesalius carried out his dissections was itself a radical innovation that moved dissection into the university from the public courtyards where they had been conducted. That theater was in turn the end result of the employment of a range of crafts whose members first designed and then built the theater. And, of course, all that first required a bureaucracy that empowered universities to teach medicine, typically through royal franchise, and then paid for the university's building and its lecturers.

It was this bureaucracy, and that of the greater community, that defined dissection as something other than sacrilege. And it was only for the honorable purpose of physician instruction that bodies otherwise destined for whole Christian burial were transformed into medical cadavers, a social good that outweighed what otherwise would have been an unholy violation of the corpse. The tools of dissection were the product of metal workers specializing in the production of the clamps, probes, scalpels, saws and other tools of a medical profession whose members were constantly modifying the tools they had, or having better implements designed.

The *Theatrum* and the *Civitates* were similarly collaborative. Indeed, they were more so because the atlas makers were not themselves principally artisans, explorers, or professionals but entrepreneurs who brought together and standardized the maps of others. It was Mercator, not Ortelius, who defined the method of projection that permitted the globe to be reduced to a flat surface in which distance—the measure of travel—rather than the area of landmasses was the critical constant. Mercator is carefully cited in the Ortelius's extensive bibliography that lists not only the mapmakers whose work he used but also the more than ninety works he consulted.

In 1595, the year after his death, Mercator's own atlas was published—*Atlas sive Cosmographicae Meditationes de Fabrica Mundi et Fabricati Figura*[6]—in a manner that acknowledged the work of Ortelius's commercial and cartographic genius. We know the sources of Braun and Hogenberg's work because they were at equal pains to document their sources and their contributors. Indeed, it was in those bibliographies that the authority of their texts was lodged, a compendium of work that proclaimed its authority through the long list of contributors whose work the atlas employed in its own construction.

Whether the subject was nations in relation one to the other or the cities of nations laid out page by page, the mapping required a complex merchant enterprise that might carry a sufficient number of explorers and traders to a large enough number of places that sufficient knowledge about the world's locations could be returned. Only then, would there be a body of sufficient knowledge to produce the *Civitates* or the *Theatrum*. The atlases were thus dependent on the shipping industry as well as, of course, the paper and print industries. Trees were cut, cured, sawed, and then fashioned to make the presses on which the paper was pressed. Metallurgists fashioned stamps that could be used to engrave consistent, repeated symbols in maps through the 1600s (Delano-Smith 2005). Other metalworkers would produce not woodblock but copper plates on which engraved maps could be replicated. None of this would have happened were there not a growing and literate audience interested in these volumes and able to pay for them. Finally, there was the civil bureaucracy at large that both encouraged this work through policies advancing a mercantile society and economy and as a market for the works that were thus produced.

Maps and atlases, anatomy textbooks, and other works all depended on the urban explosion that brought artisans, craftspeople, and shopkeepers into the expanding sixteenth- and seventeenth-century cities where paying customers were to be found. To talk about maps within the printing revolution is to talk about printing as a handmaiden of the progressive urbanization that has continued from the Renaissance into the present day. One sees this in London, for example, where during the first years of the printing revolution the population exploded from approximately seventy-five thousand to two hundred thousand persons (Porter 2005, 63). Some saw this as, at best, a very mixed blessing. To slow the influx of both rural workers and foreigners seeking a London home, in 1580 the government of Queen Elizabeth I issued a proclamation prohibiting new building and limiting dwelling density in individual houses to one family. It was a proclamation violated almost from the start. In the city that evolved, disease would become common—and a subject common to the physician and mapmaker alike.

CHAPTER 4

DISEASE IN CITIES: THE NEIGHBORHOODS OF PLAGUE

DISEASE

From the Middle Ages through the Renaissance, leprosy was perhaps the most feared disease and one against which local bureaucracies labored mightily (Foucault 1973, 44–45). Terrifying as leprosy was, the numbers of those affected in any village at any one time were relatively small and the response to its presence orchestrated principally at the level of local parishes and their officials. Attempts at disease control focused primarily on the exclusion of the leper as a way to protect those still healthy. Lepers thus were cast out of their communities, ordered to finish their days on the edges of the towns and villages in which their lives had unfolded. "The relations governing the isolation of lepers were very detailed and precise. The awful finality of exclusion from the human community was symbolized by an enactment of the funeral service involving the participation of the 'Leper' dressed in a shroud and attending a solemn mass for the dead in the leper's honour" (Rosen 1993, 41). The status of those afflicted as disease carriers was assumed even if the nature of the disease and its mode of transition were not understood.

Unlike leprosy, plague was a recurrent threat that affected whole communities and entire nations rather than a scattering of individuals here and there. Epidemic outbreaks decimated populations from Europe to Africa, and from Persia to Portugal. The "plague of Justinian" ravaged Constantinople in AD 541–542, a historical horror still not forgotten in the thirteenth century, when the Black Death reigned. Sixteenth-century scholars and officials remembered those earlier histories as smaller, but new, still-virulent epidemics raged. Plague was never wholly absent in Europe. In most years outbreaks were reported in this or that country or kingdom, in this or that province or town. In the sixteenth century, England alone experienced at least

48

nine separate epidemic visitations (Drew 1970). Nobody understood the etiology of the disease, its active agent, or local source. Physicians could diagnose it because of the swollen lymph nodes whose distinctive buboes followed the onset of a fever. Its virulence and intensity in affected populations made plague the very definition of epidemic disease.

With little understanding of its causes and less ability to treat its symptoms, officials were restricted in their ability to confront plague's challenge. In some jurisdictions special burial sites were set aside for plague victims, and for those whose illness was not yet mortal, quarantine houses were constructed. Another approach, desperate and ineffective, was preventive. Assuming that plague originated in the unclean homes of the poor, attempts were made to remove the poorest of citizens from areas where outbreaks occurred to try to stem the disease at its point of origin. Thus, in response to a plague outbreak, the Venetian senate in the seventeenth century passed decrees to remove beggars—those whose lives were assumed to be unhealthy and whose living quarters unclean—from the city (Pullman 1992, 101–2).

One idea was that plague was divine punishment for profligate and unreligious lifestyles. Another was that the piles of human and animal waste amidst the hovels of the unwashed poor created odors within which disease was generated. Plague as a thing of the unsanitary and poor had a long life as an idea, one that stretched in some quarters through the nineteenth century. Remedies based on this theory were uniformly ineffectual and often destructive. In the plague pandemic that began in the 1890s, for example, British colonial forces burned the homes of local Chinese citizens in the mistaken belief the plague originated in their close, malodorous section of the city. Similarly, the local Chinatown in Honolulu, Hawaii, was quarantined and much of it burned in a similar attempt at plague containment and eradication (Mohr 2005).

Plague was not to be contained by either eviction of social undesirables or the razing of their dwellings, however. And it was clearly ecumenical in its choice of subjects, attacking bishops and beggars alike. Another idea came forth that plague was not a disease rooted in the characteristics of place but a portable disease carried from place to place by travelers with their goods, from person to person as if by the touch of death itself. This anthropogenic, mobile plague was a central character in Hans Holbein the Younger's *Dance of Death*, first published in 1538. As Vesalius brought forth the body in all its parts, Holbein made Death's diseased visitations an equal partner in the world of men. In its forty-one woodblock prints, Death is shown to be a real and constant presence in every walk of life, at once a personal and collective threat.

We *see* the relation between plague as a disease and death as its portable in the woodblock prints Holbein produced. Disease, and especially plague as *the* epidemic disease, transmitted from place to place, was a principal tool of Death's office. The Dark Angel climbs the mast of a cargo vessel's mast as it moves from port to port. Death's hand is placed on a merchant's head as he reaches to open a bale of goods,

DISEASES IN CITIES

49

CHAPTER 4

FIGURE 4.1 A plate from Holbein's *Dance of Death* (1538) presents Death as a cargo inspector capturing a merchant with his bale of goods.

FIGURE 4.2 In this plate from Holbein's *Dance of Death*, Death rides the oxcart between towns, an unwelcome addition to the carrier's goods.

releasing mortal disease.[1] In another scene, Death rides on a cargo wagon carrying barrels down from a cultivated hill. Beside him his horse lies dead, presumably a result of Death's touch. As the waggoner looks on in horror, death opens the barrels the man was transporting. The landsman and the sailor and the merchant they supplied: all were depicted as links in the causal chain of transmissible, epidemic, mortal disease. Through the eighteenth century and, indeed, much of the nineteenth century the idea of plague as a portable byproduct of trade would remain a medical assumption. In his account of "several plagues that have appeared in the world since the year 1346," eighteenth-century physician Dale Ingram argued plague as something packed with the travelers' goods, bundled in the merchant's wares that, when opened, would release it into the atmosphere (Ingram 1755, 77).

Quarantine

There was the idea: plague traveled the human network of trade in the world Ortelius's *Theatrum* brought forth. It was transported by the people of the trading world, on ship and ox cart in bales of goods. Whatever its agent, it was a natural phenomenon made epidemic and pandemic through the commerce that was beginning to tie together the various parties of the mercantile world. Once transported to a new place it flourished among the poor. The way to impede its spread, therefore, was to stop the movement of goods.

Quarantine was perhaps the first true public health initiative: "The public health community can trace its roots to fourteenth-century Italy, when fear of the Black Death prompted the government of Venice to exclude from their ports those ships with persons reported to have pneumonic plague" (Thacker and Stroup 1998, xix). Absent a detailed knowledge of disease agency, it was the only action available to bureaucrats concerned with the health of citizens and thus one that, in the sixteenth century, became increasingly common.

The English word "quarantine" comes from the Italian *quarantenaria*; the forty-day period of isolation first mentioned in Venice in 1127. "In 1374 Venice denied entry to suspect travelers, vehicles, and ships based on the belief that plague was introduced chiefly through shipping" (Rosen 1993, 44–45). In 1564, the Duchess of Parma, Philip II's half sister, and his regent in the Netherlands, prohibited trade with London during an outbreak of plague in that city. In 1580 ships from Lisbon were quarantined in London's harbor, at the Lord Mayor's suggestion, in an attempt to prevent the plague from spreading into that city (Porter 2005, 61). During epidemics, the cities so nicely imagined in Braun and Hogenberg's *Civitates* became places not of civility and trade but of diseased, mortal danger. Merchants were unhappy with trade restrictions, of course, and typically sought either to prevent a quarantine or, if one was imposed, to find alternate trading points. During periods of London's closure, for example, Dutch ships sailed for Emden, Germany, instead of London (Porter 2005, 57).

PLAGUE IN BARI: 1690

The situation had not changed greatly from Holbein's day when, in the 1690s, a plague outbreak occurred in the province of Bari in the kingdom of Naples. The names of both the city and province of Bari are minute, almost indecipherable, in Ortelius's map of Europe (fig. 3.6). But it was here that plague returned, the sequel to a ferocious epidemic in 1630 that killed more than 30 percent of the population of Venice (Pullman 1992). The royal auditor and military governor, Filippo Arrieta, deployed troops to isolate the province from its neighbors, hoping to thus curtail the spread of the disease. Within the province more troops were sent to cordon off cities and towns where plague was active from those where it had yet to appear (Arrieta 1694). To explain his actions (and justify the expense) to his liege, Arrieta included two maps in his official report on the plague outbreak and his quarantine.

Physician Saul Jarcho,[2] long associated with the New York Academy of Medicine, commented on only one of the two maps, describing it as an "interesting and attractive" curiosity (Jarcho 1970, 132). Arrieta's maps are far more than that, however. Together they present a careful spatial argument about disease containment and, implicitly, the idea of plague as a portable disease spread by travelers. To see how

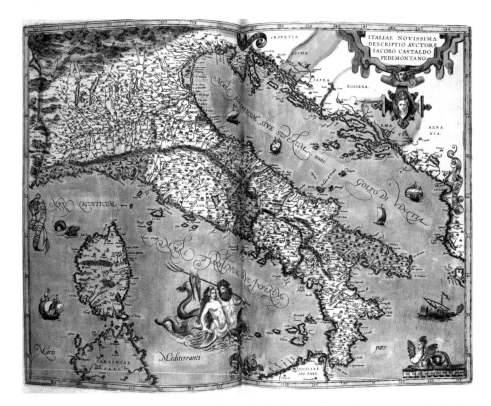

FIGURE 4.3 It is hard to see the province of Bari in Ortelius's atlas. The scale is too coarse for it to be clearly revealed. But in Arrieta's map of his military campaign against the plague the province and its capital are large, finely resolved, and show detailed theaters of operation.

self-consciously the ideas are formulated in these maps and how different they are from a map in which issues of plague and its progress are absent, one need only seek the map of Italy in Ortelius's *Theatrum* (fig. 4.3). The province of Bari is almost invisible, its capital a mere dot on the Napoli coast. In Arrieta's maps, the province becomes the whole, and the city central, in a detailed region of cities and town amid woods along a coastline that defines an administrative area bounded by neighboring jurisdictions to the south and west.

In his map Arrieta literally created the province he administered and in doing so created the field in which plague as a portable disease was to be restrained. In these maps Bari is the center of coastal Italy, the lynchpin that holds the area together and thus is the center of the fight against plague's diffusion. Arrieta's goal was first military, to design a campaign to secure the province against the invader, plague, which seemed somehow to be attacking with mortal results. Secondarily, he had to justify both the idea of his attack and the expense of those actions to his liege. Thus Arrieta's maps are at once territorial and official, establishing the political boundaries of the province over which he exercised military and civilian control as his sovereign's surrogate. To the west is Capitonata, to the south Balificata, and to the east is the province of Scala.

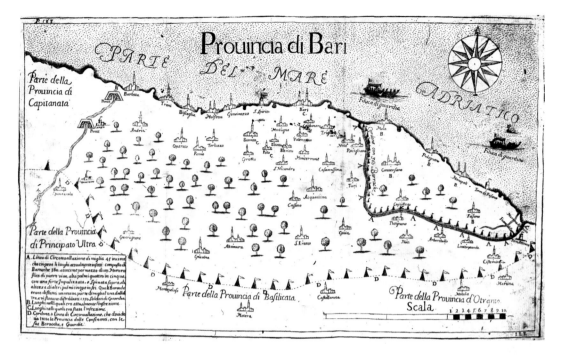

FIGURE 4.4 Arrieta employed troops in the province of Bari, Italy, in an attempt to halt the progress of plague in provincial towns during an outbreak in 1690–1692. This map describes a series of nested quarantine areas were created within a province-wide *cordon sanitare*.

Pervading both maps is a sense of objectivity in which authority and responsibility were not so much asserted as assumed. "Maps pass as descriptions of the territory when they project the sense of being unauthored or, if authored, then by a machine-like medium through which the territory passes" (Wood 2006, 8). Where landscape paintings are imbued with a sense of artistic ownership, these maps, with their repetitive symbols, seem authorless. Towns are symbolized either by buildings with steeples topped by a cross (churches), or in the case of towns with hospices, with a flag. Trees symbolize the countryside and at the western border of the province in both maps can be seen a river with two bridge crossings. Coastal patrol boats (*filuca di guardia*) at once assert Bari's coastal waters and the likelihood of plague as a portable disease brought in by maritime trading. The point is not the aesthetics of the landscape but the territory brought forth in the map: the province of Bari that Arrieta controlled.

In the first map (fig. 4.4), as the legend explains, a thick line lettered with an "A" separates the area of the province suspected where plague had been, and separately, those in danger of its outbreak, from the rest of the province.[3] An 80-mile-long military barrier is identified to create a *cordon sanitaire* supported by 350 troop barracks, each with a compliment of 50 soldiers. A shorter, second line 10 miles in length is an internal area, designated "B," where plague existed in specific towns. The tents along this length symbolize 50 troop encampments. Within these areas, the letter "C" identifies cities in which there had been infection, and the letter "D" shows where infection was active. Along the southern border of the province is a less well-guarded line that separates the province from its neighbor, Basilicata.

In the second map (fig. 4.5A), line "A" is the 45-mile-long border surrounding areas of actual infection, a cordon composed of barracks attached to each other by, the legend says, a 4- or 5-hands-high wall to which fortifications raising its height further (*impalizzata e spinata sopra*) had been added. The letter "B" identified areas in which actual infection continued, as opposed to areas marked with a "C," which identify towns where infection had been in the past. The provincial cordon line "D" divided the province at its borders with, again, a lighter compliment of barracks and their guards.

The complex of recurring signs in both maps argued a provincial, not a local or a regional, point of view. Arrieta did not himself investigate the cholera deaths in Molá, or any other town, any more than he personally counted the number of soldiers manning this or that encampment. That necessary work all flowed from the authority given Arrieta by his liege as administrator and military governor. Ecclesiastical and secular officials were charged with recording deaths occurring in their civil jurisdictions and reporting them through the existing bureaucracies—civilian and military—that reported in turn to Arrieta and his staff.

The resulting troop encampments are the "exercise of power" the postmodernists (Harley 1988) have suggested we must expect from maps. Certainly the map served "for guiding action in and across space, especially in the contexts of military activity" (Edney 2005, 91). Here the space mapped was neither simply military nor exclusively political but at once economic, medical, military, *and* political. The data inscribed in the map instantiated a theory of plague as a portable thing whose spread could be contained through travel and trade restraint. Arrieta initiated the quarantine on this basis and because he believed the safety of the citizens of Bari and neighboring provinces depended on its success. The political decision to contain the plague through a costly quarantine only made sense *if it was based on the understanding of plague as a unique and perhaps uniquely portable disease.* Otherwise there would be no need for quarantine at all.

Arrieta's quarantine program enacted a divide-and-conquer approach inherently military in its nature. In the map, sites of battles fought and won—where plague had been—are distinguished from those where fighting continued and casualties mounted. In addition, there were the towns where plague had yet to be reported but where its potential was feared.

At the level of the province the result of this thinking was a multitiered containment that modern military strategists (and public health experts) might admire. First, a cordon separates Bari, the general theater of operations, from its provincial neighbors. Second, protective barriers are raised along the seacoast to assure no new plague combatants can enter the province, no "reinforcements" to the disease already present. Third, areas of greatest plague activity are isolated. Fourth, within these areas, special garrison compliments are assigned to cities of active plague. Smaller gar-

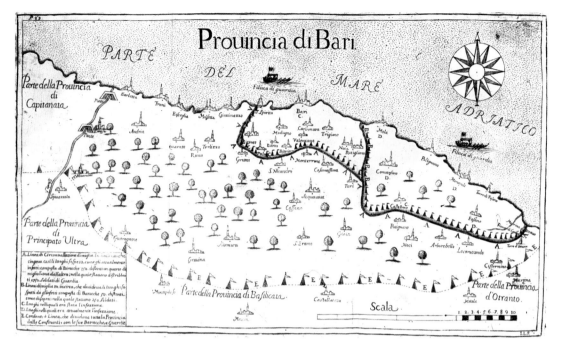

FIGURE 4.5A Fillipio Arrieta's second map makes clearer than the first the detailed containment strategy was designed to slow or halt the progress of plague within Bari.

risons are used to protect cities where plague had been active but seemed to have passed, and finally, other garrisons protected those cities at risk but without reported cases of plague.

The idea was similar to that enacted within individual city states like Genoa, although at a very different scale of action. In the city not only were persons with plague quarantined ("where the disease is"), family members of the afflicted were placed in a special section of plague houses for forty days, a period sufficient to either permit the plague to flourish (at which time they would be separated) or to guarantee they were free of the disease (Orent 2004, 160–62). Those surviving the quarantine (or the disease) then endured a period of "convalescence," still in seclusion. The provincial equivalent were the contained spaces in the map where plague had been but was no longer.

Plague is everywhere implied but nowhere seen in Arrieta's map. It is there, however, in the map's linguistic ("A," "B," "C," "D") and iconic (troop encampments) codes (flags, crosses, buildings, tents). The plague *had been* in Bari and Palo but then *moved to* the eastern towns of Eafano, Molá, Monovoli, and onward, where troops were set to impede its progress from entering Arborebelo, Turi, and Santo Spirito. Towns were not simple locations on a map—unchanging places—but elements on a timeline of plague's spatial progression and possible future. In this manner plague was argued as a temporal as well as spatial thing, one whose progress could be charted in time and space a manner that permitted predictions to be made.

FIGURE 4.5B Arrieta's containment strategy map, detail. Plague-free towns like Trigano are distinguished from others where plague had been (marked with a "C") and where plague continued to be active (marked with a "D").

Arrieta's plague was more a "where" than a "what." It was a portable killer spread through trade and travel, a temporal thing lodged in cities whose spatial progress could be seen in the map. Physicians named it by its distinctive symptoms, but the agent of the disease was unknown and really it was not Arrieta's concern. It was a "there" (in Molá, for example) to be understood by having been first here, and here, and here, on the map of Arrieta's area of administration. For Arrieta, plague was defined by its mobility and his response to it was the same as he might have mounted against any invader: his quarantine cut the enemy's supply lines and constricted the space of its operations.

The *cordon sanitaire* and its resulting restraint of trade created a dilemma for administrators like Arrieta. While it might slow or even stop the spread of the disease, quarantine created its own hardships. Interrupting the flow of goods and travelers interrupted as well the commerce that provided income for tradespeople and merchants. This in turn restricted tax revenues for the state that had to pay for the costs of the military operation. The map was at once a plan for the address of plague and in its description of plague's actuality a justification for the expensive program of containment enacted by Arrieta as an administrator.

There were other ideas of plague and its causes that avoided these burdens. If plague was a divine punishment for ungodly behavior then its answer lay not in quarantine but in prayer and appropriately pious behavior. In an attempt to impede plague's progress or mitigate the ferocity of an outbreak, in most jurisdictions (including Bari) prayers were said at local churches whose officials also extolled cleanliness and moderation for their spiritual as well as medicinal benefits. Some medical authorities argued plague was a creature of the very air, perhaps generated or at least accelerated in the malodorous as well as sinful slums from which it spread into other urban areas.

Were that the case quarantine would be not only useless but worse, prideful . . . an attempt to stop the wind. As a Royal Society physician, Dale Ingram argued in the mid-eighteenth century, "As the plague is propagated by hot and moist air . . . I would as soon imagine that the wall round Hyde Park would prevent the rooks from building in the oaks, or the small birds in the bushes for birds have wings and fly, so can Plague. The wings of the plague are the winds, and therefore it can soar far higher than the towering superstructures of man" (Ingram 1755, 130).

Choosing from among these different plagues was as much a civil and political necessity as it was a medical and humanitarian goal. Plague was financially costly: quarantine was expensive and epidemics deprived the state of revenues through its effect on taxable trade and subjects. To better understand this and other then-epidemic diseases would require a different type of knowledge at a different scale of concern from the one Arrieta employed in Bari. Studies of disease incidence would need to focus not only on regional occurrence but also on disease in the cities themselves. These were the expanding sites of both commerce and government whose populations were increasing and increasingly necessary to the economic welfare of the nation. Understanding epidemic disease would require theories not simply about this or that condition—leprosy, plague, influenza, and the like—but as importantly about the general nature of health and disease and how they might best be generally

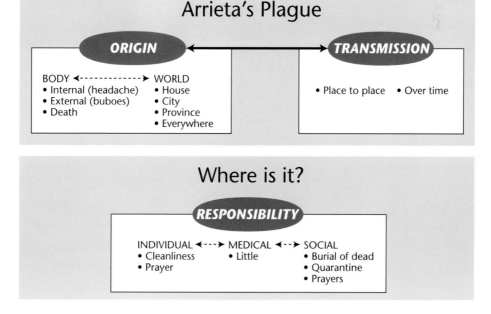

FIGURE 4.6 Arrieta's focus was on the official responsibility to impede plague spread both internally, between provincial towns, and by isolating the province from its neighbors and the commercial world.

investigated. Until that occurred, a *cordon sanitaire* was not simply the best Arrieta could do but the *only* thing to be done. In the interim, physicians might observe a disease's symptoms and speculate on its characteristics but no theory other than one invoking God's imperious will could wholly explain its presence.

DISEASE IN PLACE

Plague was not the only concern. It symbolized the general problem of disease, the host of traditional and emerging conditions active across Europe in the seventeenth and the early eighteenth centuries. There were also measles, scarlet fever, and small-pox, for example. The first smallpox pandemic occurred on the Continent in 1614 and then in England in 1628. Malaria and typhus and typhoid fever—the latter two were difficult for clinicians to distinguish—were also threats that marched through populated areas in Europe and its colonies. "English dysentery" (sometimes called "summer dysentery"), that we today assume was a type of bacterial food poisoning, was initially a sporadic and locally endemic condition which became epidemic in England around 1670 (Marks and Beatty 1976, 128–39). There were also pulmo-nary diseases like influenza and, of course, plague: at least twenty acute episodes in fifteenth-century England and more severe episodes through the seventeenth century (Marks and Beatty 1976, 139–40).

What all these health threats had in common, what made them a single subject, was a rapid clinical onset and a more or less ecumenical social pattern of attack. Rich and poor alike were affected in sufficient numbers to call for a communal response by civil and medical authorities. These were, to those affected by them, diseases of place.

A considered response first required the systematic categorization of diseases as recurrent classes of events for the purposes of diagnosis, reportage, and study. With-out that logical construction the study of disease as a general phenomenon could not proceed. Conditions with similar symptoms (including diarrhea, fever, jaundice, and cough) were categorized in a manner that permitted their differential diagnosis and hopefully an insight into their nature. This was the work of Dr. Thomas Sydenham (1624–1689), the "British Hippocrates" whose prolific writings defined what was to become a revitalized Hippocratic medicine based on observation and clinical experi-ence.[4]

Hippocrates was perhaps the first to consider, or at least write about, disease as a local phenomenon, one originating in a place, famously advising "whoever wishes to investigate medicine properly is first to consider the season in which a disease occurs and then the winds, the hot and the cold, especially such as are common to all countries, and such as a peculiar to each locality" (Howe 1963, 7). The itinerant Greek physicians who were his audience needed to impress potential patients with their knowledge of disease-in-place, their ability to diagnose and explain a patient's ill health in relation to local conditions (Rosen 1993, ix).

FIGURE 4.7 The "British Hippocrates," as he is sometimes called, Dr. Thomas Sydenham envisioned disease as a set of symptoms in much the same way Vesalius saw the body as a collection of related parts, each precisely described.

The colonists of Hippocrates' day needed to select the healthiest sites for the establishment of their colonies, and that demanded attention to the natural elements that seemed, to Hippocrates, to inhibit disease. Whatever its nature and however it might migrate, the Hippocratic cannon insisted that eventually this or that disease was to be understood as lodged in individuals living in local geographies of hamlets or towns.

This or that condition is defined symptomatically by the presence or absence of fever, of discolored stool or foul-smelling urine. It is seen as a bump on the skin or a pain in the gut. What distinguishes it from another condition with a similar but not identical set of symptoms often depends on the theory of medicine employed. Each individual condition sits within a set of disease categories that are organized by how we understand disease generally. These disease categories are inherently socially grounded, an outgrowth of the scientific, social, and cosmologic thought systems of the society in which they are applied.

In his 1676 *Observationes medicae* and the 1680 *Epistola responsoria* Sydenham created a taxonomy of individual diseases, defined symptomatically, that in the Hippocratic tradition was both locationally grounded and temporally precise. He described a five-year epidemic cycle observed over fourteen years of progressive epidemics occurring from 1661 to 1675. "Intermittent fevers," including malaria and typhus, preceded "pestilential fevers," like the plague. These were followed by small-pox, the "dysenteric constitution" of summer diarrhea, and, at the end of the cycle, a class of carefully described respiratory diseases, including influenza. The diseases Sydenham differentiated were assumed to be native to specific landscapes characterized by specific airs, geologies, and soil types. Individual habits and dispositions were agents that activated in the patient conditions that were native, and thus natural, in their origin. After Sydenham, most experts agreed that while some diseases—smallpox and scarlet fever, for example—were caused by morbid actions within a body, the majority, especially febrile diseases, were generated in and spread by "bad air," a "mal-aria" rising from odiferous animal and human wastes that were not properly disposed of (Valenčius 2002, 393–94).

Sydenham evoked a series of epistemic things in his writings. First, he created a system of classification that permitted individual conditions to be differentiated symptomatically and named, a "list of specific criteria used to decide whether or not a person has the disease of concern" (Jekel, Elmore, and Katz 1996, 45). These were grouped on the basis of their symptoms. This provided a common language for diagnosis. Diarrhea or constipation, fever, jaundice, muscle aches: the symptoms common to many diseases were ordered in a fashion that permitted a definitive, differential diagnosis of specific complaints. Plague was transformed from a portable thing without any real clinical referent into a virulent member of the "pestilential fever" category.

Second, Sydenham ordered the diseases he thus evoked as recurrent, cyclical, and typically climatic phenomena within a theory of miasmatic disease generation. Respiratory diseases came in the winter and diarrheic diseases in the summer. His record of patient care convinced him further that disease occurred over years in a predictable fashion. There was thus a natural progression of illnesses occurring over time within the annual change of seasons. Third, these patterns were to be understood in a Hippocratic fashion. Disease born in the general constitution of the environment caused imbalances in the patient's constitution. Careful attention by a person to lifestyle (cleanliness, diet, hygiene, and so forth) thus might offer at least some natural protection. The result was medicine as a natural science whose central methodology was one of close observation whose results were understood through logical organization. Disease was a thing to be understood symptomatically, chronologically (year by year, season by season) and on the basis of location.

To explain their effect on the individual, Sydenham exploited the classical theory of physical humours, whose imbalance caused disease just as health was the result of a balanced constitution. Historian of science Steven Shapin has summarized this theory succinctly: "English physician Thomas Sydenham had introduced the notion of an 'epidemic constitution of the atmosphere.' Something had contaminated the local air (possibly, he thought, noxious effluvia from 'the bowels of the earth') in a way that unbalanced the humours. The occasional appearance of these effluvia accounted for the intermittent character of epidemic disease" (Shapin 2006).

Fevers were divided into three major groups or classes, recurring across the fourteen-year period of observation, in which Sydenham recognized five periods of epidemic occurrence, "each of them being characterized by a particular *epidem-*

ic constitution or disposition of the atmosphere. It was the nature of the epidemic constitution that caused outbreaks of certain fevers" (Marks and Beatty 1976, 127; original emphasis). The seasonality of some conditions—influenza in the winter and dysentery in the summer—corresponded to the cold and hot airs of the particular season. These were amplified in some locations by "external or environmental factors such as proximity to cemeteries or sewage" where noxious, obviously foul smelling miasmas lurked (Choi 2003, 67). It was these factors that bred the "malaria," the bad airs in which this or that disease was first generated and then transmitted (Mitman and Numbers 2003, 393–94).

The result was a system that envisioned disease as a collection of symptoms that could be individually defined and that together became the diagnosis. Just as Vesalius had seen the body as a precisely coordinated series of interrelated parts operating in a specific landscape, so too did Sydenham's system envision diseases. The resulting system of classification and disease differentiation gave medical authorities a way to both diagnose more accurately and think more clearly about specific conditions and their individual (and shared) constituents. The Hippocratic underpinning gave to Sydenham's medicine a lineage that helped assure its acceptance. As William Hillary put in an eighteenth-century text on diseases: "That Father and Prince of Physicians, Hippocrates, advises all physicians to examine and duly consider the Situation, Air, and the Water, used by the People of such Cities, or Places, as they are called to, or may practice in" (Hillary 1759, 1).[5] If Hippocrates was the father then Sydenham was certainly the son or grandson.

Against this background disease was, first and foremost, a natural phenomenon arising from local airs and soils, a thing rooted in world where people lived, that land on which Vesalius posed his skeletal bodies. If disease was natural, it was also, however, anthropogenic. Those who lived lives of moderation and maintained standards of cleanliness might gain some protection through the proper cultivation of their constitution. At least some disease burden was therefore a penalty for the manner in which some lives were inappropriately lived. But that did not explain how or why diseases like plague and influenza affected everyone, bishops and poor parishioners alike. Nor did it account for the penchant of diseases like plague to appear as resulting from trade and travel. Whatever disease was, human agency played a role. Disease was somehow natural and unnatural, explicable and mysterious. Understanding them would require a different focus at the scale of urban infection.

PLAGUE: THE URBAN DISEASE

In 1593 British physician Simon Kellwaye blamed the ferocity of a plague outbreak on the odor of putrid things from "some stinking doonghills, filthe, and standing pools of water and unsavory smells" accumulating in city streets (Palmer 1993, 67). Similarly, the sixteenth-century surgeon Ambroise Paré located the origins of disease in those smells "emanating from dead animals, sewage, and the like would infect the patient and his parts with their putrid vapors." Another source of odiferous attack occurred at the scale of the sick themselves. If plague was generated in the foulness of the city, wrote Daniel Defoe in his famous, semifictional plague memoir, the "calamity was spread by . . . some certain steams or fumes called Effluvia, by the Breath, or by the Sweat, or by the stench of Sores of the sick Persons" (Defoe 1969, 197).

This plague was literally *in* the air of the city, an urban phenomenon not imported but generated in the smell of human and animal waste. "It was generally agreed through experience that filthy, stinking, and overcrowded environments were particularly attractive to the infection and that plague was more prevalent among the dirty poor" (Healy 2003, 24). Here was the Hippocratic disease rooted not in the natural place of rocks and stones and winds but the smells of people in the unnatural, manmade city. This plague had a demographic profile, its intensity defined by deaths, parish by parish, across the affected city. Its reality might best be understood, therefore, by the varying intensity of disease in different parts of the city. If plague was more violent here and less intense there, what distinguished these two parishes in a manner that could explain the variable incidence of disease?

Plague's pattern of incidence appeared to support the thesis of an odiferous disease whose origins lay in the poor areas of the city. As British historian Paul Slack writes, "The parishes least affected were in the center of the city. The worst affected were on its fringes" (Slack 1985, 156–57). The richest parishes typically were located near the center of a town; the poorest lived on the margins. During epidemic outbreaks those with money and transportation typically fled the city for country homes or residences (Tomalin 2002, 167); those without other residences were stuck in place in neighborhoods of densely settled, inferior housing where an even minimal sanitary infrastructure was unknown. That said, it has been a constant truism that epidemic "outbreaks are typically blamed on the poor and their habits even if they are the ones who suffer more severely than other economic groups" (Slack 1992, 3).

The pattern was observed again and again across the recurrent epidemics of the seventeenth century not only in London but also in towns like Bristol, Exeter, and Norwich. Evidence of plague as a disease of the odiferous (and immoderate) urban poor could be read in the parish burial records that located plague and other diseases at the level of parish streets and homes. The figures are at best rough estimates; only the births and deaths of parish members were reported, not those of resident unbelievers, Jews, and Roman Catholics (Slack 1985).[6] Still, the numbers trace the evidence on which the theories of disease were formulated. In Norwich in 1664–1665 plague mortality as a percentage of estimated parish population was 25.2 percent of total parish population in the eleven poorest parishes, 8.2 percent in the eleven richest parishes, and 12.6 percent in those in between (Slack 1985, 138). The three poorest parishes located along the River Wensum in the southeast of the district lost between 28.6 and 45 percent of their population. The inverse relation between income and plague mortality made perfect sense. *If* plague is generated by foul odors emanating from the homes of the poor, *then* the poor will be more likely to be diseased than the wealthy. The proposition gave to plague—and by extension other diseases—a geographic explanation for its pattern of occurrence within a miasmatic theory of disease.

By the late seventeenth century there were at least six different things called "plague," each argued differently and each demanding a different response. First there was a congress of physical symptoms that physicians named as plague. That thing attacked the constitution of those afflicted. Second, there was a plague of the sick house, a disease of domiciles that would have to be sanitized before they could be again inhabited. Third, plague was a recurrent disease threat with a climatological profile, one occurring not in the winter but principally in the warmer seasons. Fourth, plague was a local disease of the noxious airs that accumulated in the expanding city. Where the disease was worst was where the smell of offal and waste and illness were strongest and this, typically, was in the poorer sections of town. Cleanliness and better sanitation might decrease the airs and thus serve as a defense. Fifth, plague was a creature of the greater air that leapt the barriers placed by officials to restrain its progress.

Finally, plague was a traveler that moved from city to city and nation to nation in the holds of cargo ships and the possessions of travelers. It might seem that this plague was not plague at all but a process, a means of its transmission that was agnostic on the nature of the disease agent and its locus of generation. For Arrieta, however, whose interest was in stopping the disease's spread, the distinction would have been irrelevant. And for clinicians of the day, who believed the disease was born in foul airs (on ship, in the cities, wherever), the distinction if any spoke to the immediate source of an outbreak and not its ultimate point of origin.

To find a way to combat plague (whatever it actually was), and by inference other epidemic diseases, meant choosing between these different plague things. To the extent it was a portable condition that spread with trade and travelers it would be necessary to exclude ships and travelers coming from areas where plague was active. If plague originated in the foul airs of poor neighborhoods then officials would be obliged, for their own safety and that of their subjects, to force the cleanup of those neighborhoods. But if it was in fact simply an effect of the changing seasons, an inevitable result of the warm airs of late spring and summer, then nothing could be done. And whatever its origin, if the local homes of the patients were disease reservoirs any answer to its occurrence would be home quarantine, an attempt to slow contagion from the person's body. In some places and some outbreaks, patients were removed to a communal estate where the likelihood of general contagion it could be diminished.[7]

DEMOGRAPHIC PLAGUE

To critically consider these various plagues the disease had to be transformed from a clinical diagnosis into a demographic thing whose effect could be counted, year by year, in accurate records of the city, region, and nation. The seventeenth century was the beginning of population health as a demographic study, a method of disease consideration that relied at first on tabular statistics and secondly on graphic arguments. We see this in the work of William Petty (1623–1687). A physician and economist interested in the health of populations as well as patients, Petty coined the phrase "political arithmetic" to describe "the art of reasoning by figures upon things related to government," and especially population health as a critical barometer of the state (Rosen 1993, 87–88).

Petty's friend, John Graunt (1620–1674), advanced the idea by demonstrating a pattern of regularity in London deaths based on those reported on an annual basis over twenty years by parish officials. His goal was to uncover a "physico-mathematical relationship" between the periodicity of disease, seen by Sydenham, and elements of the local environment.[8] The medium of his concern was *Bills of Mortality*, compiled by officials to warn of threatening outbreaks and thus to permit the promulgation of measures to check an epidemic's progress (Porter 1999, 15). The earliest London bill was compiled in 1519; while the full bill reported all causes of death reported by parish officials, the focus of official concern was plague.

From our perspective the data was at best incomplete. During London's plagues, including that of 1665, causes of death were not certified medically but by the "searchers of the dead," typically "impoverished, ignorant old women who trudged from house to house and examined corpses to certify the cause of death" (Orent 2004, 150). And because the bills were based on parish burial records they did not include the deaths of Jews and others outside the official faith. Still, the results were sufficient to provide a medium in which the state of an epidemic could be understood. "In the terrible plague year 1603 the Bills were studied by officials—civil and medical—to see whether the disease incidence was increasing or had passed its peak" (Hacking 2006, 102).[9] During the plague year of 1665, Samuel Pepys, the great diarist and secretary of the admiralty, noted in his diary the rise and fall of mortality reported in the bills. If the idea of an "epidemic profile,"—the temporal face of an epidemic—had yet to be graphed its general form apparently was already understood in the early to mid-seventeenth century.

Plague was but one possible mortality condition, if, perhaps, the most terrifying of them all. Graunt's *Table of Casualties* listed seventy causes of death reported in the official *Bills of Mortality*. He transposed its figures into an evidentiary set that could be manipulated mathematically to produce, for example, the likelihood of an individual of a certain age dying over the next ten years. Population mortality and its specific risks for specific groups was born.[10] Just as Sydenham built his classificatory system

64

THE TABLE OF CASUALTIES.

The Years of our Lord	1647	1648	1649	1650	1651	1652	1653	1654	1655	1656	1657	1658	1659	1660	1629	1630	1631	1632	1633	1634	1635	1636	1629 1630 1631 1632	1633 1634 1635 1636	1647 1648 1649 1650	1651 1652 1653 1654	1655 1656 1657 1658	1629 1649 1659	In 20 Years.
Abortive, and ftilborn	335	329	327	351	389	381	384	433	483	419	463	467	421	544	499	439	410	445	500	475	507	523	1793	2005	1342	1587	1832	1247	8559
Aged	916	835	889	696	780	834	864	974	743	892	869	1176	909	1095	579	712	661	671	704	623	794	714	2475	2814	3336	3452	3680	2377	15757
Ague, and Fever	1260	884	751	970	1038	1212	1282	1371	689	875	999	1800	2303	2148	956	1091	1115	1108	953	1279	1622	2360	4418	6235	3865	4903	4363	4010	23784
Apoplex, and fodainly	68	74	64	74	106	111	118	86	92	102	113	138	91	67	22	36		17	24	35	26		75	85	280	421	445	177	1306
Bleach			1	3	7	2				1															4	9	1	1	15
Blafted	4	1			6	6			4		5	5	3	8	13	8	10	13	6	4		4	54	14	5	12	14	16	99
Bleeding	3	2	5	1	3	4	3	2	7	3	5	4	7	2	5	2	5	4	4	3			16	7	11	12	19	17	65
Bloody Flux, Scouring, and Flux	155	176	802	289	833	762	200	386	168	368	362	233	346	251	449	438	352	348	278	512	346	330	1587	1466	1422	2181	1161	1597	7818
Burnt, and Scalded	3	6	10	5	11	8	5	7	10	5	7	4	6	6	3	10	7	5	1	3	12	3	25	19	24	31	26	19	125
Calenture	1			1		2	1	1			3										1	3		4	2	4	3		13
Cancer, Gangrene, and Fiftula	26	29	31	19	31	53	36	37	73	31	24	35	63	52	20	14	23	28	27	30	24	30	85	112	105	157	150	114	609
Wolf				8																				8					8
Canker, Sore-mouth, and Thrufh	66	28	54	42	68	51	53	72	44	81	19	27	73	68	6	4	4	1			5	74	15	79	190	244	161	133	689
Childbed	161	106	114	117	206	213	158	192	177	201	236	225	226	194	150	157	112	171	132	143	163	230	590	668	498	769	839	490	3364
Chrifomes, and Infants	1369	1254	1065	990	1237	1280	1050	1343	1089	1393	1162	1144	858	1123	2596	2378	2035	2268	2130	2315	2113	1895	9277	8453	4678	4910	4788	4519	32106
Colick, and Wind	103	71	85	82	76	102	80	101	85	120	113	179	116	167	48	57					37	50	105	87	341	359	497	247	1389
Cold, and Cough							41	36	21	58	30	31	33	24	10	58	51	55	45	54	50	57	174	207	00	77	140	43	598
Confumption, and Cough	2423	2200	2388	1988	2350	2410	2286	2868	2606	3184	2757	3610	2982	3414	1827	1910	1713	1797	1754	1955	2080	2477	5157	8266	8999	9914	12157	7197	44487
Convulfion	684	491	530	493	569	653	606	828	702	1027	807	841	742	1031	52	87	18	241	221	386	418	709	498	1734	2198	2656	3377	1324	9073
Cramp			1														1	0	0	0	0	0	01	00	01	0	0	1	2
Cut of the Stone		2	1	3		1	1	2	4	1	3	5	46	48				5	1	5	2	2	5	10	6	4	13	47	38
Dropfy, and Tympany	185	434	421	508	444	556	617	704	660	706	631	931	646	872	235	252	279	280	266	250	329	389	048	1734	1538	2321	2982	1302	9623
Drowned	47	40	30	27	49	50	53	30	43	45	63	60	57	48	43	33	29	34	37	32	32	45	139	147	144	182	215	130	827
Exceffive drinking			2																						2			2	2
Executed	8	17	29	43	24	12	19	21	19	22	20	18	7	18	19	13	12	18	13	13	13	13	62	52	97	76	79	55	384
Fainted in a Bath					1																					1			1
Falling-Sicknefs	3	2	2	3		3	4	1	4	3	1		4	5	3	10	7	7	2	5	6	8	27	21	10	8	8	9	74
Flox, and fmall pox	139	400	1190	184	525	1279	139	812	1294	823	835	409	1523	354	72	40	58	531	72	1354	293	127	701	1846	1913	2755	3361	2785	10576
Found dead in the Streets	6	6	9	8	7	9	14	4	3	4	9	11	2	6	18	33	26	6	13	8	24	24	83	69	29	34	27	29	243
French-Pox	18	29	15	18	21	20	20	20	29	23	25	53	51	31	17	12	12	12	7	17	12	22	53	48	80	81	130	83	392
Frighted	4	4	1		3		2		1	1				9	1			1				3	2	3	9	5	2	2	21
Gout	9	5	12	9	7	7	5	6	8	7	8	13	14	2	2	5	3	4	4	5	7	8	14	24	35	25	36	28	134
Grief	12	13	16	7	17	14	11	17	10	13	10	12	13	4	18	20	22	11	14	17	5	20	71	56	48	59	45	47	279
Hanged, and made-away themfelves	11	10	13	14	9	14	15	9	14	16	24	18	11	36	8	8	6	15		3	8	7	37	18	48	47	72	32	222
Head-Ach		1	11	2		2	6	6	5	3	4	5	35	26							4	2	0	6	14	14	17	46	051
Jaundice	57	35	39	49	41	43	57	71	61	41	46	77	102	76	47	59	35	43	35	45	54	63	184	197	180	212	225	188	998
Jaw-faln	1	1			3			2	2		3	1			10	16	13	8	10	10	4	11	47	35	02	5	6	10	95
Impoftume	75	61	65	59	80	105	79	90	92	122	80	134	105	96	58	76	73	74	50	62	73	130	282	315	260	354	428	228	1639
Itch		1																	10				00	10	01				11
Killed by feveral Accidents	27	57	39	94	47	45	57	58	52	43	52	47	55	47	54	55	47	46	49	41	51	60	202	201	217	207	194	148	1021
King's Evil	27	26	22	19	22	20	26	26	27	24	23	28	28	54	16	25	18	38	35	20	26	69	97	150	94	94	102	66	537
Lethargy	3	4	2	4	4	4	3	10	9	4	6	2	6	4	1		2	2	3		2	2	5	7	13	21	21	9	67
Leprofy			1									1		2	2						2		2	2	1		1	3	06
Livergrown, Spleen, and Rickets	53	46	56	59	65	72	67	65	52	50	38	51	8	15	94	112	99	87	82	77	98	99	392	356	213	269	191	158	1421
Lunatique	12	18	6	11	7	11	9	12	6	7	13	5	14	14	6	11	6	5	4	2	2	5	28	13	47	39	31	26	158
Meagrom	12	13		5	8	6	[illegible]	[illegible]	[illegible]	[illegible]	[illegible]	[illegible]	[illegible]	[illegible]			[illegible]					22	24	22	30	34	22	05	132
Meafles	5	92	3	33	33	62																							
Mother	2					1																							
Murdered	3	2	7	5	4	3																							
Overlayd, and ftarved at Nurfe	25	22	36	28	28	29																							
Palfy	27	21	19	20	23	20																							
Plague	3597	611	67	15	23	16																							
Plague in the Guts				1		110																							
Pleurify	30	26	13	20	23	19																							
Poyfoned		3		7																									
Purples, and fpotted Fever	145	47	43	65	54	60																							
Quinfy, and Sore-throat	14	11	12	17	24	20																							
Rickets	150	224	216	190	260	329																							
Mother, rifing of the Lights	150	92	115	120	134	138																							
Rupture	16	7	7	6	7	16																							
Scal'd-head	2				1																								
Scurvy	32	20	21	21	29	43																							
Smothered, and ftifled			2																										
Sores, Ulcers, broken and bruifed (Limbs	15	17	17	16	26	32																							
Shot																													
Spleen	12	17																											
Shingles																													
Starved		4	8	7	1	2																							
Stitch				1																									
Stone, and Strangury	45	42	29	28	50	41																							
Sciatica																													
Stopping of the Stomach	29	29	30	33	55	67																							
Surfet	217	137	136	123	104	177																							
Swine-Pox	4	4	3																										
Teeth, and Worms	767	597	540	598	709	905																							
Tiffick	62	47																											
Thrufh																													
Vomiting	1	6	3	7	4	6																							
Worms	147	107	105	65	85	86																							
Wen	1		1		2	2																							
Sodainly																													

Abortive, and ftilborn	335	329	327	351	389	381	384	433	483	419	463	467
Aged	916	835	889	696	780	834	864	974	743	892	869	1176
Ague, and Fever	1260	884	751	970	1038	1212	1282	1371	689	875	999	1800
Apoplex, and fodainly	68	74	64	74	106	111	118	86	92	102	113	138
Bleach			1	3	7	2				1		
Blafted	4	1			6	6			4		5	5
Bleeding	3	2	5	1	3	4	3	2	7	3	5	4
Bloody Flux, Scouring, and Flux	155	176	802	289	833	762	200	386	168	368	362	233
Burnt, and Scalded	3	6	10	5	11	8	5	7	10	5	7	4
Calenture	1			1		2	1	1			3	
Cancer, Gangrene, and Fiftula	26	29	31	19	31	53	36	37	73	31	24	35
Wolf				8								
Canker, Sore-mouth, and Thrufh	66	28	54	42	68	51	53	72	44	81	19	27
Childbed	161	106	114	117	206	213	158	192	177	201	236	225
Chrifomes, and Infants	1369	1254	1065	990	1237	1280	1050	1343	1089	1393	1162	1144
Colick, and Wind	103	71	85	82	76	102	80	101	85	120	113	179
Cold, and Cough							41	36	21	58	30	31
Confumption, and Cough	2423	2200	2388	1988	2350	2410	2286	2868	2606	3184	2757	3610
Convulfion	684	491	530	493	569	653	606	828	702	1027	807	841
Cramp			1									
Cut of the Stone		2	1	3		1	1	2	4	1	3	5
Dropfy, and Tympany	185	434	421	508	444	556	617	704	660	706	631	931
Drowned	47	40	30	27	49	50	53	30	43	45	63	60
Exceffive drinking			2									
Executed	8	17	29	43	24	12	19	21	19	22	20	18

FIGURE 4.8 This is a recreation of the 1667 Bill of Mortality used in early quantification of deaths over a twenty-year period.

upon observation over time, Graunt and Petty used mortality counts to built their image of disease as a recurring event that might be quantified and then differentiated as a demographic exercise.

The work of men like Petty and Graunt led to Christian Huygens's (1629–1695) early tables of life expectancy, and in the eighteenth century, the population-based analysis of specific diseases like smallpox by, for example, Daniel Bernoulli (1700–

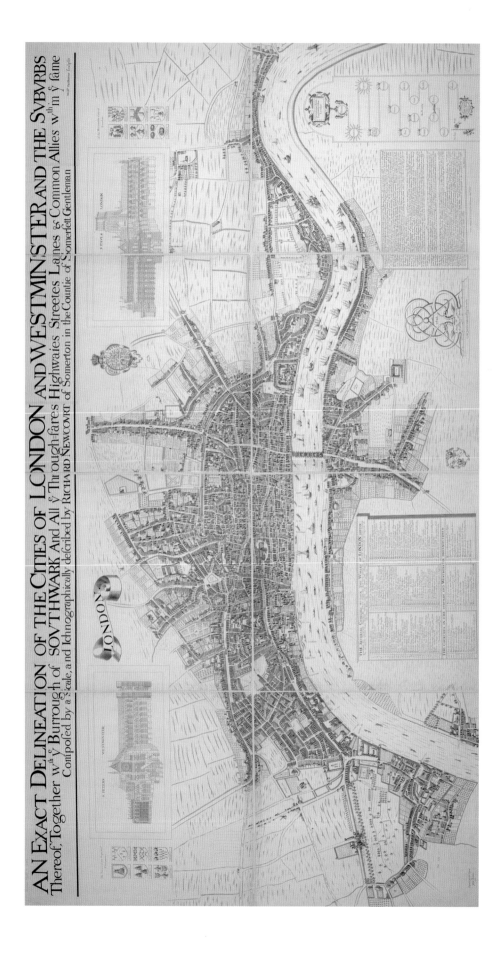

AN EXACT DELINEATION OF THE CITIES OF LONDON AND WESTMINSTER AND THE SVBVRBS Thereof. Together wᵗʰ ȳ Burrough of SOVTHWARK And All ȳ Through-faŕes Highwaies Streetes Laines & Common Allies wᵗʰⁱⁿ ȳ ſame Comſpoſed by a Scale, a̅d Ichnographically deſcribed by RICHARD NEWCOVRT of Somerton in the Countie of Somerſett Gentleman

1782).[11] The end would be to quantify as a national concern the presence of specific conditions whose symptoms had been categorized by medical experts like Sydenham but whose existence became, through quantification, the responsibility of not only the physician but also of the state itself.

All this brought forth a new idea, that of the quantifiable disease grounded in "its specific phenomena and is peculiar variables: birth and death rates, life expectancy, fertility, state of health, frequency of illness, patterns of diet and habitation" (Foucault 1978, 25–26).[12] It was at once too much and not enough. There was too much data to be easily understood, too much to be held in memory. For the extent of disease to be calculated and applied in urban areas Graunt required a means of calculating the size of the city, of the areas within which people lived and disease sometimes flourished.

For this he used Faithorne and Newcourt's *Exact Delineation of London*, "The Map of London set out in the year 1658 by Richard Newcourt, drawn to a scale of yards" (Marsh 2007, 7).[13] The map presented an enabling methodology (fig. 4.9A) in which precise areas were carved out of the city in a manner permitting area to be measured precisely and thus disease incidence to be seen locally. The map's accuracy, and thus its general utility, was proudly asserted in its title, an "exact delineation of the cities of London and Westminster." The scale bar and compass at the bottom left of the map underscored the idea of the map's precision and correctness.

The city imagined in the map remains, as it was in Braun and Hogenberg's day, one dominated by the Thames, the city's essential thoroughfare. But this London is one whose *raison d'être* is business and trade. At the top of the map on both left and right sides are the coats of arms of the principal trades that worked in the city (cloth workers, ironworkers, tanners, and so forth). They are emblematic of the daily businesses that were transforming seventeenth-century London from a royal capital into a mercantile power, from a city of religion and royalty to a city of commerce. Ecclesiastic and bureaucratic London are similarly called forth at the top of the map with, on the left, a beautifully rendered drawing of St. Peter's and Westminster Abbey, and beside it the coat of arms of the city itself. To the right is St. Paul's Cathedral.

These were, as writes Antoine Picon in his discussion of Paris maps, elements of a portrait tradition that was "linked to the ambition to depict not only the layout of the city but also its appearance and its spirit, as expressed through its main monuments" (Picon 2003, 136). Here that spirit resides in the coats of arms advertising the many trades London supported, the commerce that was the city's lifeblood. The images of Westminster Abbey and St. Paul's Cathedral asserted the greater theological underpinning of the city and as England's capital. The streets and lanes inscribed in the maps were transformed into the locus of the trades carried out in the city. The streets exist not simply in the city, however, but in individual parishes, the bureaucratic division of the city symbolized by the churches, each church with a higher roof, a distinct outline, and a number that in the legend would give the parish its name.

FIGURE 4.9A (*opposite page*) The Faithorne-Newcourt map of London (1658) created the spatial context within which both mortality statistics were to be collected and understood.

If the coat of arms trumpeted the commercial city, divided into parishes but ready for business, the self-importance of London is underscored in the text at the bottom right of the map and the genealogical table beside it. The text asserts London as a city not of mercantilism but of antiquity grounded in the "third period of the world" with the progenitor Brutus, grandchild of Aneas of Trojan. Beside the text, the chart links Brutus quite literally to the stars, a history whose provenance is if not divine at least celestial. This to use fanciful mythology needs the chart to emphasize the accuracy of the history as part of the general care of measurement with which the city, historical and modern, was constructed.

The Faithorne-Newcourt map and Graunt's calculations based on the official bills of mortality were perfect compliments. It is not simply the accuracy of the map's area calculations that permitted Graunt's work but, more importantly, the map's content. The city is conceived as a population center divided into parishes numbered in the map and named in its legend. It was in these parishes that disease was reported and that data collected to create a calculated argument about disease in the city, calculations that in part rested upon the spatial analytic of the mapmakers. The result was a city within which one might think seriously about disease and its relative location, about the conditions that might promote its spread or restrain its progress.

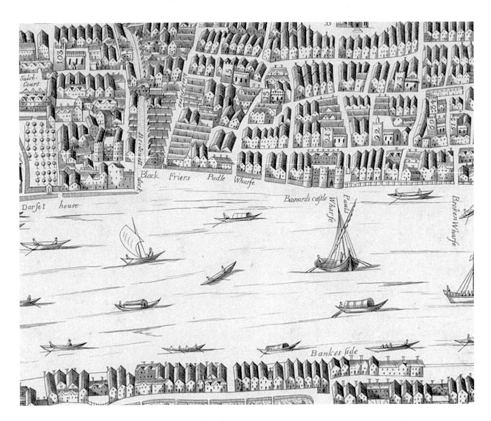

FIGURE 4.9B This detail of the Faithorne-Newcourt map shows watercraft, numbered churches, and the wharfs that were critical way stations on the Thames.

GRAPHS AND TABLES

The attempt to forge connections between mortality data and spatially grounded health studies within a population also demanded another thing, the graph of population data that in the nineteenth century would be a common addition to maps of specific diseases (White and Hardy 1970). These distilled the rows of mortality tables into comprehensible arguments (for example, on life expectancy) and would permit sophisticated arguments about the nature of disease occurrence. There was Christian Huygens's *ligne de vie* (the line of life) that plotted the expected number of survivors as a function of age. Included in a letter written in 1669, Huygens's graph used data from Graunt's 1662 mortality tables to craft a continuous curve of mortality, and thus life expectancy, the number of persons surviving to age "x" based upon a hypothetical initial cohort of one hundred.[14] This brought to population and disease studies a type of graphic statement that others were then applying to physical studies, for example, Edmond Halley's 1686 graph of the relationship between altitude and pressure (Spence 2006).

Both the graphs and the tables were part of the processes of spatialization and visualization that began the previous century with Vesalius and Ortelius; both were a means of seeing the aggregate when the mass of particular data was too large to be easily grasped. Halley's graph of altitude and pressure was the description of a physical relationship without any temporality, one that was natural and constant. Huygens's analysis argued populations within Sydenham's progressive time frame in which years and decades marched one after another, from individual birth to timely or untimely death. The methodology would assert, as it developed, population as a thing divisible into specific, geographically located age groups in a manner defining different levels of disease risk. Disease would be a set of physical symptoms not apart from but embedded within demographic and geographic environments. Put another way, disease would become a thing of tables, charts, graphs, and maps at once.

HOBBES, BOYLE, AND SCIENCE

One final component needs commentary here. In the second half of the seventeenth century, Robert Boyle emerged as the major proponent of a method of systematic experimentation and its importance in natural philosophy. He was not a "scientist"—the idea as we understand it had yet to be advanced—but a philosopher who proposed a method of experimental knowing in which arguments based on a kind of replicable or evidence was to be judged by the agreement of one's peers. "Boyle relied on a parajuridicial metaphor: credible, trustworthy, well-to-do witnesses gathered at the scene of the action can attest to the existence of a fact, the matter of fact, even if they do not know its true nature. So he invested the empirical style that we still use today" (Latour 1993, 19).

In this Boyle was opposed by another natural philosopher, Thomas Hobbes, who rejected Boyle's idea of the laboratory as an arena of choosing between contested truths through the judgment of a knowledgeable jury. Hobbes believed Boyles's experiments served nothing, that real understanding was not constructed in a vacuum tube or other apparatus, but rather, came from close observation and a careful logic that could discern the real. Their dispute over what would later be thought of as science and its construction of truth as a socially agreed upon and scientific thing is a turning point in the formation of what would today we think of as science (Shapin 1994). "Boyle's air-pump experiments have a canonical character in science texts, in science pedagogy, and in the academic discipline of the history of science. . . . It is an oft-told tale and, in the main, a well-told tale" (Shapin and Schaffer 1985, 3–4).

Steven Shapin and Simon Schaffer (1985) retold this tale in their *Leviathan and the Air-Pump* in a way that has become canonical, a new basis for our understanding of the birth of what would later be the scientific perspective and its reliance on the experimental method. As Bruno Latour put it: "Before Shapin and Schaffer, other historians of science had studied scientific practice; other historians had studied the religious, political and cultural context of science. No one, before Shapin and Schaffer, had been capable of doing both at once" (Latour 1993, 20). They did this, he continued, "in a quasi ethnographic way what philosophers of science now do scarcely at all: they show the realistic foundations of the sciences. But rather than speaking of the external reality 'out there' they anchor the indisputable reality of science 'down there,' on the [laboratory] bench" (Latour 1993, 21).

What was a relatively simple narrative about the triumph of the new experimental scientific espoused by Boyle over the old natural philosophy of Hobbes became, in the hand of Shapin and Schaffer (and others like Latour), a complex allocation of things scientific and political. In the crucible of this dispute was created the notion of observable "scientific fact" as an outcome of replicable testing accepted by knowledgeable peers. As Latour interprets Shapin and Schaffer's *Leviathan and the Air-Pump*, however, so too the notions of power, political interest, and the general polity were creations lodged in Hobbes's critique (Latour 1993, 26). They were opposite sides of the same coin, both necessary for the science that would result, both revolutionary in their way.

None of this is about disease as such. To the extent that disease studies would become an experimental science, however, the politics and science of Hobbes and Boyle would together define the systems of knowledge production that would result in the often-contested sciences that became medicine in the eighteenth, nineteenth, and twentieth centuries. We saw this general idea presaged in Vesalius's challenge to physicians to test the anatomy of the ancients against the reality of the corpse they dissected and then examined. We see it awakening in the London map's promise of the precisely delineated city. The dividers measure divisions on the rule that promises a scale of three feet to the yard and 1760 yards to the British mile. It was

FIGURE 4.9C The scale bar and dividers of the Faithorne-Newcourt map argue an accurate, precise knowing of natural phenomena symbolized the wings and horns entwined in the dividers and the mapmakers curving line.

thus that the parts of the city might be drawn in precise relation, one to the other, just as Vesalius had drawn the body parts in their correct relation, one to another. And conceptually, this was not distant from Sydenham's belief that exact observation of symptoms might distinguish them, one from another to create a reasonable taxonomy of disease itself.

The map was a principal workbench in which the practice of disease studies continued to evolve, a space in which disease incidence among populations could be collected and considered. It was in the map, and the tables of data used in its construction, that ideas about specific diseases were to be tested as part of an evolving disease science. The outcome would be a formulation in which politics and science were indivisibly joined. A consequence would be the creation of civil boards of health and boards of inquiry into specific disease that would seek in the map, and the tables containing mappable data, answers to disease incidence and occurrence.

None of this happened quickly or in a simple, linear fashion. One sees its beginning in the sixteenth-century anatomies and atlases, in the expanding cities of seventeenth-century mercantilism and concern for the diseases urban expansion appeared to promote. The promise is there in Arrieta's map of plague as a progressive condition and in the numbered parishes of Faithorne and Newcourt's London, the locus of the data that was collected into *Bills of Mortality* analyzed by John Graunt and those who followed his work with their own. While messy and uneven in their development, all these things would bring forth the disease studies of the eighteenth and especially nineteenth centuries.

CHAPTER 5

THE YELLOW FEVER THING

All the pieces were in play—but not in place—by the end of the seventeenth century. Hobbes's realpolitik was in the mix, a way of knowing and of organizing worlds of knowledge that did not just fade away. There was as well the emerging experimental science of Robert Boyle, whose perspective was presaged in medicine by Vesalius's challenge to physicians that they see for themselves, *test* their anatomical preconceptions against observed reality. To that was added the pathology of Morgagni in which Vesalius's healthy anatomy was a standard against which diseased tissues were compared to create a pathology of disease (Nuland 1988, 145–70). Symptoms in the living became pathologies of the dead, defining the physical affect of diseases whose origins were rooted in the air, climate, and soils of local geographies. The nature of those diseases were known through Sydenham's taxonomy and its neo-Hippocratic perspective of disease as something environmental whose elements would be known through careful observation.

Slowly, all these things would be filtered through the tables of public data on birth, mortality, and morbidity that were increasingly being collected by medical researchers and political agents for whom disease was a civil as well as medical concern. Bureaucracies of city and state would come to focus on the nature of disease and its effect on the economic and social enterprises of mercantile trade and the urban development. The early mortality statistics of Petty and Graunt would become more complex, serving as a foundational resource for bureaucrats and professionals in their struggle to understand both the nature of this or that disease and its affect upon this or that population. All this was made exigent by the globalization of seventeenth-century mercantilism. The biological transfers that began in 1492, what Alfred Crosby called "the Columbian exchange," accelerated in the eighteenth century as diseases like measles; typhoid, smallpox, and syphilis were exported across the trading world (Crosby 1972; Carrell 2003). In return, unfamiliar diseases would be imported into Europe from the expanded trading world, conditions for which treatments would need to be found.

It was in the eighteenth century that disease studies would begin a shift from observation toward experimentation, from Hobbesian natural science toward the juried, experimental perspective of Boyle. None of this happened over night. Disease would continue to be understood by most through the logical, inductive consideration of closely observed phenomena. Others, however, would gravitate toward a more deductive approach with an experimental method that was still a relatively new idea. Through much of the nineteenth century it was not either/or, however, but more commonly both/and. Consider, for example, medically trained Christopher Packe's 1743 map of the countryside within a sixteen-mile radius of Canterbury. The goal of his "Philosophico-Chorographical Chart of East Kent" was to describe fully the valley and its drainage system through precise measurement and careful study (Friendly and Palsky 2007). Packe's methodology was grounded in the language and approach of medicine and the anatomic realities it promoted.

"His comparison of the draining of the land with the circulation of the blood was natural to one trained in medicine" (Campbell 1949). Local geology was described using medical terms, with "anastomoses or inosculations" affecting the "capillary or extreme vessels by intermixing or indenting them very artificially with one another" (Jarcho 1978, 49). Rivers were veins and mountains became spines. For all its splendid cartographic detail and its production of the natural, Packe's map was a medically grounded imaging of the physical world. It applied to nature the lessons of anatomy and thus argued that the visual and experimental methodologies of anatomy would also serve in the greater world.

In Packe's master text, *Ankographia*, the black-and-white map (printed in a limited edition in green) was divided into four sheets (Jarcho 1978, 48). Concentric circles indicated miles from Canterbury, and compass bearings from the city were shown as radii mapped at the scale of the Canterbury city and countryside. The scale was necessary to accommodate the detail inscribed in the map, to do justice to the elevation of its hills, and the related, circulatory system that was Packe's streams. The natural geography thus created served to relate through a single methodology both the anatomic figures in Vesalius's landscape and the landscape itself. How the observational might be joined to the experimental, and how either or both might best serve the understanding of disease is in large part the story of the eighteenth century struggle to come to terms with a then emerging disease, yellow fever.

YELLOW FEVER

As epidemiologist Robert Desowitz tells it, sometime around 1647–1648 a "concatenation of conditions" led to near simultaneous outbreaks of a previously unknown disease in Havana, Barbados, Guadeloupe, St. Christopher, and Mexico's Yucatan peninsula (Desowitz 1997, 99). We now know that those conditions included the importation of African slaves to work plantations and presumably the accidental inclusion of the *Aedis aegypti* mosquito in the holds of ships whose sailors and cargos of slaves carried the flavivirus that causes yellow fever. Once in the New World the virus prospered, as did its mosquito carrier, in European settlements across the tropical and subtropical latitudes.

Mild cases were flu-like, with symptoms of diarrhea, fever, headache, and joint and muscle ache. In more severe cases the chills and pain were so intense it could feel as if the bones themselves were being broken (Spielman and D'Antgonio 2001, 58). In fatal cases hemorrhaging caused blood to ooze from the mouth and nose; blood filled the stomach, only to be vomited as humoural "black bile." Yellow fever's name comes from the jaundiced color of those with moderate to severe cases in which the liver and its system of enzymatic production are attacked. The collection of conditions suffered by patients and observed by physicians was, in Sydenham's taxonomy, a remittent bilious fever generated in hot moist airs. Some called it a "hemogastric pestilence," a disease of the blood and the gastric systems whose symptoms were its name.

Whether it was in fact a new disease or one previously seen was not clear from its symptoms, however. In 1740 a slave ship physician insisted it was just another name for black water fever, a malarial complication with which it shared many symptoms.[1] A Jamaican physician, Dr. Park Bennett, insisted it was obviously a new, distinct disease. Neither yielded to the other's authority, and in the acrimony that resulted both men died together in a duel defending their opposing definitions (Desowitz 1997, 96). Whatever its precise nature, what was clear was the disease or diseases clearly favored specific geographies: "It appears to be a fever that is indigenous to the West Indian Island and the Continent of America Which is [in] the tropics, and Most Probably to all other Countries within the Torrid Zone" (Hillary 1759, 14). There were reports of outbreaks in other tropical and subtropical areas. The Spanish reported it in their colonies, for example, and the French called it *la maladie de Siam* because that was where they encountered the same complex of symptoms.

Dale Ingram stated authoritatively in 1755 that, "this distemper is *epidemical*, that it springs from a corruption of the air, and that its violence is in proportion to the continuance of the *heat* and *moisture*" (Ingram 1755, 120; original emphasis). It was also, he continued, recurrent, "shewing its periodical appearance to be similar to the plague." Plague was the exemplar of epidemic disease, the thing that rampaged through populations. Invoking it in a treatise on yellow fever defined this new disease as a similar threat to white populations in colonial areas that were clearly at risk.

Both yellow fever's periodicity and its observed, climatic characteristics were legacies of Sydenham's system of long-term observation and naturalistic investigation. If disease was defined by its symptoms, its nature was to be discovered through a careful monitoring of air, climate, and environment. Physicians studying yellow fever therefore kept careful records of all these things in hopes of uncovering a correspondence that might better reveal the precise nature of the disease. Dr. William Hillary

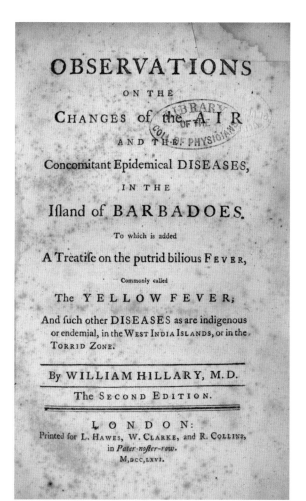

FIGURE 5.1 William Hillary carried out careful measurements of air pressure, rainfall, and temperatures in Barbados in an attempt to describe the conditions most conducive to yellow fever and other recurring fevers.

described the general tenor of these studies in his *Observations on the Changes of the Air and the Concomitant Epidemical Diseases in the Island of Barbados* (fig. 5.1): "I mean such Changes in Diseases, as arise from the Variations in the Weather, and either produce different Symptoms in the same Disease, or such as may determine the morbid Matter to a different part of the Body, or lastly to be carried off by a different critical Discharge, than it was before; all of which should be carefully observed by the attending Physician, and Nature should always be assisted by him, to effect such ways and Methods as she indicates and endeavours to do, if it can be done; these I have endeavoured to observe" (Hillary 1759, 2).

The disease occurred at once in the patient's body, the body of a local population, and the physical environment within which that population resided. To understand the disease as environmental required a careful recording of daily climatic changes: "*The Degrees of the Heat, or the Coolnesses, of the Air, were observed by* Fahrenheit's Mercurial Thermometer, made at Amsterdam, every Morning at or before the Rising of the Sun, and again between the Hours of Twelve and One o'clock. *And the Height of the* Mercury *in the* Barometer *was observed at the same Times, tho' I have only recorded its station at Noon*" (Hillary 1759, 2; original emphasis).

This made perfect sense to those physicians who saw themselves as inheritors of the Hippocratic tradition advanced by Sydenham. If disease was born in the airs and soils of specific places then understanding an outbreak required a close study of the location where it appeared. As Hillary put it, "That Father and Prince of Physicians, Hippocrates, advises all physicians to examine and duly consider the Situation, Air, and the Water, used by the People of such Cities, or Places, as they are called to, or may practice in. It therefore is necessary that I should say something concerning the Situation, Air, Water, Etc. of this Island, before I give an Account of the Observations made on the epidemical and endemical Diseases in it" (Hillary 1759, 1).

If yellow fever was a natural result of hot moist air and its affect on white colonialists there was little to be done. The argument for that yellow fever lay in the climatological records of temperature variations kept by researchers like Hillary and Mathew Carey (fig. 5.2). But others argued that while moist, warm tropical airs might be "exciting agents" that the source of outbreaks lay in the odiferous wastes of the city and cargo from ships stored in warehouses and later distributed in the city. This was an *un*natural yellow fever, one anthropogenic in its construction. Where epidemics occurred there were therefore concerted attempts to identify malodorous sites that created a local environment within which yellow fever could advance. These could be dockside bags of spilled coffee beans, the offal from slaughtered animals, or pools

METEOROLOGICAL OBSERVATIONS,
MADE IN PHILADELPHIA, BY
DAVID RITTENHOUSE, Esquire.

AUGUST, 1793.

	Barometer A.M.	Barometer 3 P.M.	Thermometer 6 A.M.	Thermometer 3 P.M.	Wind 6 A.M.	Wind 3 P.M.	Weather 6 A.M.	Weather 3 P.M.
1	29 95	30 0	65	77	WNW	NW	cloudy,	fair
2	30 1	30 1	63	81	NW	SW	fair,	fair
3	30 5	29 95	64	82	N	NNE	fair,	fair
4	29 97	30 0	65	87	S	SW	fair,	fair
5	30 5	30 1	73	90	SSW	SW	fair,	fair
6	30 2	30 0	77	87	SW	W	cloudy,	fair
7	30 12	30 1	68	83	NW	W	fair,	fair
8	30 1	29 95	69	86	SSE	SSE	fair,	rain
9	29 8	29 75	75	85	SSW	SW	cloudy,	fair
10	29 9	29 9	67	82	W	SW	fair,	fair
11	30 0	30 0	70	84	SW	WSW	cloudy,	cloud
12	30 0	30 0	70	87	W	W	fair,	fair
13	30 5	30 0	71	89	SW	W	fair,	fair
14	30 0	29 95	75	82	SW	SW	fair,	rain
15	30 0	30 1	72	75	NNE	NE	rain,	cloud
16	30 1	30 1	70	83	NNE	NE	fair,	fair
17	30 1	30 0	71	86	SW	SW	fair,	fair
18	30 1	30 0	73	89	calm	SW	fair,	fair
19	30 1	30 1	72	82	N	N	fair,	cloud
20	30 1	30 1	69	82	NNE	NNE	fair,	fair
21	30 15	30 25	62	83	N	NNE	fair,	fair
22	30 3	30 85	63	86	NE	SE	fair,	fair
23	30 25	30 1	63	85	calm	S	fair,	fair
24	30 1	30 1	73	81	calm	calm	cloudy,	rain
25	30 1	30 1	71	66	NE	NE	rain,	great r
26	30 15	30 2	59	69	NE	NE	cloudy,	cloud
27	30 2	30 2	65	73	NE	NE	cloudy,	cloud
28	30 2	30 1	67	80	S	calm	cloudy,	clear
29	30 16	30 15	72	86	calm	SW	cloudy,	fair
30	30 1	30 1	74	87	calm	SW	fair,	fair
31	30 0	30 0	74	84	SW	NW	rain,	fair

FIGURE 5.2 A table of climatologic data used by eighteenth-century disease investigators like Mathew Carey to describe conditions in which yellow fever was generated in the atmosphere.

of human wastes washed from the city to its docklands, any place anything might generate disease-causing odors.

If only the city could be made cleaner, and thus less odiferous, then this yellow fever might be contained. Failing that, colonialists at risk could take protective measures. Carey described these nicely: "Those who ventured approach, had handkerchiefs or sponges impregnated with vinegar or camphor, at their noses, or else smelling bottles with thieves' vinegar. Others carried pieces of tar in their hands, or pockets, or camphor bags tied around their neck" (Carey 1793, 29).

A brisk trade in powders grew up in affected areas, all promising protective scents that even if they did not protect their user against yellow fever at least masked the foul odors whose inhalation was assumed to cause it. The physician Robert James, for example, described one of these nostrums in a 1764 report on yellow fever in Jamaica. It was packaged in a special paper "to prevent counterfeits" and marked as "by the King's patent" to assert its official provenance. Indeed, manufacturers boasted the product was so efficacious it was "lately ordered to be used on board His Majesty's Navy" to protect its sailors (James 1764, 87).

To medical practitioners yellow fever was a subject of compelling professional interest, to settlers and sailors a matter of personal concern. For governments increasingly dependent on revenues generated by international trade, yellow fever threatened the very structure on which that trade was based. Epidemics, even without quarantines, diminished the workforce and restricted colonial production and the financial benefits that accrued from it. The continuing importance of yellow fever, and the threat it presented to the emerging trade of nations, is underscored by the location of Carey's *Account of the Malignant Fever Lately Prevalent in Philadelphia,* a critical port and a center for medical research. But as its full title makes clear (fig. 5.4), the focus was not just the localized outbreak of Philadelphia but the yellow fever pandemic in eastern United States, generally of which Philadelphia's experience was one example. The continuing importance of yellow fever studies, and the recurrent epidemics experienced, was underscored in an 1811 edition that included notes by Dr. Benjamin Rush, perhaps the most important physician researcher of his time in the United States.

If yellow fever was in the patient and in the cities where patients resided it also seemingly infected the trading links that joined cities in nations to the world. In the language of modern medicine, it was, at least potentially, a "connective tissue disorder" whose local effect could damage the broad commercial body. The extent of trade between England and its colonies in the West Indies, and its importance to both, made yellow fever a subject of economic and political—not simply medical—urgency. That trade is shown in a chart from William Playfair's *Commercial and Political Atlas and Statistical Breviary*, first published in 1786 (Playfair 2005).[2] Playfair's distillation of data tables into graphics arguing commercial and geographic relations was an important innovation (fig. 5.3). As Playfair wrote, "Figures and letters may express

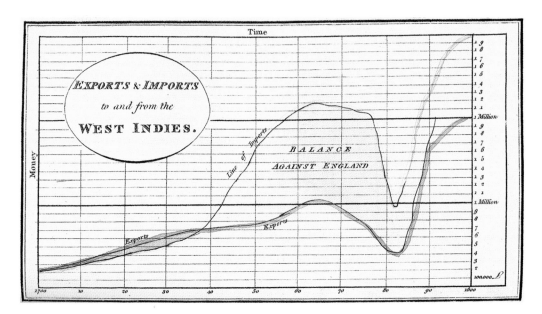

FIGURE 5.3 Playfair's chart of trade over time presented a graphic argument of trade growth between the West Indies and England.

with accuracy, but they can never *represent* either number or space" (Wainer and Playfair 2005, 3). It was that sense of presentation that he sought to enhance in his treatment of official data.

Playfair's chart proposed two distinct areas, England and the West Indies, as separate entities whose members joined over time in communal trade. It argued a timeline along which the value of trade relations measured in pounds of cargo could be measured. The shape and physical relation of the trade partners were implied; a reader could imagine the map. Constantiated in the map was trade as its own space. What was produced was a landform between the traders called the land of "trade deficit." The result was a new way of understanding aggregate public and professional data: large collections of economic statistics were widely available—and had been since the time of Graunt and Petty—more than a century before Playfair thought of publishing such data in pictorial form (Playfair 2005, 11). The graphics Playfair created were not mere representations but intellectually grounded arguments, the exposition of ideas inscribed in the colored charts he developed. As Playfair said, "As the *eye is the best judge of proportion* . . . it follows, that wherever *relative quantities* are in question, a gradual increase or decrease of any . . . value, is to be stated, this mode of representing it is peculiarly applicable . . . giving *form and shape to a number of separate ideas*, which are otherwise abstract and unconnected" (Playfair 2005, 30; original emphasis).

Playfair's atlas and breviary gave shape and form to the statistics that governments increasingly collected about things important to them. Understanding the tables required a method by which tabulated returns could be fashioned into arguments grounded in officially acknowledged evidence. One could see in Playfair's chart the growing importance of West Indian trade to the British, and by inference, the degree to which epidemic disease there affected England and its increasingly vital colonial trade. Like tendons on bones, the elements of the chart knitted together the parts of the world that on the map were distant and appeared unrelated. In the nineteenth century, charts like Playfair's would be increasingly inserted into maps in a marriage of media encouraging the exploration of data that was at once spatial, temporal, and statistical.

YELLOW FEVER: NEW YORK

Yellow fever did not remain safely tucked away in the Caribbean but migrated to the America colonies, where outbreaks occurred in the Gulf Coast port of New Orleans (Colten 2006), whose climate presumably favored it. It also became epidemic along the Atlantic Coast, first in Charleston, South Carolina (Lining 1799) before appearing in Boston, Philadelphia, and New York. "Yellow fever struck New York City in 1702, 1732, 1741, 1743, 1745, 1747, and yearly from 1893 to 1805. Boston was similarly visited by the virus as were other maritime sites as far north as Halifax, Nova Scotia" (Desowitz 1997, 100).

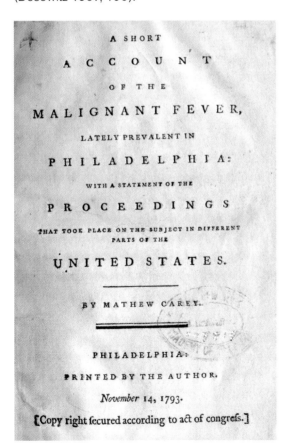

FIGURE 5.4 Philadelphia bookseller and publisher Mathew Carey's 1793 report on the "malignant fever" in Philadelphia attempted to describe the outbreak and the distinguish theories about the origin of the epidemic disease.

These virulent, recurring epidemics threatened the very cities in which they struck and thus the entire colonial and mercantile enterprise. Noah Webster put the problem succinctly when in 1796 he asked, "Why should cities be erected, if they are only to be the tombs of men?" What made these epidemic outbreaks especially frightening was they were not *supposed* to happen in cities like New York and Philadelphia, whose climate was temperate rather than tropical. This temperate-zone yellow fever was something decidedly *un*natural, and thus especially dangerous: "Distempers of very hot southerly countries, and natural to those climes, are unnatural to other countries situated in a northern latitude, and that the north is never attacked but when the atmosphere abounds with heat and moisture, the prerequisites of the disease" (Ingram 1755, 100).

The warm, sticky heat of the temperate climate's summer months was one answer. Another was the odiferous local condition in the mercantile cities where yellow fever occurred. Perhaps the disease-inducing odors of animal and human waste were the source of this tropical disease in a temperate region. If that were shown to be true then the logical response would be cleanliness, not just of person but also of place. "Too much cannot be said in favor of cleanliness," wrote Benjamin Rush in his *Medical Inquires and Observation*, first published in 1797. "Offal matters, especially those which are of a vegetable nature, should be removed from the neighborhood of the dwelling house" (Rush 1809, 199). For the individual, prudence demanded as well as personal hygiene the avoidance of disease generating airs: "So, too, the healthy person would avoid breathing both chill evening and early morning airs, at least until the body has been fortified with solid aliment, or a *draught* of bitters" (Rush 1809, 272; original emphasis).

Following the 1793 epidemic in Philadelphia, which saw 10 percent of the city's population of fifty thousand killed by the fever, Mathew Carey offered *A Short Account of the Malignant Fever, Lately Prevalent in Philadelphia*. The first step in this account was to detail the mortality that occurred during the outbreak. For this he used burial records kept by local religious officials who supervised the individual cemeteries and their burials (fig 5.5). The number of burials was tallied and their distribution among individual cemeteries hints at the social (and thus geographic) division of the decedents. In the days before nineteenth-century central public health registries were created, burial records were the only count of deaths available besides the individual casebooks of local physicians. Still, for Carey, they were enough to provide the data he required to emphasize both the severity of the epidemic and the rough ethnography of mortality in this period of malignant fever.

Carey was a critical figure in Philadelphian medicine. As a bookseller and publisher he assured the publication of a number of books of science and medicine.[3] In a treatise on yellow fever presented to the American Philosophical Society, Carey sought to deliver "plain truths in plain language," a scientific account and explanation accessible to all. While "the general [professional] opinion was, that the disorder

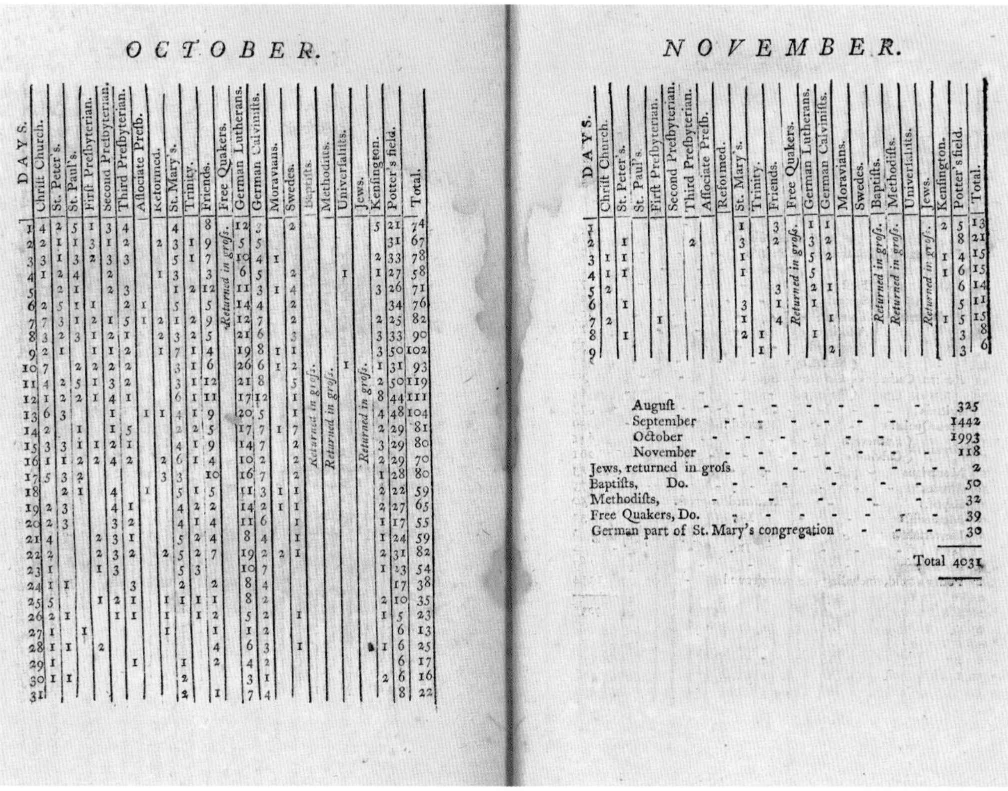

OCTOBER.
NOVEMBER.
DAYS.
Chrift Church.
St. Peter's.
St. Paul's.
Firft Prefbyterian.
Second Prefbyterian.
Third Prefbyterian.
Affociate Prefb.
Reformed.
St. Mary's.
Trinity.
Friends.
Free Quakers.
German Lutherans.
German Calvinifts.
Moravians.
Swedes.
Baptifts.
Methodifts.
Univerfalifts.
Jews.
Kenfington.
Potter's field.
Total.
Returned in grofs.
August - - - - - - - 325
September - - - - - - 1442
October - - - - - 1993
November - - - - - - - 118
Jews, returned in grofs - - - - - 2
Baptifts, Do. - - - - - 50
Methodifts, - - - - - 32
Free Quakers, Do. - - - - - - - 39
German part of St. Mary's congregation - - - 30
Total 4031

FIGURE 5.5 To document the mortality of the severe 1793 yellow fever epidemic in Philadelphia, Mathew Carey used burial records collected from local religious officials in charge of different cemeteries.

originated from some damaged coffee, or other putrefied vegetable and animal matters" at a wharf a little above Arch Street in the waterfront district of the city, another very different, popular theory was offered by citizens untrained in Hippocratic medicine and medical theory. "That it is an imported disorder is the opinion of most of the inhabitants of Philadelphia. However there is much diversity of sentiment, as to the time and manner of its introduction" (Carey 1793, 18).

Thus nonmedical professionals saw yellow fever as not indigenous but portable, traveling the world in cargo ships. Like plague, it took root in different cities and countries irrespective of local climate and conditions. "Some believed it was carried on the ship *Il Constante*, which had arrived in port in May; others argued it was more likely born aboard the ship *Mary*, which arrived August 7 carrying French emigrants from the Cape" (Carey 1793, 18). In this theory, local geographies might *contribute* to the intensity of a yellow fever epidemic but its origin lay elsewhere. Like the plague, outbreaks were in this reading an effect of trade and not local airs and odors alone. This was not a new idea. It was, for example, embedded in Arrieta's map of plague prevention in 1690. But in the expanding mercantilism of the late eighteenth century it had immense ramifications.

Historians have tended to see as irreconcilably opposed these two views of disease as either unnatural, introduced from elsewhere (contagious), or as the sole natural result of local conditions (anticontagionist). As L. G. Stevenson put it, "This was the era of the great debate between contagionists and miasmatists (anticongationists). The apparent beginning, and the subsequent development, of the mapping of disease were conditioned by this debate" (Stevenson 1965, 228). It was rarely a matter of either/or, however. In an article on nineteenth-century cholera, George Davy Smith argued "the distinction between contagionist and anti-contagionist thought is too binary with many theoretical varieties containing elements of both . . . being widespread" (Smith 2002, 926). More frequently two or more of these disease theories combined in a multifactoral argument in which origin and source were sometimes distinct and sometimes assumed to be identical.

Within Sydenham's classificatory system of traditional diseases, yellow fever was certainly a fever but one whose origin and source were unclear. The protocols of careful observation Sydenham championed were insufficient to reveal its nature. Some believed it was the result of the warm, muggy summer airs that, with the foul odors of rotting urban garbage, created a disease-generating miasma. To test this theory Carey and other contemporary researchers collected meteorological records, recording urban temperatures in an attempt to correlate them with disease mortality (fig. 5.2). Within the miasmatic theory of recurring fevers advanced by Sydenham the idea of a seasonal disorder occurring because of temperature and rainfall variations made perfect sense. And, better, it was something that could be checked by comparing temperature and burial records. If higher temperatures resulted in increased deaths then the positive correlation would seem to prove the disease thesis. Indeed, there was a general correspondence, although today we know that the connection, while real, was not causal.

At least five different yellow fevers, each distinct in its etiology and at least partially distinct in their treatment, were therefore proposed. First, yellow fever was the name given to a congress of physical symptoms repeatedly described by the medical authorities attempting to treat affected populations from Barbados to Boston to Siam. The diagnosis was made difficult because those same symptoms were shared with a range of other, typically tropical conditions like malaria, most of them believed to originate in odiferous, miasmatic airs. Second, this was a tropical disease rooted in local geographies. The proof was that in Barbados and other outposts white settlers were far more severely affected than locals and African slaves.

Third, the "torrid zone" symptoms either migrated to the temperate regions of the coastal United States or were independently produced there. This yellow fever was clearly a seasonal affliction whose local source if not its origin likely lay in the odiferous airs generated in the refuse of the mercantile city: Boston, New York, or Philadelphia. Fourth, this pestilential yellow fever was a portable condition that traveled along trade routes just as plague had before it. Fifth, there was, finally, a sense that these different yellow fevers were not irreconcilable. They might somehow be combined into a multifactoral disease in which two or more distinct disease descriptions were joined. To sort through the multitude of yellow fevers and definitively understand their natures required an experimental approach. Only that would permit choices to be made between the options of trade quarantine, local sanitary initiatives, and the like to be made by the officials of the cities effected.

NEW YORK: YELLOW FEVER

In a landmark study, New York physician Valentine Seaman argued that "no yellow fever can spread, but by the influence of putrid effluvia" (Seaman 1798, 324–25). This was not proposed as a conclusion but advanced as a testable proposition argued in a monograph (Seaman 1796), as a chapter in a book on bilious fevers edited by Noah Webster (1796), and in a 1798 journal article that included two maps. Seaman was a medical luminary, a surgical pioneer who authored the first journal article in the United States on a surgical topic (Stevenson 1965, 234) and the author of a critical study of water quality in Saratoga Springs, New York (Seaman 1793).

A practicing physician and surgeon (the two were not as easily separated then as now), he was also a member of the New York Health Committee created in 1795, a forerunner of the nineteenth-century Board of Health that later became the New York Department of Health. A familiar of Noah Webster's, his research carried weight both within medicine and the evolving city bureaucracies whose citizens had been hard hit by the yellow fever epidemic of 1976.

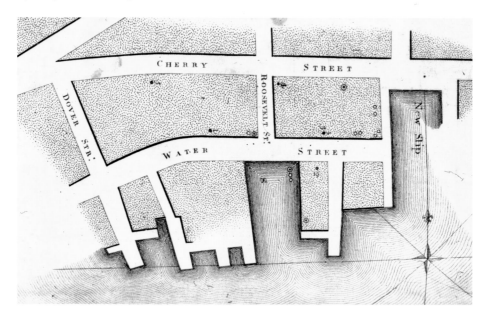

FIGURE 5.6 Valentine Seaman mapped the progressive location of cholera deaths in the New Slip area, numbering cases in his map, to prove the proposition that proximity to noxious airs caused yellow fever.

In proposing that yellow fever originated in the "putrid effluvia" of the malodorous city Seaman sought at another level to test the Hippocratic idea of disease as a locally generated condition. He had to do this in a way that he believed would demonstrate conclusively that yellow fever was not an inevitable result of the city's summer miasms. To make this case Seaman had produced two copperplate "spot maps" to accompany an article published in the inaugural issue of the journal *Medical Repository* (Shannon 1981). These are among the earliest known maps published as evidence *testing* a propositional disease theory (*if* cholera is caused by odiferous waste, *then* the greatest density of cases will be closest to waste sites), although it is unlikely they were the first to have been made. In his article Seaman makes no claims about the innovative use of mapping which suggests he had seen other, prior attempts to map disease even if those earlier examples have not survived. Further, mapping was already a commonplace adjunct to urban life if not necessarily to medical investigation. "Taken together, the New York *penchant* for drawing municipal maps for all sorts of purposes and the medical and sanitarian penchant for sketching out disease maps verbally seem to point to an inevitable conjunction—the actual drawing of an epidemiological map of disease" (Stevenson 1965, 241). What appears most innovative, perhaps, is the publication of such maps in a medical journal, a triumph of general publishing rather than scientific technology.

By arguing his case in the inaugural edition of one of the first medical journals originating in the Americas—and one published by Benjamin Rush—Seaman also was making an argument for a type of science whose results would be judged by a jury of professional peers. That, after all, is what such journals were designed to do: present arguments and conclusions in a forum where they could be widely disseminated and generally judged by knowledgeable professionals. Undoubtedly, Seaman's prominence as an important New York physician gave a certain authority to the new journal. Conversely, this publication legitimated Seaman's cartographic and experimental rather than more traditional approach of natural observation (barometric pressure, rain fall, temperature, and so forth).

Seaman's cartographic argument was grounded in three propositions. First, that the area in which yellow fever was reported in New York City could be clearly and cartographically described in a meaningful way. He brought forth its neighborhood in the same fashion that, in the first chapter, Vancouver was constructed for the diarrhea outbreak study. In the copper plate maps the city's streets and New Slip Harbor are instantiated as a part of a whole recognizable as being in New York City. Both of Seaman's maps are of the docklands where the discharge of common sewers and street washings came to rest along the city's wharfs and riverbank. There it might sit at low tide for hours or days until the tide carried it out to sea (Seaman 1798; Stevenson 1965, 234).

Second, Seaman's mapping proposed that yellow fever could be identified as a single disease whose symptoms were sufficiently well recognized within the medical

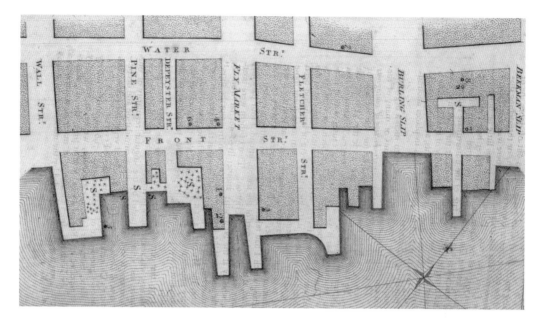

FIGURE 5.7 Seaman mapped the sources of the "furry-fostering miasmata" washed to the dockside from the city. An "S" marks filth and general garbage; an "x" marks areas of human refused called by Seaman "common convenience."

community to permit a single class of differentially diagnosed cases to be located. This permitted Seaman to reference in the map individual cases as if identical except for intensity. In his first map (fig. 5.6) Seaman engraved five numbered cases, each indicated by a small period-like dot, to locate the sequential deaths of the first reported cases of yellow fever. Small, open circles identify a further nine cases he described as either likely yellow fever, or a mild case from which the patient recovered. In the map are two circles in the water, the cases of seaman onboard ship. Finally, three near-fatal cases—one onboard a ship—are mapped using a circle with a dot inside.

These were a fraction of the total yellow fever cases reported in the city. Seaman explained the partial mapping of the available data was a result of "the want of proper marks to identify [the disease] where it is slight" and his inability to map all relevant data while maintaining legibility (Seaman 1798, 317). The index case, a seaman who arrived on the ship *Polly* from the Carolinas, is located not onboard his ship but on land, at the house where he died. It was mapped this way because Seaman did not think his voyage relevant even though, as he reported in the text, another *Polly* seaman with similar symptoms died at sea.

Third, in a separate map (fig. 5.7), Seaman referenced a set of odor-generating sites whose commonality was asserted as members of a malodorous class. Different symbols distinguished several classes of odor generation. An "x" symbolizes areas of "common convenience"—human or animal excrement—and an "S" sites where garbage and other odiferous "putrid matter" discarded by the city accumulated. These sites also shared another characteristic: dense swarms of mosquitoes in a quantity "never before known, by the oldest inhabitants, to have been so numerous as at this season" (Seaman 1798, 331). In the eighteenth century mosquitoes were assumed to spawn in the same "unwholesome exhalations" as disease (Colten 2006, 37). Seaman's odiferous waste sites thus were characterized by foul odors and the mosquitoes bred by them.

While each map is distinct they are not unrelated. Each draws from the other. We trust the map of the New York area because the elements that bring forth the New Slip area conform to our personal experience and our knowledge base and we therefore we believe it to be true. Because we believe the city map to be true we are already inclined to accept as true the georeferencing of individual cases to this or that street address. That is, we trust these locations exist and therefore that the decedent may be located at an address. Accepting this the mapping of odor sights is assumed to be true as well. The mapping of individual cases and odor sites affirms simultaneously the idea of the street network in the map. The evidentiary context is mutually supportive, each adding authenticity to the other.

Implied but not stated was Vesalius's challenge to those who doubted his anatomy: go see for yourself. Those who doubted the presence of foul wastes could visit the sites and smell for themselves. Those who questioned the case reportage were free to knock on the family's door. The yellow fever cases were collected and located because Seaman wished to test a theory of disease generation *in situ*. The odiferous sites were mapped because the argument he sought to explore was that of a relationship between yellow fever incidence and proximity to foul odors. The map made of these different propositions a meta-argument. *If* yellow fever was caused by the foul odors of human and animal wastes, *then* the incidence of death should be strongest in areas where the stench was greatest.

Each map proposes a constituent of the argument that *if* yellow fever is caused by miasmatic odors, *then* disease incidence should be greater the closer one came to odorous sites. If the evidence was accepted as accurate then the classes of facts (these deaths, those malodorous sites, these streets, this city) could be confidently used to test the theory. As facts, they were *only* important if Seaman's theory was accepted as at least provisionally tenable. If there were no reason to propose a relation between odiferous sites and yellow fever cases then the conjunction of the data sets in the maps would be meaningless. In the mapping Seaman thus legitimated the theory his mapping was designed to test. Confidence in the theory grew with the totality of the mapped sets that together created the maps of yellow fever and odor in the New Slip area of New York City.

Seaman did not have to demonstrate the correspondence he proposed: the maps did the work. In both maps the density of, in one case, yellow fever deaths, and in the other, sites of odiferous waste, were sufficiently dense to argue the mapped incidence of each as a class. The maps when seen together effectively created a new event class, one that said: "These sites of miasmatic odor are causally related to those deaths on the basis of proximity." This proved, Seaman argued, to "every unprejudiced mind that in the city there appears to be an intimate and inseparable connection between the prevalence of yellow fever and the existence of putrid effluvia" (Seaman 1798, 324–25).

The mapping demonstrated, Seaman wrote, an "intimate and inseparable connection" between yellow fever and locally odiferous airs in the Old Slip area. "Most of the patients infected with dangerous fevers, were either such as resided in the neighborhood of slips (which were or lately had been cleaned out) or whose employment led them to frequent such places (Seaman 1798, 317). In mapping the odiferous sites of the area that correlated visually with the incidence of yellow fever he had, he thought, proven his case.

Seaman did not deny the possibility that another form of contagion might have contributed to the incidence of yellow fever. The form of testing he applied only spoke to the relationship between disease and odor cities the miasmatic theory predicted. Seaman therefore could insist only that any other form would be secondary to the relationship seen in the maps. In speaking of the index case Seaman wrote: "It may be that a partial principle of death lurked in his [the *Polly* seaman's] system, during the whole time after the death of his comrade, and most likely would never have seriously acted upon him had he not immersed himself in this or some such like furry-fostering miasmata [in New York City]" (Seaman 1798; Stevenson 1965, 236–37). Yellow fever *might* be something that traveled on ships but even if that were so (and how might it be proven?) it would require local airs suffused with the odor or rotting waste to be activated.

Seaman mapped not simply a local outbreak but a disease theory (miasmatic) applied to a specific yellow fever outbreak. In his argument he assumed all yellow fever outbreaks were the same disease, and further, he implied yellow fever everywhere (Barbados Boston, Philadelphia, New Orleans, and on) were the same. Thus the testing of a specific outbreak spoke to a general theory with wide applicability. In the mapping, Seaman argued a methodology by which relationships between environment facts (physical and/or social) and disease incidence in local populations could be tested. By publishing his findings in a new professional journal, Seaman endorsed in this publication a method of argument whose results were not inherent in the logic of the presentation but instead to be judged by professional peers, readers of the journal *Medical Repository*.

Pascalis

A generation later another physician, Felix Pascalis, mapped another of New York harbor's recurrent yellow fever outbreaks (Stevenson 1965, 243–44). It was his third and most definitive attempt to assert a miasmatic theory of yellow fever. In 1796 Pascalis published his *Medico-Chymical Dissertation*, regarding yellow fever as a miasmatic disease amenable to "the best antinomial preparations" for its treatment. In 1798 he won a prize from the Medical Society of Connecticut, where he was a "corresponding member," for a report on a 1797 epidemic in Philadelphia, where he was a member of the local medical society. That study promised answers, its lengthy subtitle said, to "the questions of its causes and domestic origin, characters, medical treatment, and preventives" (Pascalis 1798). Then in 1819 Pascalis published *A statement of the occurrence of a malignant yellow fever, in the city of New-York,* condensed the next year as an article in the *Medical Repository* that included a reissue of his monograph's map. Its full title serves as an abstract:

> A statement of the occurrences of a malignant yellow fever, in the city of New-York, in the summer and autumnal months of 1819; and of the check given to its progress, by the measures adopted by the Board of Health. With a list of cases and names of sick persons; and a map of their places of residence within the infected and proscribed limits: with a view of ascertaining, by comparative arguments, whether the distemper was engendered by domestic causes, or communicated by human contagion from foreign ports. (Pascalis 1819)

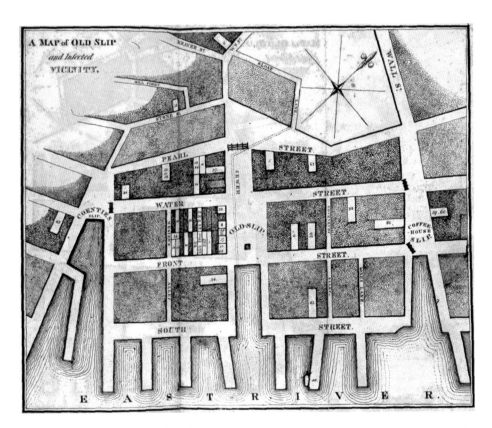

88

FIGURE 5.8 Felix Pascalis mapped a yellow fever outbreak in 1819 in New York using an approach similar to Seaman's but with a more extensive and detailed list of cases. The theory and argument—proximity implies causality—remained the same.

Yellow fever remained a subject of medical debate because nobody could identify the agent that caused the recurrent outbreaks that continued to plague New York City and especially its harbor area. And despite concerted effort, the disease was resistant to the efforts of medical officials seeking to treat it and civil authorities hoping it could be prevented. The object of Pascalis's monograph and article was to evaluate measures instituted by the New York Board of Health, the descendent of the Health Committee on which Seaman served, in the context of still contending theories of the disease. Pascalis had created for his monograph a map of the Old Slip neighborhood in which reported cases of yellow fever cases were most intense. In it are the different wharfs, including one for coffee, and the end of a sewer line that extended from the field near Sloat Lane past Pearl Street—where a cattle barrier can be seen—through the Old Slip area. In the map are sixty-six cases of yellow fever, the vast majority occurring in a single block between Front and Water streets bounded to the east by New Slip itself. The cases of yellow fever reported to health officials are numbered chronologically, each reflecting "the successive dates of sickening" (Pascalis 1819, 25).

In the accompanying text Pascalis asserted that the area between Front and Water streets was also the site of "perishing and fomenting materials," including decaying matter and animal wastes "emitting in weather an offensive smell, and no, doubt, also deleterious miasmata" (Pascalis 1819, 19). Like Seaman, Pascalis believed these odors were generative, disease origins. That was not the map's argument, however. Pascalis's map only argued the proposition that a dense concentration of cases occurred in the New Slip area. Unlike Seaman, Pascalis did not map proximate areas of offensive smell. The greater argument was made only in the text where he declared that the outbreak was indeed "engendered by domestic causes," the miasms emanating from waste sites near the mapped epicenter of the outbreak.

Seaman focused on a small set of yellow fever cases whose details he knew personally in an attempt to determine a relationship between disease incidence and environmental causes. Pascalis's far larger mapped data set established the intensity of the outbreak but did not advance an understanding of the disease itself. But then, its purpose was to assert the intensity of the outbreak in a manner that would justify sanitary measures proposed by the Board of Health in the face of strong opposition by local businessmen. The board, Pascalis noted proudly, "remained unappalled in the path of their duty against a respectable portion of their fellow-citizens, who represented the whole commercial interest, and that of the moneyed men" (Pascalis 1819, 43). The real function of Pascalis's study, and its one map, therefore, was to insist that the interests of "the moneyed men" should not hold sway over the demands of public health. Where a relation between disease and infrastructure could be shown to exist, public health was more important than commercial interests. That relationship was for Pascalis indisputably miasmatic and the area where civil authorities had to concentrate their efforts was clearly shown in the map.

Evidenced in Pascalis's map and texts were the growth of bureaucracies, professional and public, dedicated to disease studies. Pascalis's position on the Board of Health was important and may have been crucial in his provision of the data mapped in his study. The cases inscribed in the map were not his but those reported to health officials whose records he could access. Further, Pascalis's membership in medical organizations attested to his professional standing in a manner legitimating his appointment to the Board of Health, a position that added prestige to his work. Finally, in the text Pascalis argued the different perspectives of, on the one hand, persons of medicine concerned with public health, and on the other, of "moneyed men" who put commerce and profit over public health and disease containment. To stand against their influence required strong rhetoric backed by a detailed argument and here the map served as both an evidentiary statement (people died here, and here, and here) and a rhetorical device through which the civil and medical argument could be made.

THE MATRIX

At the end of the eighteenth century, there were two ideas in contention. disease was either a natural artifact of the landscape or an unnatural artifact of human settlement and trade. Depending on the researcher it could be either exclusively or, perhaps, some combination of natural predisposition within a human-made environment of trading cities. Naturalists like Carey and Hillary sought to understand yellow fever through a recording of climatic conditions that heralded its visitations. They assumed it was a natural and recurring element of the local landscape, a geographic reality to be observed. Others, like Seaman and Pascalis, believed yellow fever, whatever its origins and agents, had its local source in the decidedly unnatural conditions of rapidly expanding cities where the lack of adequate sanitation created malodorous areas that medical theory asserted would breed disease. This idea was testable and its mapped consideration resulted in confirming studies reported in official publications, professional journals, and increasingly, in daily broadsheets.

Two transformations, both in place by the first decade of the nineteenth century, made the growth of this form of disease testing possible. First was the growing sense of a professionalism that included not simply disease treatment but disease study. Medical professionals were becoming a class who would carry out studies, present papers, and perhaps publish them so that a jury of their peers might judge their arguments. This growing professionalism was encouraged by the growth of local medical associations where papers were read and journals were published. Two notable examples in North America were the New York Academy of Medicine and the College of Physicians of Philadelphia, both of whose libraries remain open today.

Second, a civil bureaucracy concerned with disease as at thing lodged not in the individual alone but in the population at large was required. These bureaucracies served both as centers for the collection of disease-related data and as authoritative bodies which could issue edits to inhibit disease generation where needed. As the work of Seaman and Pascalis demonstrate, both civil and professional arguments depended on a spatial ontology permitting the incidence of disease to be considered as a single phenomenon. The evolution of printing technologies assured ever more efficient systems of print production that permitted the reports of individual researchers to reach an increasing audience across an every larger geographic area. Printing empowered mapping as a means of study that was efficient in its use of space on the printed page as well as rhetorically compelling in its method of argument.

While physicians in the Caribbean and North America were focused on yellow fever, another disease was being generated elsewhere that would require a more sophisticated application of this evidentiary matrix. In 1781–1782, what we now know as cholera killed thousands of Calcutta citizens; and later, in 1783–1784, more than twenty thousand pilgrims died of it at Hurdwar (Morris 1976, 23). In 1817, this local Indian disease attacked British colonial forces, killing thousand of garrisoned troops in Jessor. Through most of the nineteenth century, cholera would be the disease around which medical researchers and public health bureaucrats struggled to understand disease generally. It would be within the study of this disease that a range of disparate but ultimately related epistemic things would coalesce into a more sophisticated methodology capable of more rigorously considering problems of disease and public health.

PART II

CHOLERA:
THE EXEMPLAR

CHAPTER 6

"ASIATIC CHOLERA": INDIA AND THEN THE WORLD

The idea of health as a broadly public enterprise was propelled in the early days of the industrial revolution by a recognition that the economy of the nation was at least in part dependent on the health of its workers. In 1714 John Bellers, a British Quaker, argued for the establishment of a British public health service on the grounds that untimely death wasted otherwise productive human resources. The health of the nation's workers was too important economically to be left to individual initiative or purely local oversight. Therefore, public hospitals were needed to serve not only as sites of charitable (or remunerated) care but also as teaching and research centers where investigations into the causes of disease might be pursued (Bernstein 2002).

The ideal of public responsibility for the care of those poor and ill was not new. Hospitals were the "great public health achievements of the Middle Ages" (Rosen 1993, 541), and in plague years, edicts calling for appropriate care, civil order, and efficacious prayer were common. As a public health measure civil authorities in many European countries would acquire land for "pest-houses," buildings where symptomatic citizens could be housed, and "pest-fields" for the burial of those who did not survive. Independent Italian states were leaders in this area in the sixteenth century, the first to experiment with public boards of health that had overall authority for regional health and sanitation. That model was debated but quickly rejected by Tudor officials in England.

Toward the end of the eighteenth century, however, in Western countries the idea of public centers of not only care but also research began to merge with the idea of a civil health bureaucracy capable of collecting, analyzing, and disseminating disease-related data. Seaman's New York City Board of Health, for example, became in the early nineteenth century Pascalis's Department of Health. Metropolitan health departments, especially those embedded in national systems of health concern, grew across the nineteenth century into essential loci of data collection and distribution.

With them the nature of data would change from an individual physician's collection of ten or twenty cases to public databases of hundreds and then thousands of cases.

At the same time, associations where professionals could meet to read research papers and hear the papers of others became important secondary research foci, even as advances in printing encouraged the dissemination of published findings in a burgeoning professional literature. The new lithographic printing process developed in Munich by Alois Senefelder in 1798 was not only faster but also less expensive (Twyman 1998, 47–49). It permitted the increased publication of charts, maps, and other graphics in texts whose distribution encouraged the construction of cholera as an international, pandemic experience subject to international study.

BENGAL: 1817-1819

Among the first reports on what would be the most studied and debated disease of the nineteenth century was prepared for the Bengal Medical Department by Dr. James Jameson and published by the colonial government press in 1819. There were other studies, for example one by Steuart and Phillips also published in 1819, but Jameson's *Report on the Epidemic Cholera Morbus as it visited the territories subject to the presidency of Bengal in the years 1817–1819* was the most frequently quoted in subsequent decades by British researchers. This is explained not simply by the innovative nature of its data but also by its very official pedigree. As its subtitle makes clear, the report was "drawn up by order of the Government, under the superintendence of the Medical Board." It was an early study of cholera and one that carried the imprint of the colonial machinery in India and thus of English officialdom.

Jameson's description of the disease he called cholera owed much to Sydenham's system of observation and classification. Cholera was, he wrote, a bilious fever that while described by neither Galen nor Hippocrates was characterized by a number of symptoms well considered by the ancients: diarrhea, vomiting, spasms, and in advanced cases, an erratic pulse. This new thing, while similar to the endemic, summer dysentery familiar to English medical practitioners (we now know it was caused by tainted food), was distinguished by its greater virulence and broad pattern of occurrence: "The disorder, as it lately visited India, was new in this alone, that there it, for the first time, assumed the epidemical form" (Jameson 1819, xiv).

Importantly, Jameson's report did not rely solely on his own clinical experience. When first confronted with this virulently diarrheic condition he wrote to 218 other garrison physicians in India, creating one of the earliest known physician surveys in disease studies and public health. He reported himself discouraged that only 125 of his fellow medical officers responded to his questionnaire (he apparently had hoped for a 100 percent return); and of those who did respond, 25 said they lacked first-hand knowledge of this new diarrheic disease. The 100 respondents who admitted to experience replied to a series of questions: what was the clinical presentation of the

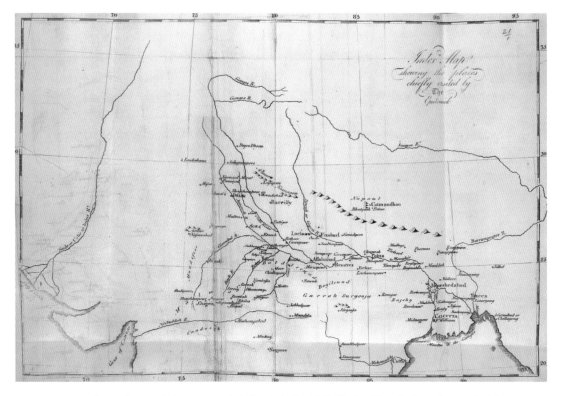

FIGURE 6.1 James Jameson's index map of cholera in India details the location of cities whose colonial officials reported cholera among the largely expatriate, colonial population in 1817–1819.

disease? When were the first cases observed? What was the epidemic's duration and how many fatal and nonfatal cases were treated? What treatments were tried and were any of those treatments successful? The responses returned added not only to Jameson's thinking but also gave a sense of communal authority to his writing. Here was not simply a lone colonial physician's ideas but the shared thinking of a coterie of British military physicians all facing a single set of epidemic symptoms.

With his report Jameson included a black and white "Index map shewing the places chiefly visited by the Epidemick" (fig. 6.1). It is neither a map of India and its population nor of cholera in India, but instead, a map of British garrisons in which the cholera had been reported in India. In the map those colonial placements were synonymous with cholera locations. Across a roughly configured country composed of coastline, some native cities, major rivers and northern mountains are imposed garrison towns whose symbology—a circle with a dot inside—is distinct from the simpler symbol for major Indian cities. The map thus argued the commonality of British military and political outposts as a single field of epidemic infection distinct from any occurring in the local populations of Indian cities and towns.

While Jameson's report made occasional reference to local "Hindoo" populations there was no attempt to consider cholera's affect on them. As occurred in the Caribbean with yellow fever, the assumption was that this epidemic was uniquely threatening to foreign colonials constitutionally at risk to conditions natural to native, nonwhite populations. In effect, there were two choleras, one that could be mapped and one that could not. The first affected Caucasians living in an environment that was unnatural to them and made them mortal prey to an indigenous disease. The second might affect local inhabitants, but because they were native to the land where the disease was spawned, and their constitutions were assumed to be different, its affect on them was thought to be minimal. The presence of the diarrheic condition among local populations was therefore assumed to be clinically irrelevant to its study in British troops and, administratively, its effect did not diminish native labor pools to the extent colonial administrators would worry about the health of Indian nationals.

Jameson looked for but could not see any systematic pattern of disease occurrence in his map. All he could conclude was the inexplicable presence of a "singularly erratic and destructive malady," a variant of well-known bilious fevers principally distinguished by first its intensity within a population—more people got sick at one time—and secondly the higher incidence of mortality of those diagnosed among British civilian and military expatriates in India (Jameson 1819, 33). As to the agent of the disease (what caused it) or exciting factors other than local climate, "No theory yet proposed will stand the test of scrutiny," Jameson wrote. Attempts to better understand its etiology "must be abandoned as placed beyond the reach of human curiosity" (Jameson 1819, 68). The question of causation could not be abandoned for long, however. Just as some Caucasians had shown themselves adept at adapting to the tropics, this torrid zone disease appeared capable of adapting to the temperate homelands of the colonial settlers.

POLAND: 1831

In the 1820s Asiatic cholera, as it began to be called, spread along established travel routes from India into first Asia and then Russia, where it stalled in 1823, only to return again several years later. In this second diffusion the disease reached St. Petersburg in 1829 (Hawkins 1831). As it spread the disease was transformed from either a geographically limited variant on a well-known diarrheic disease (British, or summer cholera) or an exotic colonial disease (Asiatic cholera) into an increasingly international, pandemic condition, cholera morbus. Alexandre Brierre de Boismont (1832) described it as "*la march du cholera-morbus dans L'Indie et Dans l'Asie Central*," adding central Asia to what Jameson and others had earlier identified as uniquely Indian. In its reports just prior to cholera's arrival in England, the British Central Board of Health—created by the government in part to prepare for cholera's likely arrival (Marks and Beatty 1976, 197)—described the disease as one "now prevailing in Northern Europe" (Central Board of Health 1831a). As such it became a general threat to the health of European populations and a subject of broad medical and political interest across Western Europe.

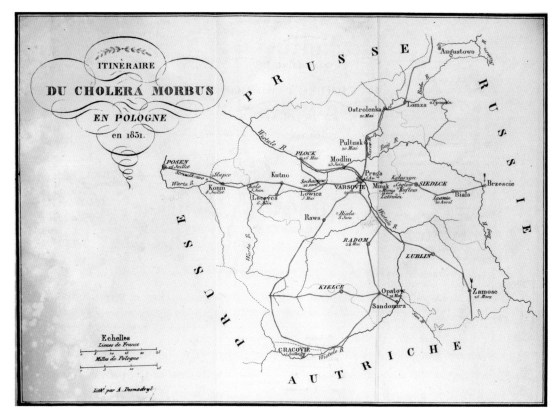

FIGURE 6.2A In his map of cholera in Poland (1831), Alexandre Brierre de Boismont argues cholera's progression from a central point through the country along a red line beside which the names of principal towns and the date of cholera's first report in them are noted.

Implicit in these studies of this particular cholera, and explicit in several, was the argument that the myriad appearances of cholera were not independent but the result of an orderly, predictable diffusion process that could be tracked. A map by Brierre de Boismont made this argument in its tracing of cholera's progress across Poland in 1831, for example. This map was included in Brierre de Boismont's book to argue a temporal progression of cholera across Polish cities in 1831. This newly named cholera morbus, the map argued, traveled in a systematic fashion from town to town along the country's roads and rivers.

Poland's topography was minimally sketched, mostly the rivers along which towns were located within well-delineated national boundaries that placed Poland in rela-tion to Russia and Prussia, where cholera had also been reported. To this Brierre de Boismont added major towns, for each of which he added, in a smaller font, the date of the first appearance of the disease. Joining these towns and dates was a red line signifying interurban routes. The map thus incorporated four propositions: (a) within a geography that was Poland (b) cholera morbus existed in towns at a specific time (c) and that its temporal progression could be mapped (d) along the nation's transit routes. As the disease moved from here to there, it could be seen to move from earlier

FIGURE 6.2B In this detail of Brierre de Boismont's 1831 map of cholera's progress in Poland, the red line links cases of the disease while dates of disease occurrence demonstrate its temporal progression.

to later. In March the disease was reported in Osterlinka to the north, Zamosc to the south, and Plock along the Weistale River. In April the disease had moved into Prague and Vasovie, and in the following months it advanced toward Posen in the west and Cracovie to the south.

Poland's cholera, the map argued, was a dynamic disease that spread contiguously (from Opatow and Zamosc to Radom and Lublin) and hierarchically, from Praga and Varsovie toward Posen. Its temporality was not simply seasonal—a "summer illness"—but also geographic, traveling in an understandable and perhaps a predictable way. Whatever else cholera morbus might be, the map argued, it progressed from town to town in a fashion that was similar to plague and dissimilar from traditionally understood, localized diarrheic conditions.

DANTZICK: 1831

In addition to studies of cholera's broad geographic progress there were others that attempted to investigate its nature in the cities where it took root. Assuming "nothing could prevent its spread over Europe" in the late 1820s, British officials dispatched physicians to study this new disease as it took hold in various cities (Morris 1976, 23). Officials needed to know whether cholera's importation could be prevented and, if not, how might its effects best be mitigated. For that they needed firsthand, medically educated data. Among those dispatched was Dr. John Hamett, who traveled to Dantzick on the Baltic coast of western Prussia to study cholera's effect. British officials chose not to publish his report, Hamett said in an introduction to a private printing, because of its description of cholera as a local rather than pandemic disease differed from official assumptions. "In opposition to the position of the government," Hamett concluded, "the disease has not been imported into Dantzick and that it did not prove contagious in that city" (Hamett 1832, x, viii).

That the British officials who had dispatched him to Dantzick chose not to publish Hamett's report was a tremendous blow to his prestige. The title of Hamett's publication included, therefore, the phrase "as transmitted to their lordships" lest the original authority of his efforts be lost upon readers. To further assert the legitimacy of his work, despite its rejection by British officials, Hamett included three pages of subscribers to the text he privately published. This was not mere pique but what today would be seen as a reaction to bureaucratic censorship. At issue was not the quality of Hamett's work but the theory he believed his data supported. He therefore independently published his findings to assure their distribution.

Hamett's argument that cholera was local rather than imported was based in part on the judgment of local medical officials, including a hospital physician, Dr. Baum ("In my experience the disease certainly proved not to be contagious"). Secondly, it was a conclusion resulting from the statement of the British consul in Dantzick who told Hamett that cholera morbus appeared "at a time when it was not known to be within a hundred miles of the place, and without there being the slightest trace of communication with any foreign means of infection." If there was no evidence of cholera's importation by sailing ship or overland cargo transportation, no mapped point of introduction, Hamett reasoned, "It must therefore have originated in the place" (Hamett 1832, 15).

If this cholera was locally generated then the secrets of its origin and source must be in the city itself. To explore this idea Hamett analyzed two tables of 841 cases of cholera reported at local hospitals. The first, organized by date, included for each patient a name, age, date of diagnosis, occupation, date of admission, and outcome (death or recovery). The second table included the same data but was organized by location, the street name, and number of each case. On the basis of the first table Hamett created a temporal portrait of the Dantzick outbreak, an epidemic curve in which a few cases rapidly built in number until the outbreak peaked and then declined as the susceptible but as-yet unaffected population grew smaller. The second table permitted Hamett to reference all the hospital-reported cholera cases on a map of the city and its streets (fig. 6.3A).

The map describes a walled city beside the Vistula River that extended along the Mollau Flus canals. Outside the wall are agricultural fields, and just north of the wall is a secondary settlement inhabited, according to Hamett, by the city's poorer citizens. In both the city and the northern settlement streets are laid out in neighborhoods identified by letters referred to in the text but not in a map legend. City dwellings are inscribed along the mapped streets and in these Hamett carefully symbolized the homes of cholera patients using a thick black bar for each hospital admission. Multiple cases at a single location are shown through a deepening of the symbol, in some cases partially or wholly inking the buildings (fig. 6.3B).

6.3B

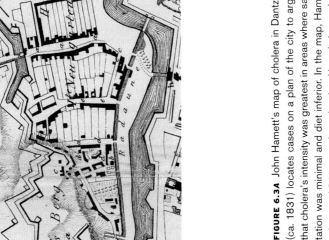

FIGURE 6.3A John Hamett's map of cholera in Dantzick (ca. 1831) locates cases on a plan of the city to argue that cholera's intensity was greatest in areas where sanitation was minimal and diet inferior. In the map, Hamett referenced 841 cases based on local hospital records to their street addresses.

FIGURE 6.3B John Hamett's map of cholera in Dantzick, detail. Deaths from cholera are symbolized by black bars, whose depth represents the number of reported cases. Letters in the map refer to places discussed in his text but not included in a legend.

6.3A

PLAN
OF
DANTZICK
In Illustration of
Dr. HAMETT'S ANALYSIS
of the EPIDEMIC CHOLERA in that City,
in 1831.

A Scale of 150 Rhenish Rods
Lithographed by Dudingston in Leicester Sq.

This is the first publication I have found of hospital admission tables and the first map based on hospital records of disease incidence. In the map Hamett identified clusters of hospital-certified cholera that originated in "close, low, and dirty alleys or places in which the air is penned up" (Hamett 1832, 132). From these sites cholera "continued to spread, irregularly, without any marked order from personal context or proximity, in low, damp, and dirty, or close and offensive situations all over the city, among the destitute and poor who are, in general, ill-clothed; ill-fed, uncleanly in their persons and dwellings; never wearing flannel next the skin; subsisting chiefly on indigestible and unwholesome foods, and in the habit of using pernicious drinks" (Hamett 1832, 23).

The map does not evidence the income or dietary habits of the denizens of areas where cholera was most concentrated. Hamett argued that on the basis of his and his local informants' knowledge of the city. To the pattern of disease distribution in the city Hamett applied a Hippocratic sense of disease as something necessarily generated in local airs and the personal habits of patients. Because he could identify no single point of disease importation and no obvious pattern of disease progression, Hamett concluded that cholera in Dantzick was a natural result of local odors and foul personal habits. "The progress of cholera is not that of the contagion [like plague], but is quite dependent on certain atmospheric states, modified or aggravated by unfavorable localities" (Hamett 1832, 160). These atmospheric states centered over poorer districts where "the *effluvia* from their persons" generated "deleterious infected air and personal communication with the sick" (Hamett 1832, 155). Dantzik's cholera was, in effect, like Seaman's eighteenth-century yellow fever in New York City.

Hamett did not deny the symptomatic similarity of Dantzick's hospital-reported cholera and the cholera reported in the British garrisons in India, the *suqs* of the Middle East, and the towns of central Asia. He agreed that the congress of symptoms identified by Dantzick hospital personnel were similar to those of the disease then being reported elsewhere in Europe. But because neither he nor his local informants were able to identify a single, obvious point of importation or a logical pattern of diffusion—one obvious in the map—Hamett was compelled to invoke a thesis of spontaneous local generation.

The presumed relation between disease intensity and the unhygienic habits of the poor was certainly nothing new. It was the living legacy of the Hippocratic perspective, advanced with Sydenham's modifications, and one that seemed to explain the data that Hamett collected from local civil and medical informants. As one historian put it, "Early writers on the disease constantly reiterated the common bourgeois belief that drunkards, layabouts, vagabonds and the idle, 'undeserving poor' were those most affected" (Evans 1992). The idea of odiferous, urban wastes generating foul, disease-causing airs was no less historical and no less useful in Dantzick than it had been for seventeenth-century plague writers in Norwich, England, or eighteenth-century yellow fever researchers in New York and Philadelphia. The spread of this new cholera was too extensive, however, and its local sightings too numerous for Dantzik's locally generated cholera to long endure. This was manifest in other mapped studies of its spread.

THE *LANCET*

In 1831 the *Lancet* published an unsigned report, what we might call today a meta-analysis of cholera studies, based on the published reports of Hamett, Jameson, and other contemporary researchers. Written before the first cases appeared in England but published as they were being announced, this study sought to synthesize what was known about the disease as it disembarked on British shores. The article included a map of "the progress of the Cholera in Asia, Europe, Africa, etc.," one somewhat limited by the reproductive abilities of the journal and the technology of printmaking it employed. At the time a real distinction appeared between the simple black-and-white maps increasingly included in professional journals and the larger, more detailed, often colored maps produced in books and official reports (Twyman 1998, 43–45).

The *Lancet* map (fig. 6.4*A–B*) asserted the progressive, international nature of Asiatic cholera. No dates were attached to the affected cities; no tables of incidence were included in the article it accompanied. Instead, in the map and text, the *Lancet* authors painted the disease as occurring simultaneously across the trading world from "Hindostan" at the southeast edge of the map to England in the extreme northwest. Within this geography circles around dots symbolized cities where the disease had been reported. The absence of a dot in a city symbol indicated cities that were cholera-free. In the context of the map these were simply cities in which the disease had yet to appear. The suppression of political borders reduced the "map clutter" but more importantly suggested the world was an open, borderless field across which cholera moved freely from city to city. The result was rhetorically powerful, letting readers make the connections. As David Turnbull observed in a different context, "internally, through the spatial arrangement of information, maps allow for enhanced connectivity" (Turnbull 2000, 91). The profusion of circled dots created their own connections implying not that Asiatic cholera had moved from here-to-there but more importantly that, as anyone could see, it's *every*where.

This *is* one disease, the map insisted, and it is present in every country, every region in the mapped world. In the article its authors boasted, "We have traced the pestilence through 700 irruptions, and shown it ravaging nearly 2000 towns. We have seen it cutting off in Hindostan one-sixth of the whole population, in the cities of Arab and in Persia a sixth, in Mesopotamia a fourth, in Armenia a fifth, in Syria a tenth, and in Russia, Poland, and Germany, a number not yet estimated with sufficient accuracy" (*Lancet* 1831, 252). Of course, the *Lancet* authors saw little of the disease personally. They did not themselves visit 700 different outbreaks, let alone 2,000 different towns. What they saw were reports of these events in the medical literature,

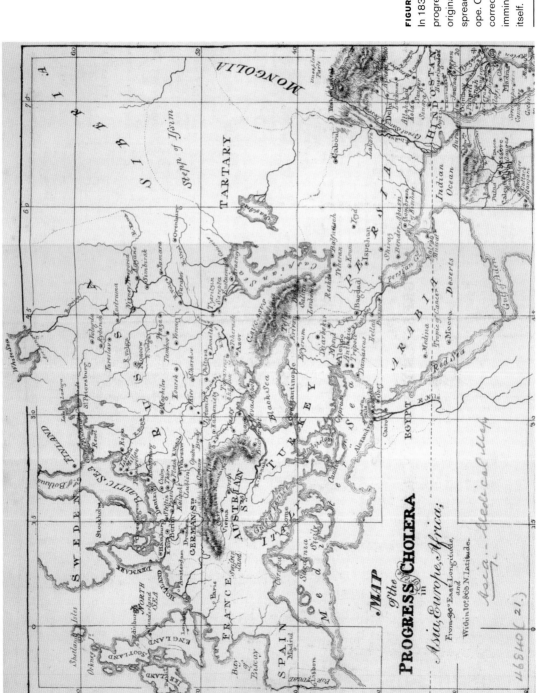

FIGURE 6.4A

In 1831, the *Lancet* argued a progressive, dynamic cholera originating in India that had spread through western Europe. On its basis the authors correctly predicted cholera's imminent arrival in England itself.

governmental reports, and in the popular magazines and news pages of the day. Reading had become the same as seeing, the literature a reality as certain as day.

The wide-ranging sources lent enormous authority to the *Lancet* authors' generally miasmatic explanation of cholera's origins, in the lower right of the map, "from the exhalations of the [sewage] tanks and morasses of Jessore (*Lancet* 1831, 267). "Never perhaps was there, in the history of the world, a more close and abundant concatenation of the causes, which transmute the decay of a vegetable life into the pestilence of the living animal, and never perhaps was a malady thus produced which then swept the world with more destructive virulence than that which we are about to consider" (*Lancet* 1831, 242).

That was the locus of the disease, the article said, the rotting refuse whose odors were generative, spawning this new disease thing. The mapped argument was that the disease born in Jessore had taken root everywhere *irrespective* of local climate or geography. "Elevation of territory possesses little influence on its [cholera's] severity," the authors argued (*Lancet* 1831, 244). That cholera "marched directly against monsoons," spreading in the direction from which monsoons blew, proved to the authors that "currents of air are not the [sole] agent in question, the disease" (*Lancet* 1831, 260). Whatever else it might be, this disease in its freedom from local geographies was not easily fit within the confines of a Hippocratic or neo-Hippocratic framework of knowledge.

The agency of its transmission, the authors argued, was evidenced in the "geographical notice we gave of the progress of cholera" that demonstrated "with singular clearness, the constancy with which the disease followed the track of ships, armies, pilgrims, caravans, and individuals, from one country to another" (*Lancet* 1831, 261). In other words, the *agent* of disease could not be seen but its progress, and thus its method of travel, could be traced in the map along the trade routes of humankind. There was no simple geography to this cholera *except* the geography of human travel. "We can only suppose the existence of a poison which progresses independently of the wind, of the soil, of all conditions of the air, and of the barrier of the sea; in short, one that *makes mankind the chief agent for its dissemination*. Such is the only conclusion which strict induction, and the argument by a motion of causes, permit any thinking mind to adopt" (*Lancet* 1831, 261; emphasis added).

Cholera was thus at once natural *and* unnatural. It was spontaneously generated in the foul odors of unsanitary Jessore and perhaps other cities. But it was carried along human networks of migration and trade by humankind itself. The *Lancet* authors speculated that cholera might be "regenerated in great quantities by those who suffer its influence," and thus that each city to which the disease was transmitted might become a new and independent center of exportable infection. To argue disease transmission along human networks was not, however, the *Lancet* authors cautioned, an argument for quarantine, a "savage system" politically odious in its restraint of trade and individual travel "which when tried on the continent had invariably been found productive of evil."

FIGURE 6.4B The *Lancet's* 1831 map of a dynamic cholera (detail) identifies cites where cholera had been reported with a hand drawn circle around a dark dot. Cities like London and Copenhagen, still free of disease at this point, are symbolized with an open circle.

Worse, quarantine had been shown to fail, the authors insisted, in the countries where it had been imposed. Here the map served a second evidentiary function. *If* quarantine were an effective barrier to cholera, it would not have spread across the continent but would have been stopped in places like Malta where quarantine was practiced. *Because* the map showed no country to be cholera-free, even those where quarantine had been attempted, the likelihood that isolation would stop its progress *anywhere* was doubtful. One might as well try to stop the wind, as Dale Ingram had argued a century before in relation to the plague (chapter 4), or the seasons themselves.

The Central Board of Health, created by royal proclamation in England in 1831 in anticipation of the arrival of cholera, had drafted a set of rules and regulations for the containment of ships arriving in English ports from cholera infected countries (Central Board of Health 1831c; Marks and Beatty 1976, 197). The board abandoned those proposals, however, in the face of fierce opposition from those who feared the economic consequences of diminished trade and travel. The *Lancet* authors "rejoiced at this salutary change [in official policy], and congratulate the British nation on its escape from that visitation, worse than pestilence,—the enforcement of a Maltese code of plague regulations" (*Lancet* 1831, 284). As a prophylaxis the authors advocated basic sanitarian measures—"ventilation and cleansing"—to diminish the foul local miasms that might serve as exciting agents of the disease in place. For the treatment of cholera patients in the slums of industrializing London the authors recommended Christian charity and "the organization of a system of effectual medical attendance for the poor inhabitants of a given district."

SUNDERLAND: 1831

The first cases of cholera in England were reported as the *Lancet* article was being set in print, a fact the authors had inserted at the end of their article. "Up to the moment of going to press we remained in anxious expectation of obtaining private information on the subject, but in this we have been entirely disappointed. By the official report, we learn, however, that eight new cases of the 'malignant' disease has been reported within the two days preceding of the 15th [October] instant. A considerable number of milder cases has occurred" (*Lancet* 1831, 284). The first to die was Sunderland keelman William Sproat who, according to his physicians, "became violently ill, had a severe shivering fit and giddiness, cramps of the stomach and violent vomiting and purging" (Morris 1976, 11). In the *Lancet* map, Sunderland is the only British town identified by a circle with a dot inside—a last-minute update to the data.

More than thirty thousand Britons would die of symptoms similar to Sproat's in the pandemic years 1831 through 1833, while more than fifty thousand others would contract cholera but survive. "One of the most terrible pestilences which have ever desolated the earth," the London journal *Quarterly Review* wailed in 1832 (Morris 1976, 23)—strong words in a city where the cultural memory of the seventeenth-century plague epidemic remained fresh. Plague, however devastating, was old and familiar. Cholera was a virulent new disease for which neither the medical community nor civic health officials could offer either effective protection or treatment.

And certainly this cholera, whatever its true nature might be, was terrifying. Some died within a day of the first symptom. Others lingered for several painful days of cramps, diarrhea, and vomiting. Hand-drawn illustrations of the day show pitiful figures spent from the physical battle the disease presented, darkened in pigmentation from the violence of its bleeding (figs. 1.1–1.3). Bloodletting, emetics, and opiates were the most common medical response by physicians whose patients were no more or less likely to survive an attack than those who faced the disease without medical care.

Both physicians and politicians were painfully aware that the new disease came upon England and its cities at a time in which urban populations were exploding. The size of the population at risk was therefore greater than it had been in previous epidemics. Also greater were the economic (lost wages, lost production) and social costs (orphans and widows requiring some form of support). As had happened in the plague year 1665, the number of local physicians was insufficient to treat all those who were ill. And again, the disease appeared to be most ferocious in the warrens of the urban poor.

The Central Board of Health submitted sixteen pages of proposed regulations to the Privy Council, most based on the assumption that cholera, if not bred like Seaman's yellow fever in fetid local airs, then its incidence was certainly promoted in the foul miasms of unsanitary urban slums (Central Board of Health 1831a). A public advisory informed citizens of cholera's symptoms: "LOOSENESS of the BOWELS is the Beginning of CHOLERA." As a prophylactic it cautioned citizens against inhabiting "crowded rooms" where close, fetid airs might encourage cholera's attack (Central Board of Health 1831c). Hamett must have taken a very personal pleasure in this. Here was his cholera, the Dantzik variant born of close rooms and bad food, argued implicitly in official public cautions.

OTHER WORLDS

Others joined the *Lancet* authors in trying to define the geographic range of this new disease. In the way of science, old arguments were returned to repeatedly, each new work trying to tease a deeper response out of the arguments of earlier author-researchers. In 1831, for example, Germany's Friedrich Schnurrer produced an international map of cholera using a ten-degree rectangular grid to map its presence in his *Die Cholera morbus* (1831). Schnurrer's study of cholera followed the 1825 publication of his great study, the *Chronik der Seuchen* (Chronicle of Epidemics), whose quixotic goal was an *enunciatio facti*, a collection of data on epidemic disease so complete as to render the need for speculative theory irrelevant (Brömer 2000, 183).

Appointed in 1814 as district public health officer in Enz, a few miles from Stuttgart, Schnurrer exemplified the "natural scientist" (today we might say ecologist) whose central argument was that "diseases do not exist autonomously, as do other products of organic life, but are instead bound to the distribution of the latter, and especially that of man, registers of disease are dependent on maps of mankind" (Brömer 2000, 185). In other words, disease could not be understood outside the reservoirs of human activity. To understand epidemic disease, one had to identify the human environments that promoted it.

Schnurrer's map (fig. 6.5) appeared similar to that of the *Lancet* authors but made

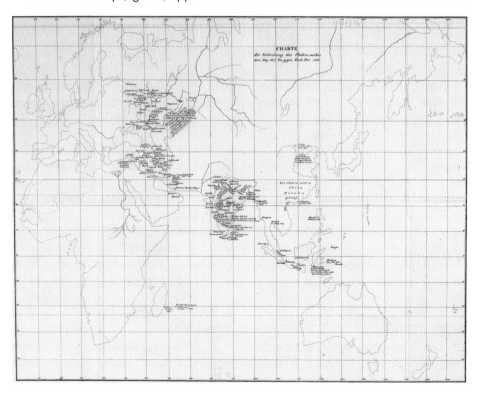

FIGURE 6.5 Friedrich Schnurrer's map of cholera (ca. 1831) argues a relationship between human settlement and disease as necessarily conjoined and interdependent realities.

a very different argument. In *Chronik der Seuchen*, Schnurrer insisted Asiatic cholera was not unique but that, as an anthropogenic disease, it shared characteristics with other epidemic conditions. The point of the map, to Schnurrer, was this: *where* there are cities there *will be* cholera. The argument was startling and profound. *Whether* it was born in the airs of the urban poor or in the trading patterns of nations, like other diseases elsewhere cholera was generated and then transmitted not by nature but by humankind itself.

In *Chronik der Seuchen,* Schnurrer used colored maps to present and conceptually link the catchments of various recurring diseases to argue their locational commonality: where some diseases prospered, others would find a welcome home. While Schnurrer believed that "their connective areas can be outlined on a map," he did not demonstrate it in his mapping (Brömer 2000, 185), nor did he attempt to map a temporal relationship that would link cholera cases in various parts of the world over time. But then, his cholera was a creature of human habitation, an environmental disease that arose in different cities where environmental conditions were favorable. If this cholera was a creature of human settlement he did not need to map the links between the cities—only their location.

In 1833 another map of cholera added a more temporal dimension that came close to testing the idea of cholera as an international disease progressing along trade routes from city to city (fig. 6.6A). Alexander Turnbull Christie was a British physician who reviewed reports from India and elsewhere to produce a 116-page monograph, *A Treatise on Epidemic Cholera; containing its histories, symptoms, autopsy, etiology, causes and treatment*. Christie's map filled in the blank spaces in Schnurrer's map including, for example, Malacca (1819), Macao (1820), and Siam (1820). For each city Christie inscribed in this map the date of cholera's first appearance. Tracing its geographic passage over time in the map argued that: "The disease has generally followed the course of great rivers, or of frequented lines of communication between different countries but this is more in appearance than reality, and is owing to various causes" (Christie 1833, 3–4).

One needed to see the geography to assert the link between cases, see the dates in the map to understand the order of disease appearance. Because cities and large towns were typically situated on rivers or on great roads, the progress of cholera noted in those towns (which amounted to Schnurrer's argument) would naturally suggest an association with travel routes (the *Lancet* argument). This was therefore a cholera of travel and not simply of local incidence. Because smaller towns were not reported and thus unavailable for mapping, the impression of cholera's progressive spread was possibly illusory, Christie cautioned. The implication was that finer, more complete data would be needed if the thesis of disease transmission along trade routes were to be proven.

In an 1832 study of cholera in India, Frederick Corbyn did connect the dots in his *The Epidemic of Cholera as it prevailed in India*. His data set was based on reports

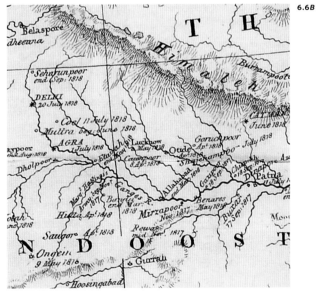

FIGURE 6.6A Alexander Turnbull Christie's map of cholera from 1817 to 1830 is centered on India and includes, with dates, those cities in which epidemics had been reported.

FIGURE 6.6B In Christie's map of cholera (detail), the dates on which cholera epidemics began in cities are noted under their names. Towns without dates are presumably included for the sake of geographic completeness.

of outbreaks in regimental hospitals in the country. Like Jameson in 1819, Corbyn ignored its behavior in the non-British, indigenous population. In his map were a number of Indian cities that served not as disease foci but as reference points for cholera-afflicted British garrisons in the map. The legend gives for each garrison the month and day cholera was first reported and the map square in which each can be found. A red line connects garrisons along local roads. Corbyn's grounding in classical disease theory prevented him from seeing a progression of the disease along those routes, however.

Wherever it might be and however it might travel, for Corbyn this was a Hippocratic illness of the type Sydenham would have understood. "The cause of cholera," he wrote confidently," is "to consist in inequalities of weather producing sudden check of respiration, a determination of blood from the surface to the center, and consequent inflammation internally" (Corbyn 1832, vi). In other words, the heat and humidity of the torrid zone country taxed the European respiratory system, causing changes in the blood and inflammation. This in turn resulted in cholera's manifest symptoms.

In effect, Corbyn mapped two very different Indias. The first was native and tropical (or subtropical), its geography carefully rendered and its population centers carefully located. Roads were lightly drawn to join those centers, many located along well-described rivers in principalities and kingdoms. Second, Corbyn mapped a nation of cholera-infested British garrisons linked by local roads distinguished by their color in the map. Living in these garrisons were British soldiers and officials affected by a disease born of local climate and geography. Cholera traveled from garrison to garrison along the roads in a manner that could be seen in the map. There was no need to map its incidence among native populations, because the disease, as Jameson had argued, uniquely affected the delicate constitution of the colonial white population.

Other researchers argued the anthropogenic nature of cholera's passage along trade routes. Dr. Amariah Brigham's 1832 cholera map, for example, published in his *Treatise on Epidemic Cholera*, was, like Schnurrer's, the graphic instantiation of tables of data listing by date of onset the world cities in which cholera had been reported. Dates for some of those cities were noted on the map, waypoints of cholera's progress. Unlike maps by Christie, the *Lancet* authors, or Schnurrer, Brigham's included the Americas, where cholera first appeared within months of its first occurrence in England. More importantly, Brigham's map linked the cities where cholera appeared with a red line that argued its progress city-by-city and month-by-month from India around the trading world.

Like the map by the *Lancet* authors, Brigham's map (fig. 6.8A) did not support the idea of a simple climatic determinism. Cholera was everywhere, equally evident in tropical and temperate climates and in both mountainous regions and flatlands. At the same time, Brigham's map argued concretely what the *Lancet* authors had only asserted inductively: "The poison of cholera may be conveyed and communicated by man to man." His red line defined the progress of the disease over time from city

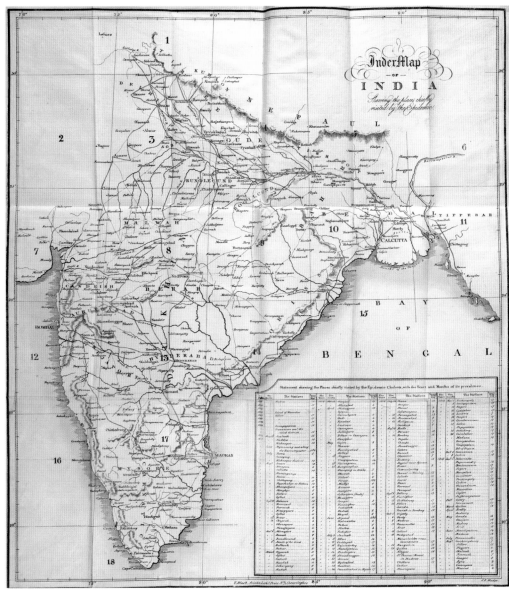

6.7A

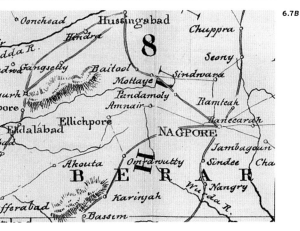

6.7B

FIGURE 6.7A In 1832, Frederick Corbyn mapped the history of cholera in England within British regimental stations. The date of each reported occurrence is included in a table of British regimental locations to describe the temporal progression of the disease.

FIGURE 6.7B This detail of Corbyn's map shows routes connecting British garrisons within the landscape of Indian towns and cities.

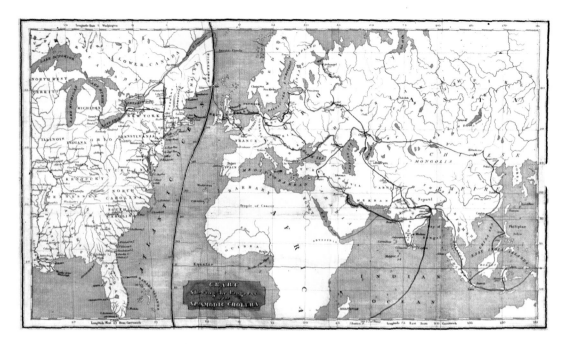

FIGURE 6.8A A red line in Amariah Brigham's 1832 map of cholera links cities where cholera was active to describe the route of its expansion over time from India to Europe and then the Americas.

to city and nation to nation along *both* land and sea routes. In this way, Schnurrer's thesis of human agency was located not in the towns and cities themselves but in travel of trade and exchange that linked them, one to another.

None of these maps would have been possible without a vast international bureaucracy capable of collecting local reports of cholera and making them available nationally and internationally. It was with real pride that the *Lancet* authors announced their tracking of cholera around the world, noting occurrences in towns and cities unknown to most of its readers, places whose mapped significance existed only in the context of cholera's diffusion. Each dot of each town made cholera real, diffuse, and pandemic. The data required for that map, crude as it appears to us, was massive. Catalogs of cholera geographies became themselves a research genre whose publication assured a common base of data for researchers from India to New York City.

In 1832, for example, Henry Schenck Tanner privately published in Philadelphia *A geographical and statistical account of the epidemic cholera, from its commencement in India to its entrance into the United States . . . Compiled from a great variety of printed and manuscript documents.* The thirty-five pages of this pamphlet were near encyclopedic, including entries for Europe, Asia (China, India, Persia), Canada, and the United States. For each nation data included, where available, the date and month the epidemic began, the total number of reported cases and the number of deaths, as well as the duration of the epidemic in each city or town. The entries for the United States, which extended from the Atlantic states as far west as Missouri,

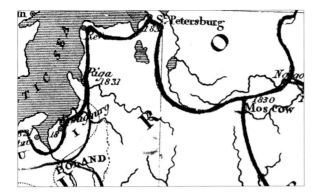

FIGURE 6.8B By joining cities to which dates had been appended, Brigham's map argues a spatial disease pattern over time that followed then existing trade routes. It was with this map that proposals of a disease that spread across human travel and trade programs was first, and best argued using international incidence data.

were especially complete. For New York City and Philadelphia the weekly report of mortality and morbidity was included on a week-by-week basis.

The collected data was described in a temporal world map; each country is colored according to the date cholera first appeared in any single country. Thus for "the first period" of reportage, 1819 to 1829, nations are green; for 1830, a year of rapid international progression, countries are yellow. Nations like England, where it appeared in 1831, are blue; and countries like the United States, where cholera first took root in 1832, are red.

Two supplemental maps detailed the diseases location and progression in India, "or Hindostan," and the United States "to give a more comprehensive view of those regions, than is afford by the general map; the character and small scale of which, prevent that detailed representation of those interesting quarters, which a larger and more complete map enabled me to give" (Tanner 1832, iv). For the United States, a small red dot was used to "indicate each of the principal towns which had been visited by cholera.[1]

In this mapping Tanner self-consciously avoided the argument of those, like Brigham, who wished to assert a spatial progression of the disease rather than a spontaneous eruption of the disease in many places. "It will be perceived that I have coloured the entire surface of each," Tanner wrote, "as represented by the map of the world, and thus avoided that common and absurd mode of delineating, by definite lines, the exact route of the pestilence, from place to place" (Tanner 1832, iv). Others might read in the temporality of his data a contagious disease that traveled the sea and land routes of commerce but Tanner refused the idea, crafting a map that argued different temporality but a more spontaneously occurring disease without human agency.

CONCLUSION

The choleras argued by different researchers in this first pandemic, which ended in Europe in the mid-1830s, would be debated through the 1870s. Four separate, and to some extent contradictory, choleras would be discussed and debated at a range of scales. Each would require a different type of study, a different avenue of scholarship. First, there was a congress of familiar diarrheic symptoms that were often difficult to distinguish from more familiar endemic gastrointestinal disorders. The commonality of these symptoms required this cholera to be distinguished from other diarrheic diseases on the basis of its virulence—more people died—and the sheer number of people sick with it at any single time. The goal was to understand the pathology of this condition and to develop better methods of clinical treatment.

Second, there was the natural, miasmatic cholera born in the warm moist airs of the torrid zone. Inimical to Caucasians if not local peoples, it was as much a part of local climate and geography as the rivers and mountains. This cholera was studied, as yellow fever had been, through a careful recording of population incidence in relation to barometric pressure, rainfall, and temperature. That this condition then somehow appeared in the dry warm summer seasons of other climates was not easily explicable. Through the 1850s, however, to study cholera would for many researchers entail at least some attention to climatic variables.

A third cholera was something unnaturally promoted, first and foremost, in the fetid airs of the industrializing mercantile city, and at another scale, in the rooms of densely packed areas of the city that housed the poor. It was a creature of poor sanitation and a response to it would require of urban officials more attention to housing and sanitary infrastructure. To study this cholera would require careful and detailed analysis at the finer scale of Hamett's investigation, and perhaps, at the even larger scale of individual neighborhoods.

Fourth, there was the poor people's cholera. Promoted by Hamett, this was a disease of bad diet, bad hygiene, and a general lifestyle that, if it did not breed disease, certainly promoted it. The argument was old and supported by evidence of increased incidence in areas where the poor lived in unsanitary surroundings. There was history here as well. The same argument had been made about plague and yellow fever with similar geographic concentrations to support them. The map made it more powerful, perhaps. And for this cholera the responsibility for its prevention lay not with officialdom at all but with those who would become ill because of their own foul habits.

Finally, there was the cholera that Arrieta would have understood, one born of international trade patterns. It somehow traveled from city to city in a manner whose agency was unknown but whose effect was evident in the map. This cholera was located in the trade routes of commercial travel, its incidence affirmed in the published reports of incidence in trading nations across the world. Studies of cholera's peripatetic travels required a literature in which international cases were reported and a means by which that literature could be accessed easily and then integrated conceptually to make an argument about the disease itself. That would require, eventually, a progressively international system of disease definition, classification and reportage.

The growth of the international as well as national disease literatures was part of a new matrix of technologies that would serve to advance not simply cholera studies but disease studies in general. It would be based upon the increasingly exact records of disease incidence in towns, cities, and nations around the trading world. Some of this data would be the result of military reportage like Jameson's, some of it would benefit from new systems of urban data collection like the hospital records employed by Hamett. Whatever its origin, however, this public data would become the essential medium in which theories of health and disease first were developed and then argued to an expanding jury of concerned officials and interested professionals.

Finally, all these choleras would be need to be considered as at once clinical, climatic, spatial, and temporal things. Whatever the individual cholera argued, all these elements were would be necessary attributes to consider at a range of scales of address. The wealth of data was too complex for a simple inductive argument, too vast for a simple statement. Mapping was becoming an essential medium for all these cholera, one in which tables of data were transformed into arguments. *If* it traveled, *then* its passage could be seen in the map. *If* it was a creature of local airs, *then* a local source could be identified in the city. *If* it was climatic, *then* its presence would be restricted in maps to areas of heat and rain. Evaluating these very different arguments would be the work of two generations.

CHAPTER 7

BUREAUCRATIC CHOLERA

Cholera was one of a number of diseases adversely affecting the body public whose causes were unknown and whose etiology was unclear. Together, these conditions attacked the working population, the factory and shop workers who were the labor base of the industrializing nation. If nothing else, that made of public disease a subject of increasing state interest. To understand them would require a matrix of bureaucratic systems facilitating the consideration of their general origins (foreign or local), immediate sources (air, soil, human, or water) and the agencies of their transmission. While cholera was not the only disease of importance, it was *the* disease of the nineteenth century, the one that—like plague and yellow fever before it—compelled unceasing concern. In effect, a new cholera came into being, one that was the subject of official attention (Newsholme 1927, 115).

A new complex of official activities involving four separate but related elements would be required for the matrix of nineteenth-century disease studies to advance. First, a system of data collection ensuring accurate details of mortality and morbidity within a carefully described population was required. Traditional reportage catchments, those relying on parish and town reports, were insufficiently precise. Only a comprehensive system of population surveillance capable of reporting precise data at a range of scales would answer. Second, for this to be brought into being, local, regional, and national maps would need to be remade in a manner permitting accurate data catchments to be constructed. Third, new methods of analysis capable of analyzing not just scores or even hundreds but thousands of disease reports would be needed. Fourth, to be of use these things would have to be joined through a system of distribution permitting the widest sharing of both the data collected and reports based on its analysis.

As these elements came into being, medicine was transformed from a study of disease in the individual or individual locales into a study of disease as a broadly public phenomenon and thus as an official responsibility. One sees this first, perhaps, in France where in 1815 physicians founded the Royal Academy of Medicine to advance the goal of disease prevention through a focus on the environmental origins of disease in the general population (Ackerknecht 1948a, 123). In 1820, the *Annals d'Hygiène Publique et Medicine Legal*, a journal dedicated to public sanitation as a clinical prophylaxis, was first published. Editorially, the *Annals* brought together an ecumenical congress of persons interested in public health: bureaucrats, lawyers, physicians, and a few independent representative workers.[1] French ideas about disease as a general and public health responsibility rather than a local, private concern were exported across Europe and to the Americas, traveling like cholera, whose expanding presence was a spur to their progress.

Underlying the whole was an argument that insisted on disease as unnatural, as anthropogenic and environmentally grounded. The motive idea was prepositional: *if* "inscrutable vapors" that arose from urban pools of excrement and decaying vegetable matter—the source of deadly and largely untreatable "intermittent fevers" (Forry, 1842, 39)—were causal, *then* the necessary response would be to attack those miasmatic sources in place through public works. *If* society cleaned up the rotting urban refuse, assured fresh water for the population, and drained the marshy areas, the sanitarians argued, *then* the incidence of cholera (and other diseases) would surely decline.

The general thesis would have been familiar to generations of previous researchers: from Hillary in Barbados to Seaman and Pascalis in the United States. The new French perspective was distinguished by a *méthode numérique* that argued for the quantification of the incidence of disease within a population group in a manner permitting theories of generation and transmission to be tested against a range of potentially causative phenomena (Porter 1998, 406–8). Seaman's limited, primarily visual spatial correlation of yellow fever intensity and proximity to waste sites in New York City in the 1790s was to become a more precise, numerically spatial argument. That in turn required maps in which suspected environmental problems could be located in a manner permitting their correlation to accurately reported disease incidence in specific populations.

This is clearly seen in the French tradition in which mapping "the distribution of mortality in cities struck by the disease" was based upon an already existing tradition of urban mapping generally (Picon 2003, 140). By 1815, "more than 3,000 partial maps had already been realized by the administration in charge of the quarries" (Picon 2003, 142). Produced for nonmedical purposes, these and other urban maps were to serve as the backcloth for studies in which disease would be mapped in place. The appearance of cholera, which in 1832 claimed 18,000 lives in Paris alone, required responses that drew upon this map library. It is not simply, as Robinson argued, that cholera stimulated an interest in disease mapping (Robinson 1982, 170). More fundamentally, mapping permitted the construction of disease events like cholera. Consequentially, spatial theories about the nature and origin of these diseases, and the pathways of their diffusion, were advanced in maps arguing environmental and social determinants.

THE POSTAL SERVICE

Resulting arguments and the theories they advanced would have had limited impact without a system encouraging their distribution. In the nineteenth century this system was postal. It was through the post that Boyle's general jury of the knowledgeable was constituted. The arguments of researchers produced in books, government reports, pamphlets, and professional journals traveled to an expanding community of readers through the mail. Each nation's postal history is slightly different; all reflect a more-or-less similar attempt to structure local, regional, national, and international exchanges (personal and professional) in a systematic and cost-efficient manner. Britain's history serves as a convenient example whose transformation is well documented, specifically its relation to disease studies in general and cholera studies.[2]

In 1837 the lithographer James Basire was commissioned by the House of Commons to create a map of England and Wales, centered on Leicester, to advance the nation's program of postal reform (fig. 7.1). Posted in the map are rail and coach routes linking cities and towns as well as horse and foot routes joining local post offices and suboffices. Using Ordinance Survey data, Basire's 1838 map created postal service areas in which service time and distance were conflated in circular zones. These zones were also economic: the more distant a city, the more it cost to post a letter to it. The reason was *in* the map's argument: *if* geographic destination was further, *then* it would take more time to travel and therefore would cost more to get there. In this manner the national map was redrawn on the basis of the speed of the post from a central postal station to everywhere else.

At another scale, the Commissioners of Post Office Management in 1837 commissioned a map of the range of local, two-penny stamp delivery in London. Cartographer James Wyld used data from a British Ordinance Survey map to create the local service delivery area in which the two-penny stamp would serve (fig. 7.2). The map's circularity and its system of projection suggest an imaging similar to that employed by microscopists who drew tumors and human cells as seen through a microscope. Transportation lines serve here as circulatory pathways across the general of the organically conceived city, a microscopic part of the national body in a manner Christopher Packe, in his medical description of Kent's local geographies, would have applauded (see chapter 5).

120

It would be a mistake to dismiss these postal maps as tangential to disease studies. If cholera traveled by carriage, ship, and train . . . so too would cholera studies. The postal system enabled both the delivery of local data on disease mortality and the national and international distribution of official reports based on that data. As importantly, the post was the medium delivering to readers the books, medical journals, and

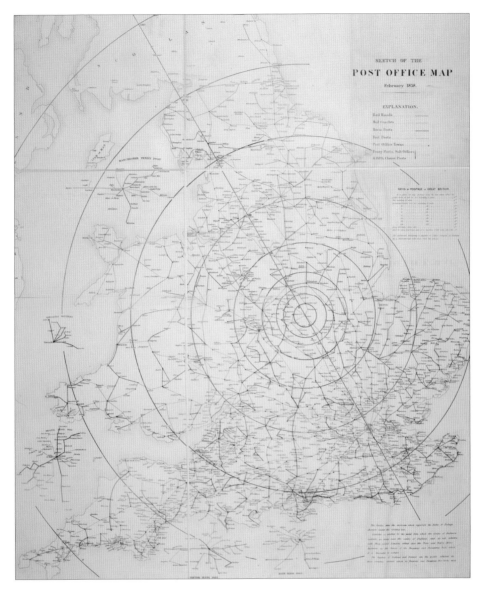

FIGURE 7.1 This 1838 postal reform map by lithographer James Basire, commissioned by the House of Commons, describes a cost-distance formula for mail delivery centered on the city of Leicester.

newspapers in which new research was reported. The British system was only one of many that together created a network of international exchange ensuring the rapid diffusion of studies, and in many cases the data on which they was based, among American, British, and European researchers. The mail also enabled the growth of an increasingly public press whose general readership was capable of exerting political pressure on public officials charged with population health and safety.

Technical advances in the collection of land survey data permitted ever more precise constructions that would serve a host of public initiatives. Certainly, it permitted a better construction of the national post. And without new topographic survey techniques drainage, embankments, and sewer projects would have been difficult to conceive and impossible to complete. In the mundane mapping of the postal system can be seen as well the development of a series of analytic tools that also would be critical to disease and public health studies. The concentric circles used to create postal zones would become the buffers used to identify disease outbreaks centered on suspected disease sources, for example. The transformation of time-distance tables for post offices into postal zones provided a mechanism by which distance and time could be together considered in disease diffusion. Finally, and no less significantly, it was in the mapping that governments created population reportage catchments in which clinical and general demographic data would be first collected and then analyzed. None of this happened overnight, of course. Again, each national history is specific but all would follow a similar developmental pattern.[3] None of it would have happened at all without the tools mapping provided.

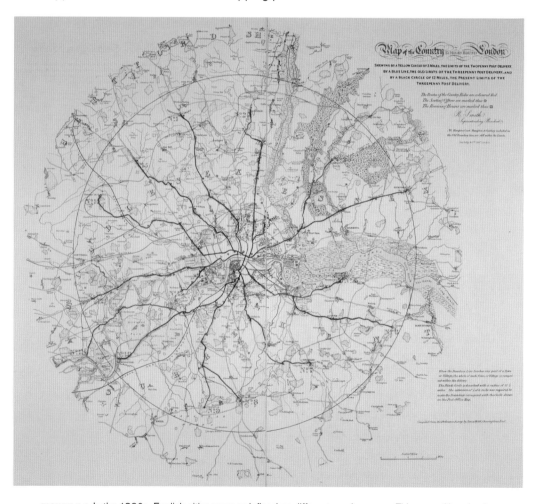

FIGURE 7.2 In the 1830s, English cities were redefined as different service areas. This map of London by James Wyld was commissioned by the House of Commons to show the limits of local (two-penny stamp) delivery. The London that results looks remarkably like the images of tissues seen under low power microscopic magnification in biology and medical texts of the day.

PUBLIC HEALTH DATA

In 1836 British Parliament brought into being the General Register Office (GRO) through, first, An Act for Marriages in England and then An Act for Registering Births, Deaths, and Marriages in England (Eyler 1979, 37).[4] Since 1539 parish churches had been required to keep records of baptisms, marriages, and burials. Bills of mortality based on the weekly report of parish clerks had been available since 1592, the basis of John Graunt's early analysis of British mortality. The data historically reported was typically incomplete, however—sometimes haphazard—and therefore unreliable.

All this would change under legislation mandating a system in which adoptions, births (including stillbirths), deaths, and marriages were to be systematically reported to a local, officially appointed and therefore publicly accountable registrar. The GRO would not only collect but also make generally available both mortality and socioeconomic data enumerated in a national census it helped organize. For this to happen maps of the nation had to be transformed, the old parish-based divisions redefined to create comprehensive and trustworthy civil reportage catchments. In the 1830s England was divided into 636 registration districts, each with a registrar whose job was the collection of relevant data and its transmission to the GRO in London. These were typically composed of two or more registration subdistricts, each often with its own local registrar. By 1837 there were in England and Wales more than 2,000 registration district and subdistrict registrars.

Other countries, most notably France, were engaged in similar national collection initiatives. These were medical, urban (sewer lines and street networks), and geographic (mineral resources, topographic surveys, and the like). In areas of social importance the idea was the same as in England, with local collection and assemblage of data transmitted to a national center for analysis and distribution. In the case of Britain one can see the transformation from parish and borough to registration subdistrict and district by comparing a map of metropolitan London's boroughs from the early 1830s (fig. 7.3) with another of registration districts created in the 1840s (fig. 7.4). In the first, traditional boroughs, each distinctly colored in the choropleth map, are listed in block capital letters. Suburban boroughs—for example Islington and Hackney—are labeled in lowercase italics. Within each can be found major streets (such as Oxford, Regent, and Piccadilly), along which are symbolized city blocks. The second map shows registration districts at the time of the 1851 national census. These administrative areas, while similar to boroughs, were part of an interlocking system of registration subdistricts and districts from which the bureaucratic metropolis was constructed (Kain and Richard 2001). Red lines also divided the metropolis into regions—northern, southern, and central registration districts—composed of adjacent boroughs.

The scale did not permit the mapping of the individual streets in which enumerators did their work. That would require a finer scale of mapping at the resolution of the registration subdistrict. The coarser maps of registration districts insisted upon the

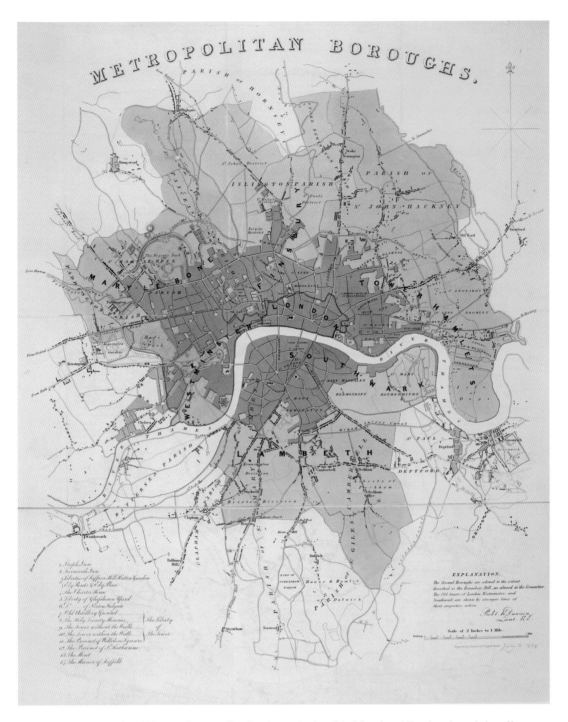

FIGURE 7.3 An 1832 map of metropolitan London at the time Britain's reform bill redrew boundaries of individual boroughs, preparing the field for what would be the creation of registration districts within them.

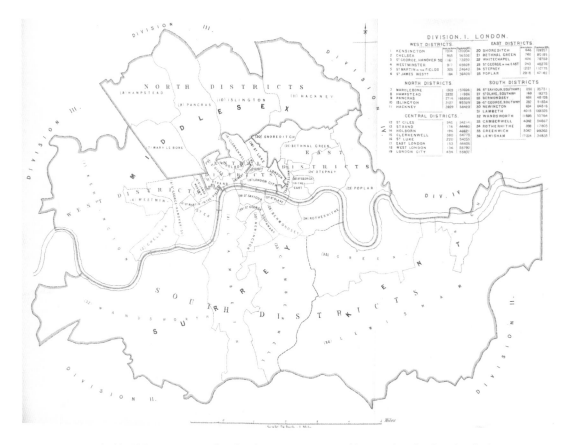

FIGURE 7.4 In this 1851 map, metropolitan London was reconstructed into a series of registration districts and registration subdistricts that, beginning in the 1840s, were used as catchments for demographic, economic, and health-related data.

city as an administrative entity, an existential reality distinct from the elements of any single division. Whatever their differences, the map argued, city core and hinterland were together the metropolis, the cartographic whole far greater than the sum of its constituent parts. It was this argument of separate parts unified into an organic whole that would come to inform everything from William Farr's 1852 study of cholera to the seventeen volumes of Charles Booth's monumental *Life and Labour of the People*, published in 1902.

CHOLERA IN THE CITY

New York City

It was at the scale of the city, and often the city district, that cholera was engaged not only by the British but also by American, French, and German researchers. Data from local boards of health and individual reportage areas were collected first locally, then regionally and nationally. Ultimately, the results of work at every scale creating an international base of cholera studies distributed the world over, widely available and generally shared. Increasingly, mortality reports were published in broadsheets and daily newspapers for the interest of everyone.

In city after city, the fact of official cholera, tallied by local health officials, was tested against the assumption of its miasmatic nature. Did it appear, as yellow fever had, in areas proximate to odiferous wastes? Did it appear in districts where unsanitary conditions (marshy lands or crowded tenements) prevailed? North America provided a field of inquiry as active as England's and Europe's. After cholera was reported in Montreal, Canada, in 1832, the boards of health of Philadelphia and New York sent physicians to that city to see what they might expect just as, several years before, British health officials had been dispatched to Europe, Poland, and Russia. They returned, As John B. Osborne has recently written, "convinced that cholera was not spread by contagion but was a local phenomenon, bred in filth and spread by miasma, striking victims who suffered from a predisposition to the disease. Their observations of cholera in Montreal confirmed these theories on the etiology of the disease" (Osborne 2008, 29).[5] The proof could be seen in the relationship between disease incidence and the socioeconomic and geographical profile of the city districts or wards where the disease was most intense.

Consider a map of the 1832 cholera epidemic in New York City (fig. 7.5), first published in David Meredith Reese's *Plain and Practical Treatise on Cholera* (1833). His goal was to argue that cholera was a miasmatic disease generated in the foul

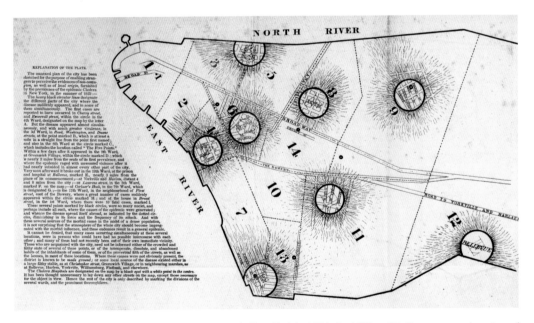

FIGURE 7.5 In an 1833 study of cholera in New York in 1832, David Meredith Reese mapped centers of intense cholera activity in lower Manhattan to argue that local environmental and social conditions were responsible for each separate outbreak.

airs of densely settled, unsanitary New York immigrant streets where outbreaks were especially severe. The *Lancet*'s contagious, spatially progressive cholera (fig. 6.4*A*) and Brigham's cholera of the trade routes (fig. 6.8*A*) were transposed by Reese into a disease more like Hamett's in Dantzig (fig. 6.3), a local disorder with local causes.

To make the case in New York City, Reese relied on weekly mortality reports released to the newspapers by the New York Board of Health, the same data that found its place in statistical compendiums like Tanner's 1832 collection of international cholera mortality reports. In his map arguing cholera's local geography, Reese inscribed an area of lower Manhattan, bounded by the North and East rivers, in which were inscribed the local ward structure of the city's administrative catchments. Numbered 1–12, the character of each was discussed in Reese's text. A more precise, local geography was suggested through the inclusion of Broadway, the Bowery, and Canal streets. Within this geography are eight circles, each a distinct locus of extreme cholera activity, each circle identified by a capital letter referenced in the text. In each circle can be seen the outline and names of local streets that appear to be magnified. It is as if each were separately under a microscope that revealed local circulatory pathways.

Hatchings surrounding each circle argue the diffusion of cholera from each locus to surrounding if unnamed streets. Circle 12 is labeled Bellevue and signified an outbreak not in the homes of local residents but at the hospital with that name, an outlier distinct from the others both in its geography and its institutional focus. Circle 4 was centered on Roosevelt Street near the Old Slip area where yellow fever had been previously so intensively studied. Its proximity to circle 6, and in it Mulberry Street, suggests a single outbreak denied by the structure of the map that argued their separateness.

Each circle identified a distinct outbreak; each outbreak was explained by the presence of local conditions Reese believed were generative. These included pools of odiferous human and animal waste, like those earlier described by Seaman and Pascalis, and at another scale, the stale, close airs of densely settled and badly ventilated immigrant apartments. The origin of cholera might have been foreign, Reese argued, but its *source* in New York's Manhattan wards was in the malodorous condition of the streets and the unsanitary homes where the disease was most severe. Were it otherwise, the separate outbreaks would have been mapped as a single occurrence. While local conditions might have explained pockets of intense mortality if cholera were solely in the air, or a general characteristic of the soil, then the whole would have been a disease surface rather than a series of distinct, diseased pockmarks. And— as Hamett might have said—if it were diffused from ships in port then it would have had a single locus of dockside introduction from which the disease could be seen to spread rather than, as Reese's map argued, eight separate independent outbreaks, each with its own localized pattern of spread.

FIGURE 7.6 As this 1830s magazine illustration makes clear, the sources of cholera were assumed to be airborne, something generated in the stench of manure and refuse in the city.

Rouen, France

A very different cholera can be seen in a map made for Dr. Eugène-Clément Hellis, chief of medicine at *l'Hôtel-Dieu* hospital in Rouen. In 1832 Hellis published a short commentary on a published report by M. E. Dubuc, who had traveled to Sunderland and Newcastle to see this new cholera and report back on it to Rouen officials (Dubuc 1832). In 1833 Hellis published his study of cholera in Rouen and the surrounding district (*Département de la Seine-Inférieure*) in his monograph *Souvenirs du Choléra en 1832 à Rouen* (Hellis 1833).[6] Engraved by A. Periaux and produced by the Paris mapmaker Pecquereau, more than one hundred cases of cholera were located in a dot map of Rouen, each dot symbolizing the home of a cholera victim. The map defined Rouen's west and east banks, separated by the Seine, in which two islands, Ille Petit Gray and Ille Lacroix, anchor bridges that join the two sides of the city and the streets inscribed in them.

128

Cholera clusters occur in three different areas: on the riverbank at the northern tip of Ille Lacroix, in the area east and south of Ille Petit Gray, and along several roads at the northwestern edge of the map. Outliers can be found in almost every map quadrant. The map thus presents an intense disease experience focused in areas of a compact city economically and geographically dominated by a river. The conclusion, the text made clear, was that its cholera was a portable disease "imported" from Sweden, Nantes, Le Havre, Rochefort, Brittany, Spain, and Eire. This was argued by the twenty-nine cases located in visiting ships on the Seine and another thirty-three cases on streets adjacent to the river's west and east banks (fig. 7.7A–B). Were it otherwise, the cholera cases inscribed on the water would have made no sense. Once introduced by trading ships cholera then spread, Hellis wrote, like "spores" in environments conducive to their growth. Those environments, the ones where clusters of mortality could be found, were the dockside areas where prostitutes lived, sailors found lodging, and the poor of the city were housed.

The map draws the reader into this urban environment, the text describing its conclusions, insisting readers *see* and consider both the maritime source of disease contagion and the areas of the city in which cholera proliferated. The map thus argued cholera as two things. First, it was an imported disease brought to Rouen on the sailing ships that visited the city from different countries. Second, it was a spore-like thing that spread from visiting ships through the poor parts of the city where sailors found

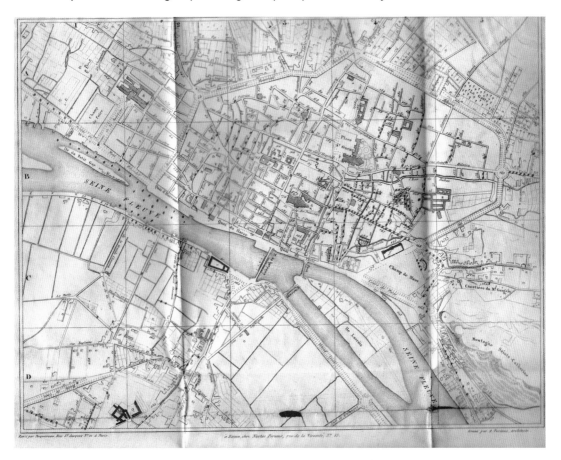

FIGURE 7.7A An 1833 map of cholera in Rouen, France, in 1832. Cases are represented by red dots. Where the disease seemed extensive, city streets are marked with red lines.

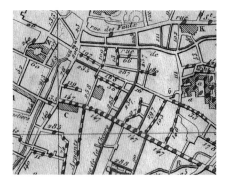

FIGURE 7.7B In this detail of the 1833 Rouen cholera map, red dots of disease are located in the river where ships were berthed and dockside, where visiting sailors found food and lodging.

temporary lodging and other, marginal elements of the citizenry offered them services. It was this social geography that provided the fertile field for cholera's expansion. Cholera was portable but it only took root in socially marginal neighborhoods.

While Hellis's map argued the experience of one city his text placed Rouen's experience within the greater epidemic context. Others had noted cholera's local source as maritime, spreading from ports into cities and from the dockside of those to neighboring towns. The Rouen experience was yet another example of cholera's ability to spread from trading ships to trading cities. And to the nineteenth-century researcher for whom plague was the paradigmatic epidemic, and portability a defining feature of its threat, this made sense. As important, the map insisted upon cholera as a general urban experience rather than Reese's mapped cholera of separate but largely unconnected outbreaks sourced to local sanitary conditions. The social bedding in which the spore-like disease flourished in Hellis's cholera did not blame the poor for their dissolute ways but instead placed the onus for cholera's spread on the greater social environment that might include but was not limited to the dietary and hygienic habits of the poor.

Hamburg

Yet another cholera, one in a different civil casting, can be seen in Rothenburg's map of cholera in Hamburg, Germany, in 1832 (fig. 7.8). It was of sufficient importance to Richard Grainger at England's Board of Health that in 1850 he included a version of it in his report on cholera to Parliament. In this map official data on cholera incidence was used to generate ranges of cholera intensity across the whole of the city surface using three shades of red, the most intense corresponding to the greatest incidence. The map, Grainger wrote, "places in a striking point of view the predominating influence of locality over the progress of the disease" (Grainger 1850, 199).

The geography argued in this map is primarily physical and not social, however. The map's authors chose to argue cholera as a natural phenomenon with at best secondary relation to the different social levels of the districts of Hamburg. By excluding the city's political division from the map its authors implied the irrelevance

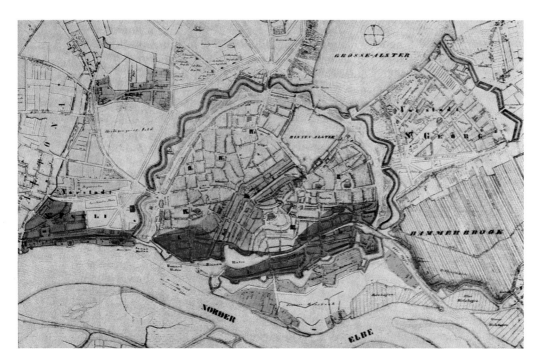

FIGURE 7.8 This 1832 map of cholera in Hamburg shows that the areas of the city most severely affected (the shaded regions) are those nearest the river where the poorer citizens lived.

of any distinction but the physical in an attempt to first document and then explain the coarse pattern of cholera occurrence. The greater intensity along the riverbanks evidenced cholera as a natural creature of lowland areas, one perhaps propagated not by the fetid airs of impoverished neighborhoods alone but climatically, in the low, dense, foggy airs of the riverbanks. This would serve Grainger well in his argument that cholera was a localized, natural condition originating in the airs of another river town: London (see fig. 7.10).

The difference between these maps lay not simply in their distinct cartographic approaches but more importantly, and more fundamentally, in the ideas they argued. Reese's cholera was born solely of local conditions, geographic and social, each outbreak its own concatenation of environmental opportunism. Hellis's dot map argued the importance of the environmental determinants of imported disease implied in the mapped clusters and argued in the text. The Hamburg map insisted on a geographical determinism across the mapped city, one in which cholera could be seen to be generally correlated as an inverse of altitude. Each of these perspectives emphasized a different etiology; each proposed a different cholera.

Nor was cholera the only disease researchers attempted to study in this way, mapping civil mortality data. In 1844, Dr. Robert Perry, senior physician at the Glasgow Royal Infirmary, created a map of an influenza outbreak based on the mortality reports of district physicians (Robinson 1982, 173–74). The question was whether the enabling geography resulted from natural variations in local climate and geography or from social variations that were most easily identified through reference to income as a surrogate for social class.

Perhaps the most famous statement of the argument that disease was a social outcome rather than the sole result of climatic or geographic determinants, on the one hand, or personal hygiene on the other hand, is found in Edwin Chadwick's map of Leeds in England. Then the head of the Poor Law Commission in Britain, Chadwick argued in the map that general conditions of poverty created environments in which disease prospered. The map was in his landmark 1842 *Report on the Sanitary Condition of the Labouring Population of Great Britain to the Poor Law Commissioners*, a seminal document in the nineteenth century literature on social welfare. The Poor Law Commission, and the conditions that inspired its creation, were at the heart of British attempts to use social data to argue a cause and effect correlation in areas of health and illness, poverty and wealth.

Chadwick was a central engine of this movement, innovatively using census data in his mapped study to distinguish three classes of city wards based upon income levels and health incidence (births and deaths by population). These were aggregated into healthy and unhealthy city ranges (fig. 7.9A). This would have been difficult, perhaps impossible, without two new innovations first made available with the census of 1841: mortality data reported by local registrars and precise income and housing data col-

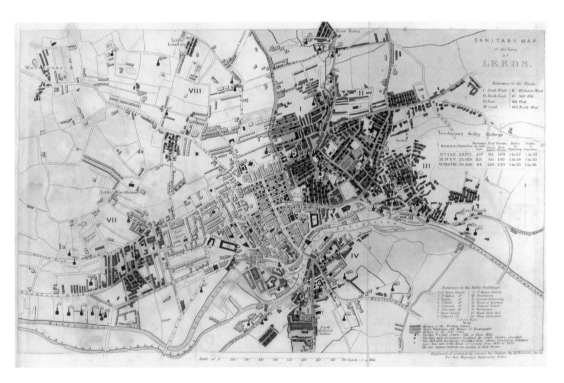

FIGURE 7.9A Edwin Chadwick mapped the proposition that income and disease mortality are inversely related. In this map of Leeds (1844), darker areas represent poorer neighborhoods in which cholera and other contagious diseases were located, street by street.

lected by census enumerators. Across the map surface, the streets of the working class, tradespeople, and upper-class citizenry are symbolized using lighter and darker brown hatchings. To this social mapping Chadwick added red dots signifying the homes of those who died in the first pandemic from cholera and dark blue dots signifying the homes of those who reportedly died of respiratory diseases. This permitted him to argue a correspondence between class, as defined by income, and mortality.

The streets of the working class and poor had more dots than those where the upper class resided (fig. 7.9B). The visual argument was advanced in an index of "good" and "bad" streets, defined by aggregate births and deaths within the mapped population, street by street. The whole visualized a statistical profile of the effect of income on fertility, mortality, and the relation of infectious diseases like cholera to mortality. Where Hamett (1832) had decried the lifestyle of the poor of Dantzick, Chadwick focused upon the conditions within which the poor found themselves: ill fed, ill housed, and ill clothed in rooms without ventilation on odiferous streets. With these elements Chadwick crafted "a forceful indictment of unsanitary living conditions in the industrial slums, as well as a severe criticism of [both] physicians ignorant of the causes of contagion and of the moribund local health boards" (Melosi 2000, 45).

Chadwick was a sanitarian, a man who believed that health was in great part dependent on the urban infrastructure that provided the population at large with housing stock, water, and sewage disposal. *If* disease was generated in the malodorous airs of uncollected human and animal wastes, *then* it was in the odiferous, waste-filled streets and close, overcrowded rooms of the poor that disease was born. At least since the plague of the 1600s people had blamed the poor for the generation and spread of disease. Chadwick and sanitarians like him inverted the argument to make the community at large responsible for the disease-generating environment of the poor.

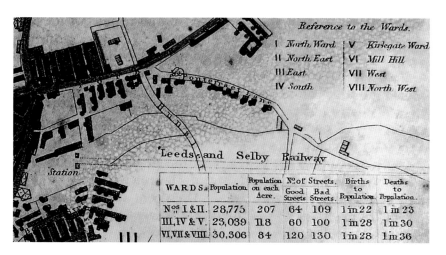

FIGURE 7.9B Map of Leeds, detail. Chadwick used basic health and income statistics to create areas of good and bad streets, healthier and poorer wards based upon population districts and disease incidence.

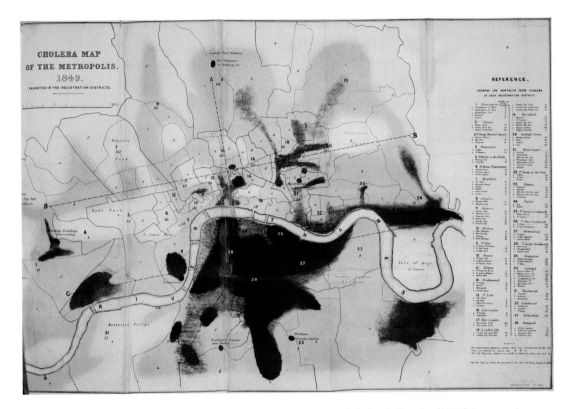

FIGURE 7.10 In this detail of Richard Grainger's density map of cholera in London (1849), he argued a negative correlation between disease intensity an altitude.

From his appointment to the Poor Law Commission in 1832 through the early 1850s, Chadwick was a critical champion of a social activism whose rationale emphasized the history of disease in a social geography of place. He was an influential advocate for the creation of the GRO as a necessity of the data required to track disease within the population were to be obtainable. It was Chadwick, too, who argued that registrars reporting to the GRO should wherever possible be medically trained and thus able to provide accurate mortality data. In 1848 Chadwick lobbied for the creation of a national Board of Health. It was under Chadwick's chairmanship of that board that Grainger wrote his report to Parliament (fig. 7.10).

That report argued cholera within the miasmatic theory of disease as an airborne thing in a way that strengthened the sanitarian call for better urban infrastructure on the basis of cholera and other disease experiences. In a map of cholera incidence in the 1830s in metropolitan London's registration districts, Grainger broke incidence classes into three rough categories of intensity. The relative intensity of cholera mortality was inscribed in darkest blue in the districts south of the Thames. This was explained by proximity of residents to the famously polluted central river, its stench a matter of periodic public and parliamentary concern. These wastes often fermented for days on the riverbanks until rain and tide could carry them to sea.[7] Adjacent to these areas of intense cholera were those of moderate mortality evidenced in a lighter blue ink wash. Finally, localized outbreaks elsewhere, principally in northern London registration districts, were symbolized by light blue inking.

It is easy to forget what a marvel Grainger's map was. In its dense ink washes the map posted thousands of deaths reported in the 1830s to local officials (Kearns 1985, 128). Collected and then located in the map of registration districts, a creation of the 1840s, Grainger created a density surface of at least three classes of disease intensity. With the geography of the river and registration districts Grainger also posted a precise series of altitude measures drawn from official land survey documents. These were small red dots to which were added altitude above Trinity high-water mark. This permitted an argument to be made inversely relating altitude as a surrogate for air quality and incidence of cholera (the higher the altitude the less likely cholera would occur). Conversely, cholera activity was most intense along the river, where altitude was lowest. Where outliers occurred (limited outbreaks at higher altitudes), explanations were posted in the map: a sewer here, a contaminated well there. In the Hamburg map the argument was simply that cholera incidence increased in correlation to proximity to the river. Here that argument was modified to insist that it was not simply proximity to the river but rather altitude, a surrogate for air quality, were also at play. Higher altitudes away from the river had better, cleaner airs and less cholera than one breathed along the malodorous banks of the river.

CITY AND NATION

Urban disease studies were the backcloth against which national studies of cholera were pursued simultaneously. These sought clues to the character of the epidemic disease through a careful consideration of national patterns of disease incidence rather than neighborhood studies of local occurrence. In 1848, for example, Augustus Petermann, a German cartographer who had worked on Heinrich Berghaus's *Physikalischer Atlas* before immigrating to England, produced two maps of two distinct choleras in England using data from the 1830s pandemic (Petermann 1848). The first was a profoundly unnatural disease, imported from abroad and spread through the nation with little respect to local geographies or social conditions. The second was, like Grainger's, a thing grounded in the local airs of the city.

To show the ubiquity of the disease and its pattern of diffusion Petermann employed a shading technique developed in 1831 by Adolphe Quetelet to present statistical data (Robinson 1982, 180–81). This was not "dot mapping," in which each dot symbolizes so many deaths but a shading technique whose application was more impressionistic than numeric. "The object, therefore, in constructing Cholera Maps, is to obtain a view of the Geographical extent of the ravages of this disease, and to discover the location conditions that might influence its progress and its degree of fatality" (Petermann 1848, 2).

For Petermann, official data of mortality collected in the first pandemic provided an opportunity to demonstrate the utility of a methodology that gave shape and place to spatial statistics, one that brought forth associations not necessarily evident in the raw tables of the data. "The masses of statistical data in figures, however clearly they might be arranged in Systematic Tables, present but a uniform appearance, the same data, embodied in a Map, will convey at once the relative bearing and proportion of the single data, together with their position, extent, and distance, and thus, a Map will make visible to the eye the development and nature of any phenomenon in regard to its Geographical distribution" (Petermann 1848, 2).

On the map Petermann included a simple graph, the already-familiar epidemic curve (fig. 7.11A). Here its subject was not the number of cholera cases but instead the number of cities and towns in which active cholera was reported in Britain from 1831 through 1833. The argument proposed in the map was that cholera was introduced first through the nation's ports and then spread in a more or less contiguous but slower fashion to inland cities across the nation. "From Sunderland the disease spread but slowly along the inland lines of communications, but the more rapidly to the different seaports, so that all the coastline round the Isles was first attacked, and from thence the disease penetrated the inland parts of the country. Thus frequently the disease broke out in places at opposite coasts almost simultaneously, while the intervening places were attacked later . . . This proves that the disease progresses more rapidly by sea-communication than by land" (Petermann 1848, 3). This was a thing whose agent was unknown but that clearly traveled, like Arrieta's plague, with trade and travelers, favoring sea over land routes but capable of moving along both.

As if to emphasize the point, the coastline of Petermann's Britain is shaded with a light blue, a halo effect that sets off the land and draws attention to the port cities where cholera activity was most intense. But because cholera also traveled inland, across the nation, the map argues that cholera was not bound by any single geography. Some cities near low and marshy areas were free of the disease; and others, like Birmingham, at an altitude of five hundred feet above sea level, were among "the most severely attacked districts in all the kingdom. The conclusion was clear, at least to Petermann: "Of all the local causes of the spread of the disease altitude is one of little comparative influence, and that it is much more affected by the density of the population" (Petermann 1848, 4–5). And if altitude did not affect the spread of cholera it was unlikely that miasmatic airs were involved in its generation since airs at higher altitudes were generally assumed to be less odiferous and thus healthier.

Most who have commented on Petermann's map have done so based on a partial facsimile published by Jusatz (1940), a section of the map reproduced in Robinson

FIGURE 7.11A (*opposite page*) Augustus Petermann's 1848 maps of cholera in England and in London in the 1830s argued that the disease entered the nation through its seaports and diffused without regard to geographical difference.

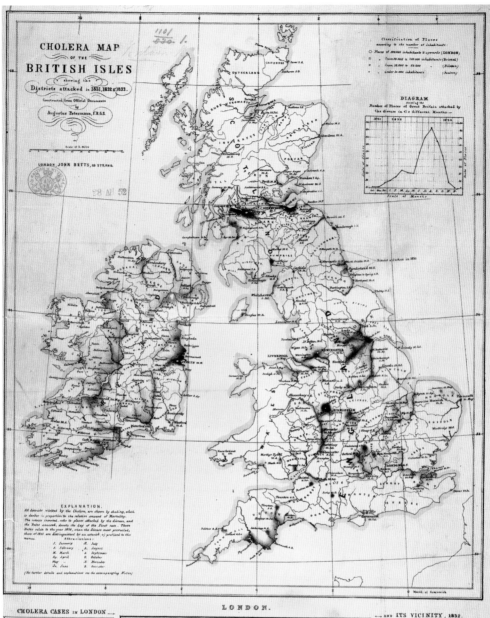

CHOLERA MAP
OF THE
BRITISH ISLES
showing the
Districts attacked in 1831, 1832 & 1833.

Constructed from Official Documents
by
Augustus Petermann, F.R.G.S.

Scale of 5 Miles

LONDON, JOHN BETTS, 115 STRAND.

DIAGRAM
showing the
Number of Places of Great Britain attacked by
the disease in 6 different Months

EXPLANATION.

All districts visited by the Cholera, are shown by shading, which is darker in proportion to the relative amount of Mortality. The names inserted, refer to places attacked by the disease, and the Date annexed, denote the day of the First case. These Dates relate to the year 1832, when the disease most prevailed; those of 1831 are distinguished by an asterisk (*) prefixed to the names.

Abbreviations:

J.	January	H.	July
F.	February	A.	August
M.	March	S.	September
Ap.	April	O.	October
May		N.	November
Jn.	June	D.	December

(For further details and explanations see the accompanying Notes)

Classification of Places
according to the number of inhabitants

Price 1s

LONDON.

CHOLERA CASES IN LONDON

	Registration Districts	Cases	Deaths	Population	Proportion Deaths to Population 1 in:
1.	Rotherhithe (without Abbey)	175	100	3,423	35
2.	Tower and Liberty	30	10	723	71
	St George Martyr				
3.	St Olave's	1050	690	75,790	94
	St Saviour's				
4.	Whitechapel	421	263	30,720	217
5.	Bermondsey	146	110	23,761	142
6.	City	626	369	53,790	125
7.	Ratcliffe	128	62	3,741	157
8.	Poplar	133	101	16,849	366
	Shadwell				
	Wapping	156	79	23,196	366
9.	Spitalfields	186	97	17,949	182
10.	St Giles	632	280	50,895	180
11.	Limehouse	103	80	18,681	
12.	Newington (Surrey)	474	300	44,644	222
13.	Lambeth	265	397	87,856	360
14.	Camberwell	127	107	28,331	366
15.	Strand	84	37	3,237	362
16.	St George in the East	256	158	20,016	515
17.	Deptford	99	58	19,795	360
18.	Bethnal Green	306	270	62,018	345
19.	Westminster	265	245	224,185	349
20.	Christchurch	87	56	13,706	309

AND ITS VICINITY, 1832.

	Registration Districts	Cases	Deaths	Population	Proportion Deaths to Population 1 in:
21.	Chelsea	186	83	32,371	395
	Brompton				
22.	St Luke, Middlesex	286	118	46,642	395
23.	Greenwich	36	24	10,400	425
24.	Putney	40	23	10,699	465
	Wandsworth				
25.	Teddington	23	13	30,282	301
27.	St Marylebone	356	224	204,304	546
28.	Clapham	37	17	9,513	394
29.	Holborn	145	65	23,926	594
30.	Rotherhithe	21	13	13,451	676
31.	Clerkenwell	143	65	47,634	733
32.	Hackney	9	5	3,866	762
33.	St George, Hanover Square	125	74	58,209	787
36.	Fulham	9	6	7,957	915
37.	Battersea	9	6	5,540	923
39.	St Pancras	126	101	103,548	933
40.	Islington	48	29	5,540	957
41.	Stoke Newington	5	3	9,480	1160
42.	Shoreditch	397	17	68,564	1303
	Bromley				
	Stratford	23	8	8,217	1343
43.	Kensington	18	14	26,902	1392
44.	Mile End	8	1	46,524	40,782

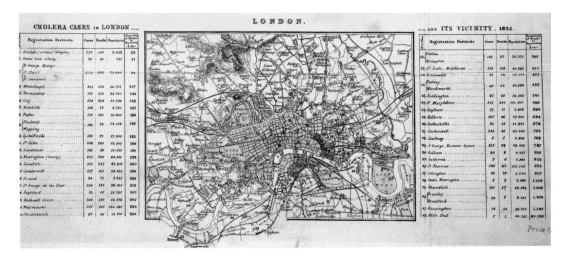

FIGURE 7.11B Petermann's map of cholera in London, based on 1830s mortality data by registration district. Population districts are likely from the 1841 census.

(1982), or a distorted version recreated by E. W. Gilbert (1958).[8] These authors ignored Petermann's cholera map of London that was originally placed beneath the natural map. In it Petermann attempted a close study of the national phenomenon that was cholera in the 1830s. But at the scale of the metropolis cholera appears to be tied to the specific geography of the Thames in central London, its intensity not generalized, but like the cholera of Grainger's mapping, geographically grounded.

In this map Petermann brought forward the city as a collection of regular urban blocks divided into forty-two sequentially numbered registration districts (fig. 7.11B). For each, the map key provides details of cholera incidence: total number of reported cases, reported mortalities, and the total registration district population based on census figures. From these Petermann calculated a simple mortality ratio, given as "one in so many persons." On the basis of these rates Petermann shaded the city in three hues of red to indicate greater or lesser areas of cholera mortality. Where the disease was either absent or sporadic, the map was left uncolored. Petermann used horizontal black bars, like those first employed by Hamett, to locate cholera cases in thinly settled areas, but in areas of dense mortality there were too many cases to attempt to mark deaths individually.

Petermann thus mapped two very different choleras, although he argued only a nongeographic, imported disease. That was the first cholera, the one he posted at the scale of the nation. It entered through ports where disease incidence was densest and then spread ecumenically, over years, across the nation. It was tied to no specific geography (climatic, physical, or social). While its origin was foreign its mode of transmission and its agency were unknown. The second cholera, the London cholera, had a distinct geographic profile that favored the lowland riverbanks over the highland regions. Both choleras were posted through population mortality figures, in the first

FIGURE 7.11C Petermann mapped the intensity of cholera in the 1830s in London registration districts by density using darker and lighter reds. Roman numerals refer to registration districts. Cases outside central districts were individually noted by a thick black line.

map using cities and towns as counters and in the second a general mortality ratio. The data for both incidence and population denominator were drawn from public records and then applied in these two very different mapped arguments.

THE NEIGHBORHOOD STUDY

What was it about riverside London, Hamburg, or Rouen that appeared to promote cholera within its boundaries? To say these were ports was to ask the same question in a different way: what was it about ports that cholera favored as a launching stage for its general diffusion? What was it about the poor quarters of Dantzik, New York, or Rouen that explained the local intensity? Seeking answers to these questions required a shift in focus and scale to the neighborhoods where cholera was active and the habits of those who lived on their local streets. An American example of work at this scale is seen in the work of John Lea in his study of a cholera outbreak in his hometown of Cincinnati, Ohio, published both as a monograph (1850) and in an article in the *Western Lancet* (1851).[9]

Briefly cited and quickly dismissed by John Snow (1855a, 114–15), Lea proposed a relationship between cholera and heavily mineralized water drawn from local wells "passing through strata of marl, producing unwholesome water" (Lea 1851, 89). After reviewing reports of cholera outbreaks in Boston, Charleston, New Orleans, New York City, and St. Louis in the 1830s, Lea was able to state definitively "that no *well attested fatal case* occurred among those who used *rain water exclusively*; and I may add boiled water" (Lea 1851, 89; original emphasis). The proposition he arrived at was simple and elegant: *if* cholera was a poison born of mineralized spring water feeding some but not all local wells, *then* not drinking well water would mean not getting cholera.

In support of his thesis, Lea quoted a number of studies, including one of an 1849 outbreak in St. Louis, Missouri, in which 90 cholera deaths occurred in 25 houses in a neighborhood whose inhabitants drew their water from a single well. To test his idea, Lea mapped the incidence of cholera during a July 1849 outbreak in Cincinnati along the 6.25° grade of Sycamore Street, where mortality from cholera was "appalling, as the annexed diagram will show, and it will also show at a glance the efficacy of rainwater as a prophylactic" (Lea 1851, 92). The map of 44 deaths among an estimated population of approximately 114 persons gave, he calculated, a mortality of about 40

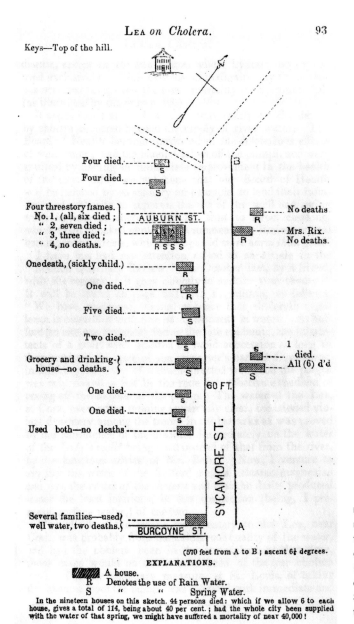

LEA on Cholera.   93

Keys—Top of the hill.

Four died.------------- [S]
Four died.------------- [S]

Four three story frames. }
No. 1, (all, six died ; | ---- AUBURN ST.
" 2, seven died ; |
" 3, three died ; | [RSSS]
" 4, no deaths. }

Onedeath, (sickly child.) ---------- [R]
One died. -------------- [R]
Five died. ------------- [S]
Two died.-------------- [S]
Grocery and drinking-} -------------- [S]
house—no deaths. }
One died.------------- [S]
One died.------------- [S]
Used both—no deaths.----------- [S]

[R]------ No deaths
[R]-------- Mrs. Rix.
      No deaths.

                    1
[S]----- died.
[S]---- All (6) d'd

60 FT.

SYCAMORE ST.

Several families—used} ----------
well water, two deaths.{ BURCOYNE ST.   A

(570 feet from A to B ; ascent 6¼ degrees.
EXPLANATIONS.

[▨] A house.
R   Denotes the use of Rain Water.
S   "   "   "   Spring Water.
In the nineteen houses on this sketch, 44 persons died: which if we allow 6 to each
house, gives a total of 114, being about 40 per cent. ; had the whole city been supplied
with the water of that spring, we might have suffered a mortality of near 40,000 !

FIGURE 7.12 John Lea mapped the relative mortality in Cincinnati, Ohio, of those who drank spring (well) water and those who drank only rainwater on Sycamore Street in 1848 to argue mineralized water was the source of cholera.

percent (fig. 7.12). Of those who died in the outbreak, 32 persons lived in 12 houses that used mineralized spring water; 2 lived in the remaining 5 houses whose water came from other sources. Lea was not concerned about those aberrant cases, assuming the decedents probably drank well water at the home of a neighbor.

There was in Lea's work a resolute scientism that might seem precious in retrospect but was impressive in his day. He knew it was necessary to calculate a mortality rate for his population. "As a test of the truth of theories," Forry had advised in 1842, "statistical investigations are of vast importance" (Forry 1842, vi). In his analysis Lea assumed six people per household to produce an overall mortality proportion of two out of thirty for those using rainwater (6.6 percent) and forty out of seventy-two for those using spring water (55 percent). The resulting ratios were a better than average attempt to quantify raw mortality data.

Lea was neither a physician nor a member of a medical academy or association. Yet his report drew on a wealth of American, British, and European published sources and compares nicely with many more professional studies of his day. By 1848 even a nonprofessional in Ohio had access to a decent sample of the national and international medical literatures as well as local mortality data sets. As importantly, that person had the ability to craft an argument on disease causation and see it published in book or journal form. No greater testimony to the advances in publishing, and the internationalization of disease studies it promoted, can be offered than this article by a Cincinnati businessman who saw himself as competing in an international field of cholera researchers with whose work he was intimately familiar.

Lea was pleased to publish his findings in the *Western Lancet*, sufficiently confident (and proud) of his conclusions that he also "made known my discovery to the British Government, referring to the especial importance it would prove to be of, to British subjects, and the army, in India, that those who used rain water *would escape the attacks of cholera*" (Lea 1851, 90; original emphasis). As collaborative evidence he included in his report a case reported in the New York edition of the *London Quarterly Review*. That cholera outbreak, occurring in Salford, England,[10] had as its apparent cause a "fermenting impurity in water" among inhabitants "smitten in rapid succession so long as they persisted . . . in drinking the water of a drain-infected well, and the plague was only stayed at last by the rude but effective expedient of taking off the handle of the pump" (Lea 1850, 95). Wells *were* a source of cholera outbreaks. Drinking only rainwater *was* a prophylactic against cholera transmitted by contaminated water supplies. The reasons were different from those Lea imagined, but our foreknowledge does not diminish the strengths of his work carried out 150 years ago.

CHAPTER 8

JOHN SNOW'S CHOLERA

Cholera returned to England in 1848, this second pandemic again traveling along sea and land routes from India through Europe, Russia, and the Middle East (*Lancet* 1848). As it rolled across the European continent so too did a series of popular uprisings in Berlin, Paris, Vienna, Sicily, Milan, Naples, Parma, Rome, Warsaw, Prague, and Budapest (Dorling 1999, 128). The poverty and malnutrition that characterized the "Hungry Forties" bred social unrest in country after country as the dislocations of industrialism—and the widening gap between rich and poor—swept Europe (Evans 1992, 158). Political reform and social change were in the air and neither was easily separated from the medicine of the day. This was also the year Dr. Rudolf Virchow started his famous journal, *Social Medicine*, and published in it a seminal essay arguing that social reforms were critical tools for combating a typhus epidemic rampaging in Upper Silesia (Nuland 1988, 314–16).

The great social statistician Adolphe Quetelet called the broadly rebellious unrest a "veritable moral cholera," a political metaphor of disease spread that ignored the etiology of social anger and its relation to the cholera that was its referent (Hacking 2006, 113). For others, however, then and now, the phrase was concretely apt, linking social justice and public health (Krieger et al. 1997, 22).

Social etiology could not be easily separated from the clinical portrait of public disease: "Who is to say what constitutes a valid 'cause'? When epidemics hit and strike the poorest hardest, is the cause miasma (reeking air corrupted by putrefied organic matter), contagion (direct transfer of a disease poison from one person to another), greater vulnerability caused by starvation and overwork, or a political and economic system that permits industry to pay starvation wages and demand 16 hour days" (Krieger 2000, 158)? Cholera was one medium in which these questions were asked, at once a metaphor and a surrogate for broad issues of public and personal responsibility in health. How one defined cholera, and by extension other diseases, would determine responses to a range of things, medical and social.

In 1848 cholera was neither the only nor the most serious disease physicians and politicians had to confront. When influenza made its annual visit to England in that year, for example, "the mortality was greatly above the average. Typhus, scarlatina [a severe scarlet fever], whopping-cough, and smallpox were epidemic in many parts" that year (Farr, 1852a, x). Compared to these traditional diseases, cholera's expected arrival was . . . less urgent. As Lord Morpeth told the House of Commons when introducing the Public Health Act in 1848, "I do not wish to lay any material stress upon the possible approach of cholera . . . it is far from any temporary evil, any transient visitation against which our legislation is now called upon to provide. It is the abiding host of disease, the endemic not the epidemic pestilence, the permanent overhanging mist of infection, the slaughter doubling its ravages on our bloodiest fields of conflict, that we are now summoned to grapple with" (Morris 1976, 205).[1]

And yet it would be cholera, and the social unrest that many feared would accompany it, that most occupied British officials. They did not have long to wait. "In October there was an outbreak of cholera among convicts crowded in the decaying [prison ship] *Justia* hull, lying off the Royal Arsenal wharf, Woolwich, in the immediate vicinity of a common sewer. Subsequent outbreaks were fatal to convicts in Millbank Prison and to lunatics in the Peckham House Asylum. In the last week of December, the epidemic broke out in Mr. Drouet's Infant Poor Establishment, Surrey Hall, Tooting, where 48 of the 61 deaths registered from cholera in the first week of January occurred" (Farr 1852a, xx).

With an eye as much to social stability as to the health of the public—British parliamentarians were well aware of continental unrest—in 1848 Parliament passed a Public Health Act, which created a three-member General Board of Health whose focus would be local water supply, sewage, and control of "offensive trades," like animal rendering, that created gut-wrenching, vile odors in the city (Krieger and Birn 1998, 1603). Limiting these trades would presumably make the air cleaner and, within the miasmatic theory of disease, inhibit the development and diffusion of diseases like cholera.

For the same reason, Parliament also passed a new Nuisances Removal and Contagious Disease Prevention Act to supercede those passed one earlier passed in 1848.[2] It required new and existing buildings to drain their refuse into the urban sewer system rather than cesspools or, as they had in Samuel Pepys's day, household cellars (Johnson 2006, 118–19). Before 1815 it was illegal to discharge raw waste into the sewer system. After 1848 it would be illegal not to. This necessitated, in turn, the construction of new sewer lines capable of serving the expanding metropolis and moving its waste away from its populated areas. For that to happen the Thames itself eventually would have to be transformed. After years of discussion, in the 1860s an embankment projection was begun to force waste in the river away from the city where, in the 1840s and 1850s, it lay malodorously fermenting on the riverbanks at low tide (Porter 1998).

The result was "a high point of the great Victorian enterprise of sanitary reform, whose central idea was that environmental circumstances—particularly pollution of air and water, defective sanitation, dampness, filth, and overcrowding—were causes of disease, particularly epidemic disease" (Paneth, Vinten-Johansen, and Brody 1998, 1545). All this made sense if, but *only* if the odiferous wastes were the source of diseases like typhus and cholera. If these diseases were promoted by something else then sanitary reforms might improve urban ambience but would not affect the health of citizens.

Advocates of the Nuisance Act included then newly appointed Board of Health chairman Edwin Chadwick, who famously insisted disease *was* smell. Among the board members was William Grainger, whose report to Parliament in 1850 exemplified the research that argued the miasmatic nature of cholera (see fig. 7.10). To enact the program of sanitary reform, and thus to protect the health of metropolitan citizens, the Board of Health worked with the London Metropolitan Sewer Commission to develop a plan for expanded service.

The commission was established to combine eight existing sewage companies into one enterprise, unifying the what previously had been separate and competing bureaucracies. This required a standardized image of the system at large to permit the planning of additional lines to the newly integrated regional system. For that to happen the very city had to be remade. This was the rationale for an 1850 sewer commission map on which a rectilinear grid was overlaid on a metropolitan ordinance survey map, dividing Greater London into sewer work areas (fig. 8.1). The forty-two rectangles were each subdivided into small, local rectangles in which the work would be carried out. Beneath the grid can be seen the boundaries of registration districts, the administrative division of the greater bureaucracy.

The focus of the work, and of the map, were in twelve darkened, central map blocks covering areas of the city in which increased service was most urgently needed to serve a burgeoning population. The existing sewer system to which new lines would be linked was included as well. Other maps included altitude points to assist in planning a sewage disposal system run by gravity (fig. 8.2A). If transportation was conceived as the circulatory system of the metropolitan body, this was its alimentary system. For the sanitarian reformers who believed, with Chadwick, that smell was disease, this *was* a health map, one that laid out a program that would decrease disease and protect the metropolis against cholera.

JOHN SNOW: 1849

It is only in this context that the radical nature of John Snow's argument about cholera can be understood. Others have suggested that his brief 1849 monograph, *On the Mode of Communication of Cholera*, is largely irrelevant since its major points are repeated with more detail in the better-known, second edition, published in 1855

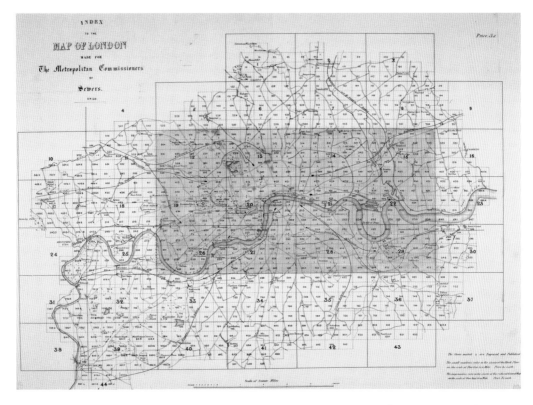

FIGURE 8.1 The London Metropolitan Sewer Commission overlaid a grid onto an ordinance survey map in 1850 to divide the city into districts in which additional sewer lines could be planned for London and environs.

(Richardson 1936, xv). But it is in this first, privately printed monograph that Snow's essential argument is laid out and here that the methodological problems it presented to his contemporaries can be most easily understood. Snow would amend this thesis, beginning with two articles in the *Medical Gazette* in November (Snow 1849b; 1849c), restating its core points and examples again and again until his death in 1858.

In a year when cholera was again epidemic in the British ports of Portsmouth, Plymouth, Bristol, Liverpool, Hull, Tynemouth, and London, Snow challenged the idea that cholera was miasmatic in its origin, source, or mode of transmission. Indeed, Snow challenged the general idea of miasmatic, airborne epidemic disease. "What is so dismal," Snow asked rhetorically, "as the idea of some invisible agent pervading the atmosphere, and spreading over the world?" (Snow 1849a, 30).[3] Snow thus challenged a large body of established medical theory that from Hippocrates to Sydenham through Seaman and his yellow fever studies had served medical practitioners. In doing so he also implicitly challenged its utility as a rationale for social reforms that had grown from the idea that miasmatic odors resulting from bad sanitation were a threat to public health. In its place Snow proposed a cholera whose source was waterborne and whose unseen agent was also transmitted interpersonally, "spreading over the world."

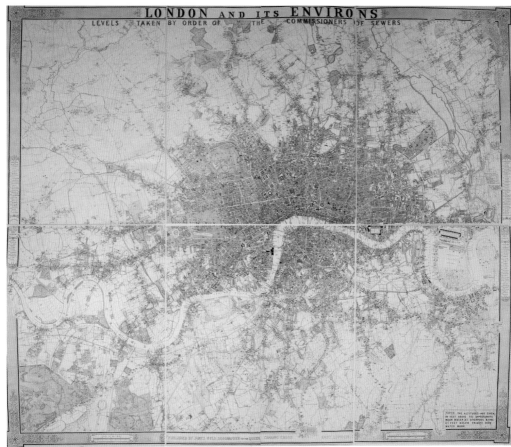

FIGURE 8.2A In another map by James Wyld of London in 1850, the sewer commission included points of altitude around the city in preparation for the addition of new lines to serve the expanding metropolis.

FIGURE 8.2B In this detail of Wyld's sewer commission map, the careful mapping of points of altitude upon the details of the city is easily seen. The result is an altitude surface that could be used by researchers like William Farr in attempting to study the relationship between altitude (and therefore air quality) and cholera.

Snow began with a view held "in common with a great portion of the medical profession, that it [cholera] is propagated by human intercourse" (Snow 1849a, 1). Cholera, in other words, was portable and not only traveled but also somehow grew (propagated) in the trade carried on by humans. It was the same thesis that had empowered Arrieta's *cordon sanitaire* and some eighteenth-century theories of yellow fever's spread from the Caribbean to the eastern seaboard of the United States.

146

Snow then asserted that cholera was something new, not transmitted "in the same way that the eruptive fevers are considered to be, viz., by emanations from the sick person into the air," but by a different route entirely (Snow 1849a, 8). In a follow-up article Snow argued cholera was in fact not a fever at all, not a disease in which "a morbid poison has entered the blood." Distinctively, the disease he proposed began with "the affection of the bowels, which often proceeds with so little feeling of general illness, that the patient does not consider himself in danger, or apply for advice till the malady is far advanced" (Snow 1849b, 745).

From Jameson through Grainger, cholera was assumed to be a fever within the existing classification of disease. Snow insisted, however, that cholera was an intense diarrhea *without* fever, a condition resulting in a "thickened state of the blood arising from the loss of fluid" (Snow 1849a, 749b). Here he accepted the general symptomatic definition, visible on autopsy, of a dark, granulated blood as a defining characteristic of cholera morbus, distinguishing it from other less virulent diarrheas. For others it was evidence of a disease inhaled and then circulated *in* the blood. For Snow, however, the pathology was the result of a dehydrated state caused by violent diarrhea and unrelated to inhalation.

Diseases of the gut are most typically caused by ingestion, Snow reasoned. Cholera was symptomatically a diarrheic disease of the gut. Therefore it was caused by something swallowed rather than inhaled. "The induction from these data is that the disease *must* be caused by something which passes from the mucous membrane of the alimentary canal of one patient to that of the other, which it can only do by being swallowed" (Snow 1849a, 745; emphasis added). That thing, he argued, was waterborne rather than airborne. Finally, Snow argued, this new waterborne thing was transmitted interpersonally. "Having rejected [airborne] effluvia and the poisoning of the blood in the first instance, and being led to the conclusion that the disease is communicated by something that acts directly on the alimentary canal, and the excretions of the sick at once suggest themselves as containing some material which, being accidentally swallowed, might attach itself to the mucus membrane of the small intestine, and there multiply itself by the appropriation of surrounding matter" (Snow 1849a, 8–9).

The evidence here was inductive and logical, a necessary conclusion based on premises that, if true, guaranteed the certainty of the conclusion. Inductive thinking is about likelihood and probabilities and Snow offered an "induction from the data" that was anecdotal and circumstantial if, to him, compelling. He did not did not dismiss entirely the airborne argument in this first work. Dried fecal matter, he reasoned, for example, might become airborne and in that medium become a secondary pathway for the disease. "The mode of contracting the malady here indicated does not altogether preclude the possibility of its being transmitted a short distance through the air; for the organic part of the faeces, when dry, might be wafted as a fine dust, in the same way as the spores of cryptogamic plants, or the germs of animalcules." (Snow

FIGURE 8.3 This is a portrait of thirty-four-year-old John Snow was painted in 1847 by Thomas J. Marker, whose family were patients of the doctor.

1849a, 26–27). Still, for Snow, logic dictated that the principal pathway was through ingesting tainted water or interpersonal contamination.

What was ingested, Snow suggested, was water contaminated by "the excretions of the sick," transferred when caregivers handled soiled products and did not wash their hands. "The bed linen nearly always becomes wetted by the cholera evacuations . . . the hands of persons waiting on the patient become soiled, and unless these persons are scrupulously clean in their habits and wash their hands upon taking food, they must accidentally swallow some of the excretion" (Snow 1849b, 746). If contaminated persons prepared food or other materials without washing their hands they contaminated those products and thus "the disease may be conveyed to a distance, and into quarters having apparently no communications with the sick" (Snow 1849a, 10). Epidemic cholera morbus therefore was a disease of poor hygiene rather than of an impoverished sanitary infrastructure.

There was something very Hobbesian about Snow's inductive approach. His argument owed much to the tradition of the old, natural science in which the world was revealed in its logical form through close observation. No agent of cholera was microscopically evident; no proof of the passage of contaminated waste from the sick could be demonstrated. No experimental evidence or test was advanced. Snow's thesis was accepted as possible but his assertion that it was a necessary conclusion was disputed.

The problem was not simply that Snow's views were eccentric but rather the absolute certainty with which Snow stated tentative, inferential conclusions as substantiated deductive fact. "The disease *must* be caused," Snow wrote. In a detailed, nineteen-page review of Snow's argument, recapitulated in the second edition of *On the Mode of Communication of Cholera*, Edmund Alexander Parkes would complain that Snow "arrived at his opinion on grounds which appear to us insufficient to warrant so grave a conclusion," one completely at odds with then current medical reasoning.[4] "We cannot admit the cogency of the must in this quotation; since we do not see that it is satisfactorily made out that the blood is "not under the influence of a poison" (Parkes 1855a, 450). Was the coagulation of the blood seen upon autopsy a direct result of cholera, one inhaled and then circulated, or simply a secondary effect of the diarrheic symptom? There was no proof either way. And while Snow sneered at the idea of an airborne agent the nature of a waterborne agent was no less uncertain.

In 1849 Snow's waterborne cholera was an unproven inference built upon a literature open to different interpretations. In 1832 outbreaks in the towns of Dumfries and Maxwell, "the inhabitants drink the water of the Nith, a river into which the sewers empty themselves, their contents floating afterwards to and fro with the tide" (Snow 1849a, 11). Similar contagion pathways were reported in Glasgow where the tidal waters of the Clyde "cannot be altogether free from contamination" and in London whose southern and eastern districts were supplied by water laden with sewage from the Thames. Snow's argument rested at every scale upon the uneven distribution of cholera cases in neighborhoods, cities, and the world.

It was, Snow insisted, "difficult to imagine that there can be such a difference in the predisposition to be affected or not by an inhaled poison, as would enable a great number to breath it without injury" (Snow 1849a, 6). And yet, as his critics would point out, cholera was not universal among those who drank apparently polluted water. And people in some areas became ill even though the water they drank was assumed to be clean. The clusters in which this new cholera appeared argued to Snow—but not necessarily to others—a very specific and in his day extremely eccentric etiology.

To demonstrate the path of cholera's introduction into local waters, Snow offered as evidence two then-recent, local cholera outbreaks in metropolitan London, the first in Thomas Street, Horsleydown, and the second at Albion Terrace, Wandsworth Road. For each he had officially reported mortality figures and firsthand knowledge based on visits to both sites. Snow's argument relied on two spatial propositions, examples of what physician and historian Howard Brody has called Snow's "map thinking" (Brody et al. 2000). First, that the outbreaks were sufficiently localized and dense that a delimited single source of infection was implied. Second, that the water source serving the affected houses was at the center of the outbreak and therefore necessarily implicated causally.

In Thomas Street "the slops of dirty water poured down by the inhabitants into a channel in front of the houses got into the well from which they [the cholera patients] got their water, this being the only difference that Mr. Grant, the Assistant-Surveyor for the Commissioners of Sewers, could find between the circumstances of the two courts" (Snow 1849a, 13). Further, the well pump had "for some time past" malfunctioned, occasionally pumping water into the street from which it flowed back again, bringing with it putrid waste. In Albion Terrace, "the water got contaminated by the contents of the house-drains and cesspools; the cholera extended to nearly all the houses in which the water was thus tainted, and to no others" (Snow 1849a, 15). This was Snow's supposition however, not something proven (and without better microscopy, it could not be so) but which seemed logically necessary to Snow.

"Snow was thinking in spatial and topographic terms, although he made no diagrams himself," write his modern biographers (Vinten-Johansen et al. 2003, 208). To remedy that lack they created a schematic of the Albion Terrace outbreak (fig. 8.4A–B) in which the houses are related to both the brick sewer drains and to the

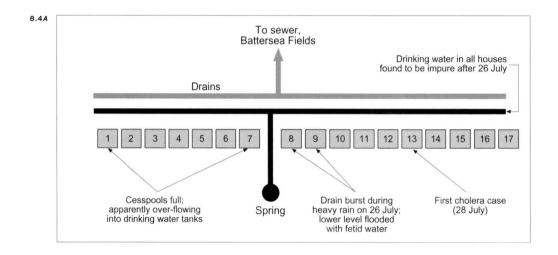

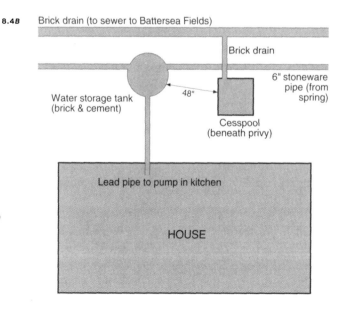

FIGURE 8.4A–B Snow argued, in this modern rendering, that the water supply was contaminated on Albion Terrace case after heavy rains caused cesspools to overflow.

local cesspools and water storage tanks. "The houses are numbered from 1 to 17 in Albion Terrace and are supplied with water from a copious spring in the road in front of the terrace, the water of which is conducted by a brick barrel drain between numbers 7 and 8, to the back of the houses, and then flows right and left to supply tanks in the ground behind each house, the tanks being made of brickwork and cement, covered with a flat stone, and connected with each other by stoneware pipes six inches in diameter. A leaden pipe conveyed water from each tank to a pump situated in the back-kitchen. There is a cesspool behind each house, under the privy, and situated four feet from the water tank" (Snow 1849a, 15–16).

When the drains behind houses 1 and 7 were opened "the cesspools at both these places were quite full, and the overflow-drain from that at No. 1 choked up" (Snow 1849, 16). It was the stopped drains that permitted contaminated materials to invade the water that served the houses whose inhabitants were stricken by cholera. On examination, Snow found that the water from one of the domestic water tanks smelt like privy soil and appeared to contain undigested food in feces (Eyler 2001, 226).

It is not the least of ironies that Snow used his sense of smell to argue against the miasmatic theory of airborne, malodorous disease. Snow's olfactory sense was not in question, but what it signified was open to doubt. Was that odor evidence of a miasmatic contamination—after all, Snow *smelt* it—or merely a symptom of the contaminated water whose ingestion Snow asserted was the source of the illness? If cholera were miasmatic, then the odors of the flooded area would be the source of the disease. And as Parkes later wrote, "We must observe, however, that in addition to the impregnation of the water, it appears that there must have been immediately before the attack of cholera, contamination of the air, also, as on the 26th of July, two days before the first case, the lower premises of two houses were flooded with foetid water from the blockage of the drain" (Parkes 1855a, 452). Snow's data was insufficient to show that his was necessarily the correct interpretation.

Snow attempt to substantiate his interpretation of the cholera pathology through an epidemiological case reports that supported his thesis of cholera as water not airborne. Absent from his report was the data that readers needed to be able to judge Snow's argument. After Snow reprised his 1849 argument in the second edition of *On the Mode of Communication of Cholera*, Parkes, who had also visited the site of the outbreak, wrote: "In this example, as in almost all the other cases adduced by Dr. Snow, we miss the very necessary information as to the number of persons resident in each house; their ages, occupations and habits, the kind of house in which they lived, etc. In six houses there were altogether twenty-four cases of cholera, in the seventh house only one case" (Parkes 1855a, 453). Without the number of inhabitants in the houses where cholera occurred the evidentiary value of the data Snow accumulated was diminished drastically. As Parkes argued, "There may have been only a single case in one of the six houses, and a greater number than the average in some of the others. If this were so, the point and force of the argument at once disappears."

Without a population denominator employed in a statistical approach, the local patterns of incidence seen by Snow were open to a range of interpretations. In an 1849 unsigned review of Snow's monograph, the reviewer, probably Parkes, noted that the pattern of concentrated incidence Snow identified as waterborne was the same as that others would expect if cholera were airborne (Anonymous [E. A. Parkes] 1849). Whatever contaminated one house might contaminate its neighbor, causing the clusters Snow observed. Snow's evidence for the prosecution thus almost equally served the defense. After all, everyone agreed that cholera's intensity was variable and that certain areas of the city were more affected than others. The question was why and Snow's audience was not convinced by his answer.

Snow admitted at the end of his 1849 monograph the limits of his analysis: "It would have been more satisfactory to the author to have given the subject a more extensive examination, and only to have published his opinions in case he could bring forward such a mass of evidence in their support as would have commanded ready and universal assent; but being preoccupied with another subject, he could only either leave the inquiry, or bring it forward in its present state, and he has considered it to be his duty to adopt the latter course, and allow his professional brethren to decide what there might be of value in his opinions" (Snow 1849a, 31). By rushing forward with a theory short of data, Snow assured his work would be received as at best a speculative statement, not a definitive discovery.

In his monograph Snow made another astonishing intellectual leap in arguing on the basis of very local, neighborhood outbreaks in Albion Terrace and Horsleydown that cholera *everywhere* in London, and by extension the world, was waterborne. Using coarse data based on registrar returns to the GRO, Snow did not map the incidence of cholera in his monograph but did include a table of cholera deaths in London's aggregated registration districts (table 8.1). The table (Snow 1849a, 24) argued both an uneven incidence of occurrence, and importantly, its concentration in districts whose water source was distinct, and therefore inferentially implicated. "The

TABLE 8.1 Deaths from cholera in London registered from September 2, 1848–August 3, 1849

District of London	Population (1841)	Deaths from cholera	Deaths per 1,000 persons
West	300,711	533	1.77
North	375,971	415	1.10
Central	373,605	920	2.48
East	392,444	1597	4.06
South	502,548	4001	7.95
Total	1,948,369	7,466.00	3.83

deaths from cholera in this district, which contains a very little more than a quarter of the population, have been more numerous than in all the other districts put together" (Snow 1849a, 23–24). For Snow, this was further proof of his thesis.

Snow's use of mortality ratios in this table acknowledged the growing requirement for basic population analytics in disease studies. The table also evidences Snow's limits in handling this type of data, a problem that would be more critical in his work in the mid-1850s. First, there is the problem of arithmetic error. The total population for the five London areas was not at the time 1,948,369 persons, but 1,945,217 persons. More importantly, there is the construction of a total global ratio of deaths per 1,000 persons. Snow derived it by dividing total deaths (7,466) by total population (1,948,369) and then multiplying by 1,000, rather than by taking the mean of the ratios for each area, recorded row by row. This eliminated the specificity based on regional difference that the table promised. The average mortality across all districts, computed by adding the mortality per districts and dividing by 5, the number of aggregated areas, was 3.48 deaths per 1,000 persons.

One can argue on the basis of these general ratios, as Snow did, but in doing so one can not also insist on a conclusion based on the specificity of local area reports. The granularity promised in his mortality by regional division disappeared in Snow's totals because of his method of calculation.[5] For Snow's contemporaries, what was significant was not the mistake in addition or his method of calculation, neither of which they commented on, but his logic and its leap from the scale of the local, neighborhood outbreak to that of region, the metropolis at large, and cholera in the world.

If his argument at the local level of Albion Terrace was unconvincing it could not be used with certainty to explain cholera at the scale of the metropolis. Doubt traveled up from the local, in other words, and this meant that in making his case Snow also had to consider what he left out, the social and geographic factors assumed by others to promote disease. How many people lived at what density in areas of variable income in a geography where altitude and air quality was variable? This would be a chronic complaint about Snow's work. "Dr. Snow does not sufficiently discuss the other conditions under which the people living in various districts of London were places, besides those of varying water supply," Parkes observed in his critique (Parkes 1855a, 439).

Nor could Snow account for the fact that across the metropolis many who cared for their sick relatives did not themselves become sick, or that many who were cholera-free drank the same water as those who became ill. Finally, Snow could not demonstrate the existence of a waterborne disease agent. As Dr. William Budd wrote, "The detection of the actual cause of the disease, and the determination of its nature, were all that was wanting to convert [Snow's] views into a real discovery" (Budd 1873, 19).

Given those limits, Snow's conclusions were certainly premature, if not arrogant, in their certitude. To stop this cholera, and by implication typhus and other diseases, Snow concluded, "It would only be necessary for all persons attending or waiting on the patient to wash their hands carefully and frequently, never omitting to do so before touching food, and for everybody to avoid drinking, or using for culinary purposes, water into which drains and sewers empty themselves; or if that cannot be accomplished, to have the water filtered and well boiled before it is used. The sanitary measure most required in the metropolis is a supply of water for the south and east district of it from some source quite removed from sewers" (Snow 1849a, 30).

Snow worked hard after the initial publication of his monograph to refine his argument. After the second edition, he almost immediately published two papers that fleshed out the theory that *On the Mode of Communication of Cholera* presented. In 1851 Snow presented papers to the Epidemiological Society of London (Snow 1851). In them he used, as an explanatory tool, a map of the South London watersheds from an 1844 report as well as Grainger's map of cholera in South London, included in the earlier parliamentary report (fig. 7.10).[6] Across this growing body of work Snow's thesis became increasingly spatial, one grounded both in the registration districts and subdistricts whose mortality reports were his principal data source, and increasingly, the water sources within those areas of study.

OTHER VOICES

Flying in the face of received truth, of dominant professional and popular opinion, the argument was astonishing. Explicitly Snow's theory challenged long-established medical theory on the basis of a disputed pathology and a suggestive epidemiological interpretation that was far from conclusive. Implicitly, it challenged the entire social reform enterprise based on the idea of a miasma with causal powers to propel an agenda of sanitary reform. That theory was all about the origins of disease in the stench of densely packed human settlements without adequate waste disposal. Snow's cholera was a thing whose source was contaminated water and whose agent was in the main . . . bad personal hygiene.

As the medical historian P. E. Brown put it, albeit without conscious irony, "All the elements of his [Snow's] theory were already in the air" (Brown 1961, 30), although none before Snow put the argument forward in so audacious a fashion. For example, Snow was not the first to worry publicly about the contamination of urban water supplies by sewage. In 1848, for example, R. D. Thomson and William Farr, then compiler of abstracts at the GRO, expressed their concerns about the wells of Glasgow in the *Lancet*; others decried as both unhealthy and wasteful the urban waste disposal system, "which consigns the excreta of the populations to rivers or water courses" (Brown 1961, 524). In May 1849, William Budd investigated a typhoid fever outbreak in a middle-class district of Bristol, Richmond Terrace, Clifton, which occurred after a local water pump was contaminated with sewage. In this local outbreak, thirteen of thirty-four inhabitants of houses that obtained their water from it, and only those in houses served by it came down with typhoid fever (Budd 1873). "The frightful fatality of the disease in particular parts of infected towns," Budd concluded, was caused by the sewer-contaminated drinking water found in that single well (Budd 1849, 9).

To prove this thesis of waterborne cholera, Snow would need something more detailed than his coarse reading of London data or his nonstatistical consideration of neighborhood outbreaks. He would need a case more definitive than a local outbreak. He would need to be able to demonstrate not simply the likelihood of cholera being waterborne, or of interpersonal transmission, but that this was more probable than any other suspected source. As Parkes put it, "We do not say that this disease may not thus be communicated to the healthy, but we consider that the facts here mentioned only raise a possibility, and furnish no proof whatever of the correctness of the author's views. The *experimentum cruces* would be, that the water conveyed to a distant locality, where cholera had been hitherto unknown, produced the disease in all who used it, while those who did not use it escaped" (Parkes 1855a).

SNOW'S CHOLERA CRITIQUED

Snow's theory of cholera was silent on the exact nature of the agent activating the new disease he proposed. The question was not simply its origin (India), its source (air and/or water), or it's path(s) of transmission. At issue was its very nature. What *was* it? Having mockingly dismissed airborne agency Snow inferred but could not prove the idea of a waterborne thing. The month before Snow's publication in 1849 two researchers reported seeing "peculiar microscopic objects" in the rice-water discharges of cholera patients. These, they believed, might be the generative, previously unseen agent of the disease. Budd believed these objects were yeast-like and alive and formulated a theory of cholera based on his conclusion. They were, he proposed, "disseminated through society in the air, in the form of impalpable particles, in contact with the articles of food; and principally, in the drinking water of infected places" (Brown 1961, 521; Budd 1849, 5).

Snow was cautious in endorsing the "discovery," one rapidly discredited by microscopists who determined that the microscopic objects in question were epithelial cells shed by the intestines in the violent diarrheic bouts that were cholera's signature symptom. But for Snow's contemporaries, cholera morbus was distinguished, if at all, solely by its intensity. There was, after all, the common "English cholera" that commonly attacked British subjects in the summer and was brought on by eating spoiled food in the days before refrigeration. Still, Snow was glad to accept any pathological support that might bolster his argument. In a reference to Budd's work, Snow wrote that his thesis "was an opinion which I thought almost peculiar to myself when I was first led to adopt it, but which, as I have since been informed, others were beginning to entertain" (Snow 1849b, 745).

In his arguments Snow relied on the published reports of others but often presented conclusions based on those reports that the quoted authors did not share. For example, Snow used Dr. Thomas Shapter's book-length report on cholera in

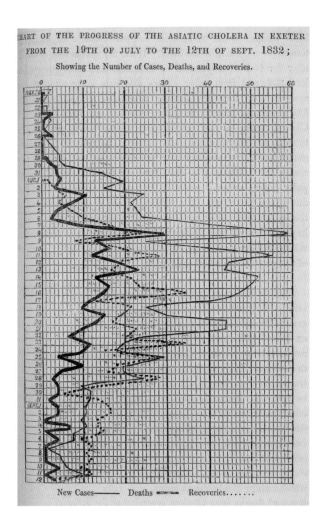

CHART OF THE PROGRESS OF THE ASIATIC CHOLERA IN EXETER FROM THE 19TH OF JULY TO THE 12TH OF SEPT. 1832;

Showing the Number of Cases, Deaths, and Recoveries.

New Cases——— Deaths ━━━ Recoveries . . . .

FIGURE 8.5 Thomas Shapter transformed official records of mortality and morbidity into an epidemic curve to demonstrate the temporal progression of the outbreak. That data was then mapped in an attempt to locate the outbreak's source.

Exeter in the 1830s, also published in 1849, to support his thesis of cholera as a waterborne disease. Shapter's text and map presented a "topography of disease" in which the incidence of cholera over three years was considered within the city of Exeter. Shapter recorded both mortality and morbidity across the outbreak in a chart of epidemic occurrence (fig. 8.5). He then used official mortality reports that included decedents' street addresses to locate cholera mortality in a map of the city. A mortality ratio based on parish population was recorded in the map legend. This amounted to a spatial description of mortality data revealing, mapped clusters of cholera appearing to occur more frequently in parishes along the low-lying riverbanks of the city, à la Grainger and Petermann, than at higher altitude away from the river where air was purer.

What Shapter saw in the map (fig. 8.6A) was a miasmatic disease, "essentially an epidemic, originated in, and chiefly due to aerial influences" (Smith 2002, 921). Snow saw Shapter's data differently, however. Shapter had informed Snow that Exeter water carriers drew their waters "almost exclusively from certain streams of water, diverted from the river in order to turn water-mills; and one of the chief sewers of the town, which receives such sewage as might come from North Street, in which the first cases of cholera occurred." (Snow 1849b, 749–50). Parishes like St. Edmund, whose water was drawn from cleaner streams, had a lower mortality rate than those in which the water came from what Snow described as polluted sources.

What could be more obvious, Snow reasoned, than simply to identify the water source as the culprit irrespective of altitude or air? For Snow no other explanation would serve or deserved exploration, even though the extent disease theory on which basis Shapter argued supported the atmospheric explanation. This was similarly congruent with an argument à la Chadwick for greater disease intensity in poorer neighborhoods, located near the river, a potentially confounding variable Snow did not consider in his analysis.

Another problem was that little was known about the water sources Snow argued where the origin of the British cholera outbreaks. Excepting, perhaps, Lea in Cincinnati (fig. 7.12), other researchers had not thought to interrogate local water sources as

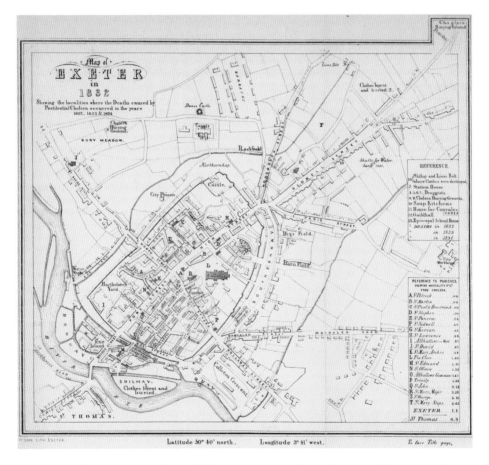

FIGURE 8.6A Shapter's map of cholera in Exeter presents parish mortality rates within a general topography that includes the location of individual cases, year by year. For Snow, it argued the proposition that incidence increased with proximity to polluted water.

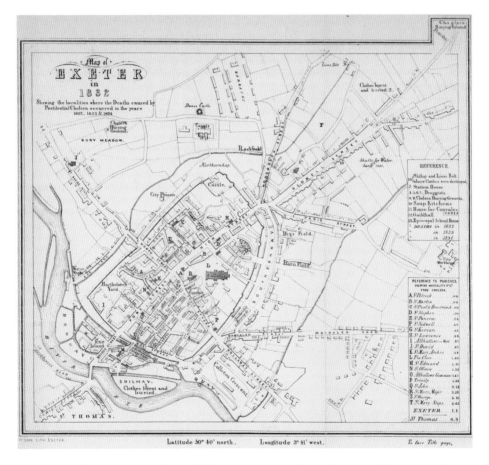

potential sites of cholera or typhus generation or transmission. "As we are never informed in works on cholera what water the people drink, I have scarcely been able to collect any information on this point, respecting foreign countries" (Snow 1849a, 926). If he were to argue a waterborne cholera Snow would need to do so in a manner that rigorously considered the nature of the urban water supply. This would not be easy.

In the mid-nineteenth century, water supply was a complicated enterprise in which varying private companies competed, sometimes on a street-by-street basis. In London in 1849, for example, ten private water companies competed for customers in the metropolitan region, most drawing their supplies from various rivers (the Thames, the River Lea, and so forth), and from various points along those rivers. In many registration districts two or more water companies competed street-by-street for customers. To complicate matters further, some businesses had their own wells and in some neighborhoods people would either send for water from other areas of the city—the water from the Broad Street well in St. James, Westminster, was considered especially tasty—or as occurred in South London, manually draw it from the nearest river.

Without knowing more about local water supplies, without being able to identify what the disease agent might be, and thus be able to test for it, Snow necessarily was limited in his argument. His cholera was a creature inferentially constructed on the basis of a paucity of evidence open to a range of interpretations. In the South London registration districts identified in the maps of Grainger and by Petermann as highly infected, Snow argued almost as an article of faith that the reason *must* be that the water supplied by the companies serving these areas was contaminated. There, however, were the beginnings of a testable thesis:

> On the south side of the Thames the water works all obtain their supply from the river, at parts where it is much polluted by the sewers; none of them obtaining their water higher up the stream than Vauxhall Bridge,—the position of the South London Water Works. Now as soon as the cholera began to prevail in London, part of the water which had been contained in the evacuations of the patients would begin to enter the mains of the Water Works: whether the materies morbi of cholera,—which, it has been shown, there is good reason for believing is contained in the evacuations,—would be sent round to the inhabitants, would depend on whether the water were kept in the reservoirs till this materies morbi settled down or was destroyed, or whether it could be separated by the filtration through gravel and sand, which the water is stated to undergo." (Snow 1849a, 23)

Snow's passing attention in 1849 to London's commercial water suppliers would serve, with additional research, to provide a better stage on which to argue the existence of his waterborne cholera. There were essentially three pathways to explore, each of which would require a different investigation. First, caregivers who did not wash their hands might inadvertently transfer cholera to others after being contami-

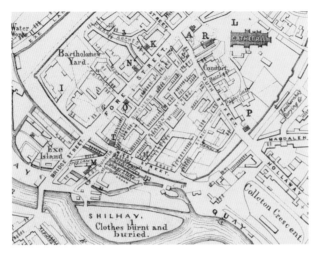

FIGURE 8.6B A detail of Thomas Shapter's study of cholera in Exeter in 1832 that shows cholera mortality in the years 1832–1834 through different symbols (bars, dots, and bars with dots).

nated by their wastes. This assumed the disease agent was in the discharge of patients and could be transferred by ingestion. Second, airborne spores of dried fecal matter might be inhaled in the close quarters of the home. This was for Snow a secondary if logically possible airborne pathway. Third, city districts were contaminated when the wastes of the sick entered the water supply. Proving any of these pathways, and thus overturning existing disease theories, and specific theories about cholera as airborne, would require proof not yet in evidence.

"I KNOW FROM HAVING SEEN THEM"

Snow did not have to publish his theory on cholera to gain public fame or notoriety. He was a physician of real renown, an expert on anesthesia with a full caseload, an M.D. with a slew of professional papers to his name. He earned money and fame from his 1847 textbook on anesthesia—the first in England—and from a bourgeoning practice as *the* London anesthesiologist (Snow 1847). Why, therefore, did Snow publish a thoroughly speculative monograph whose limitations had caused Snow to hesitate before sending it to the printer, "thinking the evidence in their favour of so scattered and general a nature as not to be likely to make a ready and easy impression" (Snow 1849a, 12)?

The answer must be that Snow published in this fashion, aggressively and prematurely, because he was compelled by a need to confront this disease. Cholera was Snow's obsession, a subject he could not leave alone. To understand this passion a brief detour into the history of the man himself is needed.[7] John Snow was born in 1813, the eldest son of laborer William Snow and his wife, Frances. We know little about Snow's early years except that, as a teenager, he was apprenticed to surgeon-apothecary William Hardcastle in Newcastle-upon-Tyne, where Snow's uncle lived, eighty miles from the family home. When cholera struck in 1831 Snow was what might be thought of as a senior apprentice charged with the care of patients under the general guidance of his superior. Snow never forgot his first experience with cholera among the miners of the Newcastle region, retelling it again and again in various publications.

> It has been found that the mining population of this country has suffered more from cholera than any other, and there is a reason for this. There are no privies in the coal pits, and I believe that this is true of other mines: as the workmen stay down the pit about eight hours at a time, they take food down with them, which they eat, of course, with unwashed hands, and as soon as one pitman gets cholera, there must be great liability of others working in the gloomy subterranean passages to get their hands contaminated, and to acquire the malady; and the crowded state in which they often live affords every opportunity for it to spread to other members of their families. (Snow 1849b, 747)

In 1836 Snow moved to London to complete his studies, qualifying as an apothecary in 1838. He then set up a practice on Firth Street in St. James, Westminster, where he attended local medical society meetings while studying for his medical degree. Soon he was first reading and then publishing his own papers: from 1838 to 1842 Snow published at least seven in the *Lancet* and the *London Medical Gazette*.[8] In 1843 Snow earned the right to put "M.D." after his name. In 1847, a year after the first London demonstration of ether as an anesthetic, Snow published his textbook on anesthesia and patented an inhaler that permitted the drug to be more reliably administered.

Against heavy odds, this son of a laborer who had dreamed of being an apothecary had made good as a London physician. Indeed, he became *the* specialist physician, the man who was the acknowledged authority on anesthesiology and its gasses. There was no *need* for Snow to go to the time, trouble, and expense of publishing on cholera. And yet, I think, Snow was in thrall to the memory of the colliers he had watched die in the mines and the working poor he saw in his Firth Street practice as an apothecary in London. Snow's passion about cholera was grounded in these experiences, and perhaps memories of the working-class people with whom he grew up.

From his first publication on cholera Snow always recognized the unhealthy effect of poverty in the overpopulated tenements of working class peoples. In what became a set piece, repeated in various publications, he used the pitmen of the Newcastle mines as an example of the unhealthy conditions in which they labored. "As the workmen stay down the pit about eight hours at a time, they take food down with them, which they eat, of course, with unwashed hands, and as soon as one pitman gets the cholera, there must be great liability of others working in the gloomy subterranean passages to their hands contaminated, and to acquire the malady; and the crowded state in which they often live affords every opportunity for it to spread to other members of their families" (Snow 1849c, 8–9).

Snow did not blame the poor for the sanitary problems they faced. He was certainly not opposed to the socially progressive agenda of reformers like Chadwick who saw in the poverty of some a reservoir of potential disease for all. But Snow's focus on personal hygiene and water as sources for cholera's spread did challenge the miasmatic theory of disease that was the rationale for many sanitary reforms. For his cholera to be accepted as *the* cholera would require a more rigorous proof, the *experimentum cruces* that Parkes called for in his 1849 review of Snow's monograph. That experiment would be carried out first not by Snow, however, but by William Farr, the compiler of statistics for the GRO in London. Farr would accept, at least provisionally, Snow's argument about an interpersonal route of transmission but would do so within the context of a hybrid thesis in which air *and* water combined to create an environment for cholera's development and transmission.

WILLIAM FARR: 1852

Farr was born in 1807, the eldest child of a Shropshire farm laborer. At a young age, Farr was sent to live with a local squire, Joseph Pryce, who effectively adopted him (Eyler 1979, 1–2). Pryce supported Farr's apprenticeship at the age of sixteen as a dresser, surgical assistant, and then student apothecary at the Salop Infirmary in Shrewsbury in 1826. Farr's patron died in 1828, leaving his protégé a legacy (£500) sufficient to fund two years of medical studies in Paris, then arguably the most advanced center of medical instruction in Europe. In his studies Farr became engaged—perhaps indoctrinated is not too strong a word—with the evolving science of "hygiene," what today we might call social medicine. It emphasized the then exploding field of medical statistics, the analysis of disease-related population statistics, at which Farr excelled (Koch 2005, 45–46). This was the training he brought back to England where he attended lectures at University College, London, briefly returning to Shrewsbury and a temporary hospital post before passing his apothecary examination in 1832. Farr then married, and after a brief stint as a *locum tenens* at a Shrewsbury city infirmary, moved to London where he established a practice on Fitzroy Square (Humphreys 1885, xv).

The city was awash with competing practitioners and in the brief years of his practice Farr was simply another struggling apothecary, and perhaps one too many. Like Snow, Farr, too, began to attend professional meetings and then submit technical articles to a range of journals, the most important of which was the *Journal of the Statistical Society of London*. Farr was a believer in the evolving *méthode numerique* that served not simply health-related studies but those of the evolving new study of criminology as well as other elements of society. Farr shared his commitment to the numerical with others of more advanced age and greater political importance who similarly sought a quantifiable vision of social and medical realities.[9] Jeremy Bentham, for example, the great legalist and social theorist, believed in statistics as a "newly cultivated branch of *Geography* having for its subject the quantities and qualities of

FIGURE 8.7 This portrait of William Farr (1870) was taken relatively late in his life.

the matter of population . . . wealth . . . and political strength" (Mack 1962, 240).

Elected a fellow of the Statistical Society in 1839, Farr described statistics as a "master science" permitting elements contributing to disease to be discovered and "natural laws" of health and disease to be formulated (Humphreys 1885, xiv). "It was not to make us hewers and drawers to those engaged on any edifice of physical science," Farr wrote in a prospectus for the society in 1838, "but it was that we should ourselves be the architects of a science or sciences." It was this science, and especially its application to the science of medicine, that would be Farr's life work and legacy.

In 1839 Farr joined the staff of the newly constituted GRO in London, where he would serve for forty years under two politically appointed registrars general, first as a compiler of abstracts and then as superintendent of the statistical department (Eyler 1979, 9). Chadwick had argued successfully for the appointment of young medical men, impecunious apothecaries like Farr as local district registrars. His goal was in part political, to bring the burgeoning medical establishment onside in support of the administrative and social reforms Chadwick was advocating. These reforms included a rigorous system for the collecting of data on births mortality and its causes as well as a comprehensive national census Farr helped supervise in 1841. Farr's view of medical statistics was little short of revolutionary, a hybrid that combined basic algebraic formulations with careful attention to disease symptomology. In the registrar general's first annual report in 1838, Farr proposed a "statistical nosology" that sought to improve on traditional disease classification systems.[10] He listed and defined twenty-seven fatal disease categories—for example, distinguishing dysentery (bloody flux) from diarrhea (looseness, purging, bowel complaint)—to be used by local British registrars in reporting causes of death.

As both bureaucrat and researcher Farr's approach to disease in general, and cholera in particular, was distinct from Snow's. His long-term contribution would be, as historian Pamela Gilbert put it, "to firmly root the connection between the already lively medical interest in statistics with a formal institutional position and a public health agenda" (P. N. Gilbert 2008, 82). It was in part the development of a mechanism for the collection of national health data, and Farr's demonstration of its application to cholera, that permitted physicians to move from treating individuals to participating in the larger question of population health.

It would be William Farr's statistically defined cholera, one based upon a mass of evidence, which Snow would have to confront. And he would do so using not simply his own observations but as well the official data that Farr's office solicited from the registrars in place around the nation. Farr would be both an early supporter of Snow's theory of interpersonal transmission and a skeptic regarding Snow's waterborne cholera theory. As the second pandemic of cholera roiled through England in the late 1840s, Farr would promote a different cholera, one that was air and waterborne at once. Further, he would do this in a manner that demonstrated a methodology that was to become the standard for research in health studies for a generation, one against which Snow's work would be measured and in his day found wanting. All this would happen at two very different scales of address, that of the metropolitan epidemic and that of the neighborhood outbreak.

CHAPTER 9

SOUTH LONDON CHOLERAS: WILLIAM FARR, JOHN SNOW, AND JOHN SIMON

In 1850, the General Board of Health produced a report on the return of cholera to England in 1848–1849 with a critical appendix authored by Dr. John Sutherland (Sutherland 1850).[1] In it was a seminal study of the possible relationship between cholera and water supply in Salford, Manchester. Sutherland also described a neighborhood outbreak in Windmill Square, Shoreditch, in which half the occupants of five houses, all supplied with water by a local well "into which surface refuse and the contents of cesspools percolated," became ill with cholera (Smith 2002, 924). Also reported was an outbreak of diarrhea in Hackney in which forty-six of sixty-three inhabitants who used a single well became ill while twenty-two neighbors using other water supplies remained wholly free of the disease.

Suggestive as these studies were, and as supportive as they appeared to be to Snow's general thesis, they were not the *experimentum cruces* Edmund Parkes had called for and Snow himself desired to produce. The scale was too limited and the methods insufficiently rigorous to be more than suggestive. Nor did these studies carefully consider the social and environmental factors many believed at least as likely as water to have contributed to patterns of disease incidence. In 1851, participants from eleven countries met in Paris for the First International Sanitary Conference. Their quixotic agenda, to create a single system of quarantine that might halt the spread of cholera, floundered upon disagreements over what cholera was, how it was spread, and therefore on how it might be stopped (Stern and Markel 2004, 1474).

Three very different studies of cholera in South London attempted to solve the puzzle that was this disease. Each argued a different theory of its nature, each employed a different approach to the official data provided by registration district and subdistrict registrars to the GRO. The first, by William Farr, sought to uncover a

FIGURE 9.1 Snow was one of a phalanx of cholera researchers in the mid-1850s. Some, like Snow, worked independently, and others, like Dr. John Sutherland (pictured here), were part of official agencies.

natural law, something demonstrably regular, describing the distribution of cholera as a recurring, physical phenomenon. The second was John Snow's attempt to argue cholera as a solely waterborne condition distributed through contaminated water and interpersonal contamination. The third, by London physician and Board of Health official Dr. John Simon, attempted to consider the nature of cholera as water or airborne in a careful study of fine-grained data.

WILLIAM FARR

In 1852, William Farr published what one reviewer admiringly called "one of the most remarkable productions of type and pen in any age or country" (Eyler, 1973, 79). Encyclopedic and authoritative, its hundred pages of dense text and three hundred pages of diagrams, charts, maps, and tables together argued for still another cholera, one that was this *and* that, at once airborne *and* waterborne. Based on data reported to the GRO by the nation's district and subdistrict registrars, and written by the GRO's expert on health statistics and their application, the work carried an immense authority. It was the standard against which work, including John Snow's, would contend (fig. 9.2).

Farr's publication emphasized the importance of the GRO as not merely a compiler of records for other bureaucracies but as a research center in its own right. From the Reform Bill of the 1830s through the acts that set up both the GRO and the national census system, promises had been made that the new bureaucracies would serve the

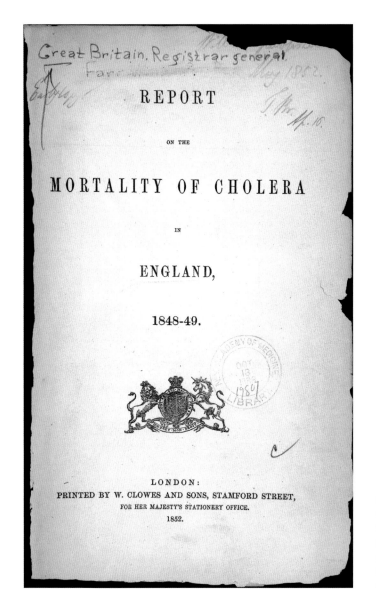

REPORT

ON THE

MORTALITY OF CHOLERA

IN

ENGLAND,

1848-49.

LONDON:
PRINTED BY W. CLOWES AND SONS, STAMFORD STREET,
FOR HER MAJESTY'S STATIONERY OFFICE.
1852.

FIGURE 9.2 William Farr's study introduced a mass of data, argued numerically and graphically, that suggested an alternate theory of cholera based on the 1849 epidemic in England.

nation, and especially its health concerns. Farr's report was . . . a deliverable. On the cover of the work they might as well have stamped: "Your government at work."

A Report on the Mortality of Cholera in England, 1848–1849 began with a letter from the registrar general, Farr's boss, commending the work to Her Majesty's Principal Secretary of State for the Home Department, the Right Honorable Sir George Grey. "In 1850 the General Board of Health had requested from the G.R.O. records of all cases of cholera in England and Wales," the letter explained, "distinguishing mortality by age and sex. After well considering the subject, I thought it desirable, not only for the use of the General Board of Health, but for circulation throughout the country, that a complete history of the late epidemic should be drawn up from the facts recorded in the register-books of deaths, distinguishing not only sex and age, but also the profession, the date of death, the place of death, and the duration of illness of all persons who died of *Cholera* or *Diarrhoea* in the years 1848 and 1849; giving not only the bare facts but exhibiting in Tables the various combinations of age, profession, locality, etc." (Farr 1852a, 2).

In his preface Farr promised "to show the effects of all the circumstances recorded in the Registers on the fatality of the greatest epidemic that has for many years infested England." The "show" was literal as well as figurative. Farr employed an unprecedented range of innovative graphic techniques—including charts, various graphs, and maps—that argued tables of data in a visually rigorous way. The sheer volume of data Farr mustered for his study was sufficiently daunting that he needed these tools to make clear what was in its raw state, an ambiguous and messy array of facts.

Farr's goal, like that of many of his contemporaries, was to uncover a "natural law," a numerical regularity that would provide a consistent description of cholera and if possible point the way to an explanation of its recurrence and pattern of distribution. In this he followed the lead of French theorists like Guerry and Quetelet who in the 1830s attempted to define social laws based on "moral" data (crime statistics, literacy rates, and so forth) in a manner that presaged more general work in medicine in the 1840s and 1850s. "Numerical regularities about disease, unknown before 1820 were commonplace by 1840. They were called laws, laws of the human body and its ailments" (Hacking 1999, 61). They were regular to the degree they described consistent patterns of relation between a quality (mortality) and a condition (cholera). Physical sciences like physics and chemistry had their laws. Farr believed that statistics was the way to reveal similar patterns of regularity in the arena of disease and health studies.

Farr's cholera

Farr first acknowledged the difficulty of distinguishing this new cholera, which originated in India, from endemic "English cholera." "The two forms are often not distinguished in the [official mortality] returns. The cases, separately considered, run so insensibly into each other, that the attempt at distinction would have been fruitless" (Farr 1852a, ix). As Jameson had first reported, because the two were symptomatically identical except for cholera's greater virulence, they were, to the average physician, often indistinguishable.

Through the 1850s mortality records submitted to the GRO included as many as fifteen different diarrheic diagnoses, all with similar symptoms, all of which were probably cholera.[2] For his study, therefore, Farr created two distinct but related classes of diarrheic disease based only on reported diagnoses. "In abstracting the cases of 1849, for the series of tables in this volume, all cases returned as 'cholera,' whether English or Asiatic, cholerine, 'bowel complaint,' and diarrhoea simply, or as a complication of other diseases, were transcribed. All the cases in which the term 'cholera' or 'choleraic diarrhoea' occurred, were referred to as cholera; about 300 cases in which diarrhoea was evidently a symptom of consumption, or some other disease, were struck out; the residue of the cases was classed under diarrhoea" (Farr 1852a, xi).

Farr then demonstrated that cholera and other dysenteric diseases had similar epidemic curves and similar temporal profiles. Both appeared in spring or early summer, peaking in September before disappearing late in the year (fig. 9.3). In this way Farr argued a commonality between the two disease sets—cholera and diarrhea—as naturally occurring, seasonal events perhaps promoted by the hot airs of summer. With the aide of records from the Royal Greenwich Observatory, other climatic variables, including rainfall, barometric pressure, wind direction and speed, were imposed upon the epidemic record. This did not mean Farr believed cholera morbus was a natural and inevitable seasonal disease, however, only that it had a climatic profile whose meaning had yet to be determined. As Jameson had first argued in India, Farr's chart demonstrated that cholera and other diarrheic conditions were to be distinguished largely on the basis of intensity (number of cases reported) and the virulence of the reported disease (its mortality).

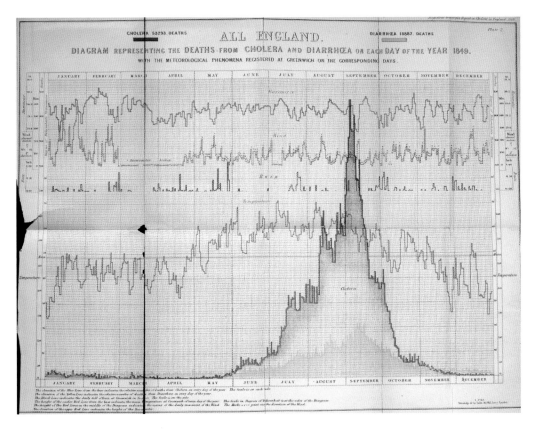

FIGURE 9.3 Farr charted epidemic curves for both cholera (blue) and other diarrheic diseases (yellow) against a range of climatological data (barometric t pressure, rainfall, temperature, wind speed and direction, and so forth).

THE MAP

Farr's study, like Petermann's in 1848, was carried out at two distinct scales. Petermann published his map of cholera mortality in London below his map of cholera in Great Britain (fig. 7.11A–B). Farr also first mapped 1849 cholera at the scale of the nation (fig. 9.4). In this map, tables of incidence that contributed to the temporal graph were located in the national landscape to argue cholera as a not only a national but also an urban health threat. In this map, unlike Petermann's, mortality in registration districts and subdistricts was calculated as a mortality ratio per ten thousand persons based on the population records returned in the 1851 national census.

The tables and the map created a demographic cholera whose distribution again could be seen to progress from the port cities inland. Mapped mortality clusters surround towns identified by name and the date of cholera's first appearance. The map's shading, rather than its use of dots per case per town, present a precise summary permitting Farr to state: "We find this striking result: 46,592 of the 532,293 deaths from cholera in the year 1849 occurred in 134 of 623 districts, or in less than a seventh part of the area of England and Wales, among four parts in ten of the population. Only 6,701 deaths took place out of 10 million people in 49, 228 square miles of territory" (Farr 1852a, xlix).

Within that general portrait of statistical incidence, Farr concluded that "the cholera was three times more fatal on the coast than in the interior of the country" (Farr 1852b, 156). The point was not simply the inference that this cholera was imported through coastal cities. Rather, the greater coastal incidence argued to Farr the necessity of a detailed analysis of cholera in the coastal area of greatest population and mortality: London. If cholera was to be understood, the national map of cholera in 1849 insisted, it would be through an interrogation of its presence in the metropolis that was at once the greatest city and the busiest port in England.

Farr next considered the general relation between mortality reported to the GRO in the 1840s and temperature recorded at the Royal Greenwich Observatory. For this he used a polar projection called a coxcomb that permitted relatively large data sets to be argued in a manner that visually proposed a relationship between them (fig. 9.5). In this type of graph the values of the sectors are proportional to the squared radii of the sectors.[3] The graphic was an important addition to the scientific language of the day, a means of presenting several different variables in a manner that was clearly understand, an argument that was easily comprehensible. In this case Farr argued à la Sydenham that *if* mortal disease was seasonal, *then* patterns of annual incidence could be observed across decades of data. Farr appears to have introduced the coxcomb projection to England, where Florence Nightingale would later make good use of it in her campaign for better support of British soldiers fighting in the Crimean War. She learned how to fashion it from Farr, with whom she carried on a detailed correspondence. André-Michel Guerry was its originator, however; he developed the analytic graphic while Farr was still a student in Paris (Guerry 1829).

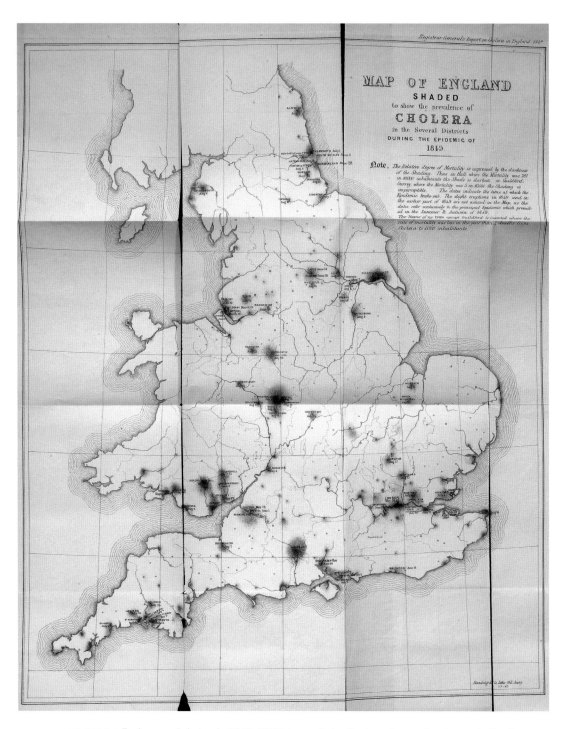

FIGURE 9.4 Farr's map of cholera in 1848–1849 demonstrates the importance of port areas to the disease's introduction and diffusion.

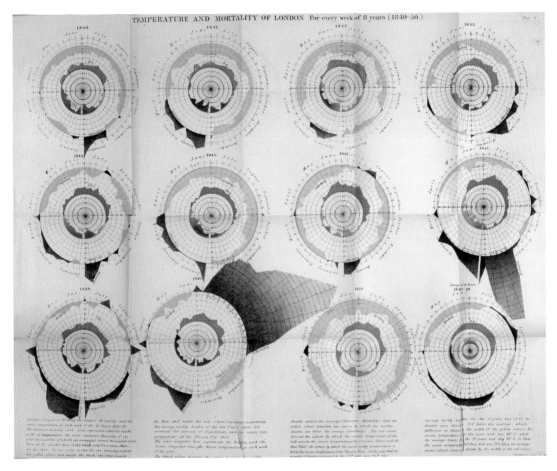

FIGURE 9.5 Innovatively, Farr used the coxcomb graph in his study of cholera in London to compare the average mortality and average temperature for the years 1840–1850.

The radii of the circle represented weeks of the year; the concentric circles in each year symbolized gradients of either 10 degrees of temperature or 100 deaths, depending on their function.[4] Black extrusions outside the circle marked periods of excess mortality, yellow figures inside the circles periods of diminished mortality. "Thus, in the year 1846 the number of deaths registered in the twenty-fifth week was 808, the average weekly number for the 10 years was 1020, the deaths were therefore 212 below the average which difference is shown in the width of the yellow colour. The mean temperature for the same week was 69.5°, while the average mean in the 70 years was only 48.4°, so that the mean temperature of that week was 21.10 above the average mean, which excess is shown by the red colour" (Farr 1852a, 163). The coxcombs laid out the unusual mortality of 1849 within seasonal mortality patterns of winter respiratory diseases and summer diarrheic diseases.

In acknowledging the seasonality of this cholera Farr was making no causal claim. Rather he was acknowledging a traditional environmental perspective without endorsing the idea of a climatic determinism that had attended earlier disease studies.

In another coxcomb projection Farr compared the incidence of cholera to that of previous epidemics, including the plague of the 1650s. Plague was the standard against which epidemic events were judged, and in that comparison, Farr's data argued, cholera was a relatively minor event. This served politically to assuage fears of a plague-like cholera's potential effect on England's economy and its population. It further located cholera within a history of epidemic disease known to Britons, minimizing without trivializing its incidence.

Metropolitan London

From the start, Farr acknowledged Snow's idea of person-to-person transmission: "Dr. Snow is unfortunately able to show that this excremental distribution—almost too revolting and disgusting to write or read—is possible to a very considerable extent" (Farr 1852a, lxxviii). Farr agreed that, as Snow proposed, cholera "multiplied in the intestines of infected people until the disease spread in this way all over the metropolis. It necessarily implies that the rice-water [diarrhea] discharges[5] of the cholera patients may, under the present system of water supply, be distributed unchanged to nearly every house in London, where water is used for drink, ablution, and washing" (Farr 1852a, lxxviii).

That begged the essential question, however: what *was* this new cholera? Why did it appear on the map, here and not . . . there? Interpersonal transmission might explain local intensity but not the greater pattern of disease incidence. The contamination of water might be a source but was it necessarily *the* source of the disease, everywhere and always? To confront that question Farr created a remarkable, semigraphic, multivariate display in which he posted a series of geographic and social variables that alone or together might serve to explain cholera's pattern of incidence.

> The relative density, expressed by persons per acre, is obtained by dividing the population of 1849 by the number of acres in each district. The number of persons to an inhabited house is for the year 1841, and has not since greatly varied. The relative wealth or poverty of the districts of the metropolis bears a certain relation to the annual value of the houses; which was obtained from the parliamentary return of the value of houses as assessed under the Income Tax for the year 1842–3. The division of this value in pounds by the number of inhabited houses, and by the population of 1841, gives the average annual value of houses, and the annual value of houses and shop room, sufficiently near for the present purpose. The poor rate in £s is added for 1842–3. The mortality of each district from all causes in the 7 years 1838–1844 is also given. (Farr, 1852a, lviii)

In addition, Farr identified for each registration district a principal water company supplier, identified by a capital letter or letters ("S" stood for Southwark Waterworks Company, for example), as well as a ratio of cholera deaths per ten thousand persons based on census population and raw cholera mortality figures for 1849 (fig. 9.6A). For comparison he included mortality by population from all causes for the years

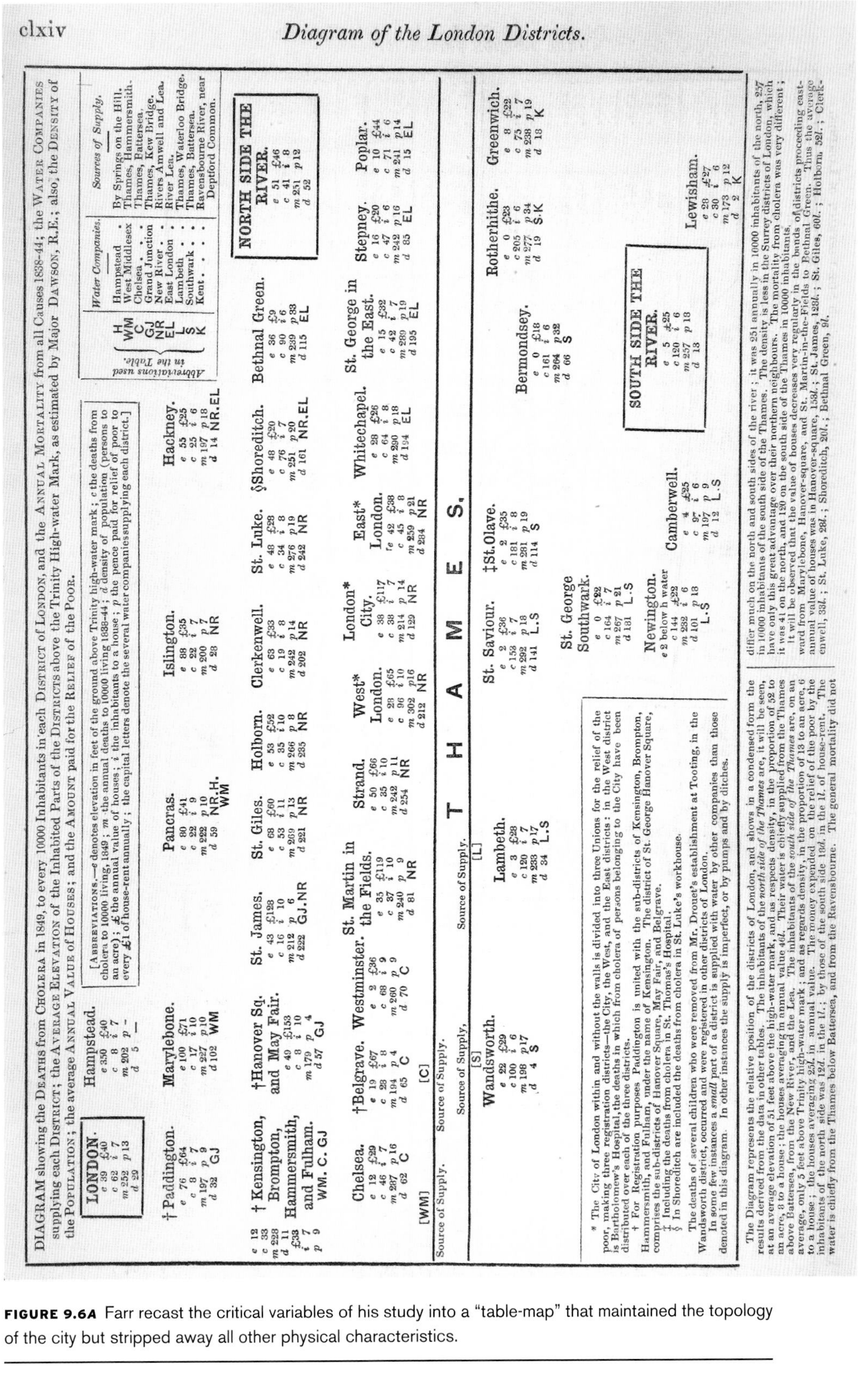

DIAGRAM showing the DEATHS from CHOLERA in 1849, to every 10000 Inhabitants in each DISTRICT of LONDON, and the ANNUAL MORTALITY from all Causes 1838–44; the WATER COMPANIES supplying each DISTRICT; the AVERAGE ELEVATION of the Inhabited Parts of the DISTRICTS above the Trinity High-water Mark, as estimated by Major DAWSON, R.E.; also, the DENSITY of the POPULATION; the average ANNUAL VALUE of HOUSES; and the AMOUNT paid for the RELIEF of the POOR.

[ABBREVIATIONS.—*e* denotes elevation in feet of the ground above Trinity high-water mark ; *c* the deaths from cholera to 10000 living, 1849 ; *m* the annual deaths to 10000 living 1838–44 ; *d* density of population (persons to an acre) ; *i* the inhabitants to a house ; *p* the pence paid for relief of poor ; *£* the annual value of houses. The capital letters denote the several water companies supplying each district.]

Water Companies.		Sources of Supply.
H	Hampstead . .	By Springs on the Hill.
WM	West Middlesex	Thames, Hammersmith.
C	Chelsea . . .	Thames, Battersea.
GJ	Grand Junction	Thames, Kew Bridge.
NR	New River . .	Rivers Amwell and Lea.
EL	East London .	River Lea.
L	Lambeth . . .	Thames, Waterloo Bridge.
S	Southwark . .	Thames, Battersea.
K	Kent	Ravensbourne River, near Deptford Common.

NORTH SIDE THE RIVER.

SOUTH SIDE THE RIVER.

FIGURE 9.6A Farr recast the critical variables of his study into a "table-map" that maintained the topology of the city but stripped away all other physical characteristics.

1838–1844. As a separate variable, Farr also included estimated elevation above the Trinity high-water mark as a surrogate for air quality, à la Grainger (fig. 7.10). This was point data developed by the Metropolitan Commission of Sewers and reviewed for Farr by assistant commissioner of the Tithe Commission in London. "The elevation and area of each district have been estimated by Major Dawson, R. E., from the maps furnished to the Sewers' Commission by the recent Ordnance Survey" (Farr 1852a, vii). Created by Parliament in 1836 to transpose the feudal remnants of the old tithe system into a more modern system of monetary rents, the commission surveyed and then mapped holdings in districts where procedural disputes occurred (Holt 1984).[6] As an assistant commissioner in London, Dawson was thus familiar with both survey mapping and with the metropolitan registration districts themselves.

While registration districts differed in area they were, Farr argued, roughly comparable administratively: "In the various combinations of the results, each [registration] district, whatever may be its extent, is considered to represent a certain state of things, and is therefore treated as of the same average extent; which is equivalent to dealing with an equal amount of the average population of the 38 districts" (Farr 1852a, lvii). In mapping his data Farr stripped away everything but a bare geometry to create what might be called a table-map permitting arguments based on data that became, in effect, the entire geography. Indeed, the table-map almost *demanded* of the reader to isolate variables—like elevation, housing density, mortality from cholera, and principal water supply company—and visually track them registration district by registration district. Mapping cholera as one datum in a set of potentially relevant elements allowed Farr to argue that cholera was a statistical fact whose relationship to other statistical facts (like income and housing density) could be tested.

In his analysis Farr tested the possible relationship between each of the variables inscribed in the table-map and the incidence of cholera mortality. There was, he noted, a weak inverse relationship at the registration district level between cholera mortality, on the one hand, and economic variables on the other. "If the 19 wealthiest districts are compared with the 19 poorest districts, the mortality from cholera is found to be inversely as the wealth, measured by the value of house income," Farr wrote. "It will also be found that the poorest population suffered most from cholera as well as from ordinary causes; but that the influence of wealth on cholera was greatest in the districts which experienced the fewest deaths in ordinary years, and were supplied with Thames water from Kew and Hammersmith" (Farr 1852a, lxvii). While real, Farr argued, the effect of income and class was too weak to be definitive and paled in comparison with other, stronger statistical relationships.[7]

Similarly, cholera mortality was more than twice as high in registration districts whose water was drawn from the central Thames, a famously polluted area, than in those districts whose waters came from the Lea and Ravensbourne rivers north of the city. What was unclear was whether water supply was the source of the difference or whether the real difference lay in geographic and socioeconomic variables

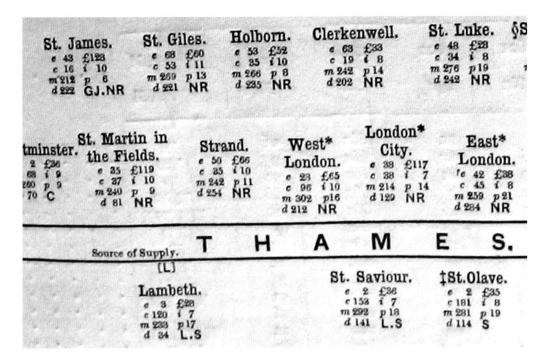

FIGURE 9.6B In his "table-map" of cholera by registration district, Farr included a range of variables including elevation (*e*), cholera deaths per 10,000 persons (*c*), annual deaths per 10,000 persons from 1833 to 1844 (*m*), population density (*d*), house value (£), inhabitants per house (*i*), and pence for poor relief (*p*). Capital letters refer to the principal water company suppliers in each registration district.

The strongest observed correspondence was between altitude above Trinity high-water mark and cholera mortality per ten thousand persons. Elevation served at once as a general environmental indicator—soils changed as one moved away from the river—and as a surrogate for air and water quality. It also served as a surrogate for proximity to the polluted river and its toxic odors. "Noxious substances thrown into the Thames are carried back and forth a number of times before disappearing below the city. Anyone who observes the same stretch of river for several days in a row can track the odd-shaped piece of wood or plastic container through this movement" (Porter 1998, 43). The result was an often incredible stench from more than two hundred tons of daily waste carried to the Thames through seventy-one main sewer outlets: "The excrement of a million and a half humans, over 100,000 horses, and perhaps 10,000 cattle and other animals mixed with rainwater runoff, manufacturing waste, tanning and slaughtering byproducts, soil erosion, and shipping spills" (Porter 1998, 51).

The potential health effect was obvious to Farr: "The effect of the river Thames and of the water supply on the health of London must be noticed. The Thames collects the waters of 6160 sq. miles of country, extending from the Cotswold Hills in Gloucestershire to the eastern coast; and the great body of this water flows and reflows through London in tides; which carry the matter below London Bridge, a mile and a half above Battersea Bridge twice a day, and ascend as high as Teddington. The contents of the greater part of the drains, sinks, and water-closets of this vast city and of the 2,360,000 people on its sides, are discharged through the sewers into its waters; which scarcely sullied by the primitive inhabitants, have now lost all their clearness and purity. The dark, turbid, dirty waters from half-stagnant sewers are agitated by the tides, but are not purified until they reach the sea" (Farr 1852a, lviii).

But if the result polluted the water supply it *also* polluted the air nearest to the river where the stench was strongest. And it was in that stench that miasmatic theorists argued cholera propagated. "The whole idea of looking at the influence of elevation was in keeping with his [Farr's] understanding of cholera's causation. The alluvial soil and stagnant water along the margins of the river contained abundant organic material for the production of miasmata. As one ascended the Thames basin the concentration of miasmata in the atmosphere dropped quite quickly, producing a regular and predictable change in cholera mortality" (Eyler 1973, 89–90).

Effectively transposing Snow's general argument about waterborne cholera into something also airborne, Farr argued that cholera was generated in the warm airs that evaporated from the polluted water in the summer months. "The Thames presents a large evaporating surface which must be taken into account, and it gives off vapours day and night in quantities which the phenomena of a 'London fog' reveal. The still air then condenses the matter which at other times enters the atmosphere invisibly, and escapes observation . . . It is a fact well worthy of attention, that after the temperature of the Thames has risen above 60°, diarrhoea, summer cholera, and dysentery become prevalent, and disappear as the temperature subsides" (Farr 1852a, lix).

Polluted water was thus appointed as a source generating airborne miasms that bred cholera and might serve as well as an agent (if not necessarily *the* agent) of its diffusion. Altitude served as a surrogate for the stench of the river, distinguishing malodorous registration districts at or near sea level and those at greater elevations whose airs were relatively free of the pollution. To nail down the relationship, Farr consulted with "a Mr. Glashner of the Royal Observatory, Greenwich" who "was requested to make an estimate of the amount of vapour raised by evaporation from the Thames in London" (Farr 1852a, lix). The estimate—"A depth of water of fully 30 inches must evaporate from the surface of the Thames annually"—was enough for Farr. "Indeed," Farr added, "the quantity must be larger than this from the circumstance of its relatively high night temperature."

Farr tested his conclusion in another table-map—a map-diagram, perhaps—that retained the general physical shape of registration districts (fig. 9.7A). In it Farr considered the relationship between altitude and cholera mortality at the finer scale of

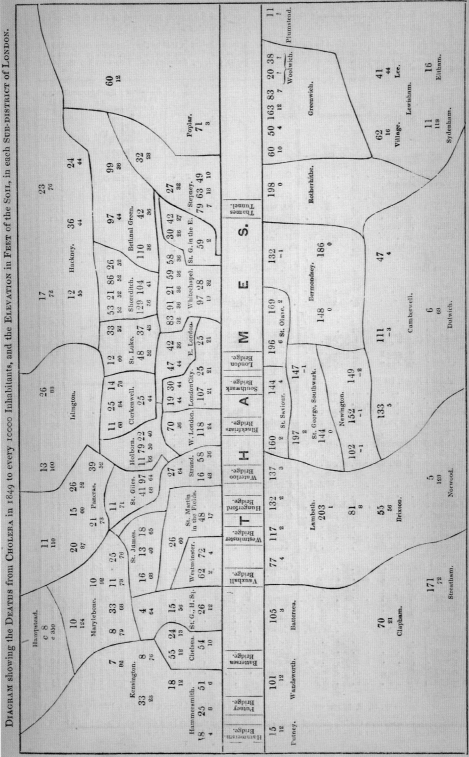

DIAGRAM showing the DEATHS from CHOLERA in 1849 to every 10000 Inhabitants, and the ELEVATION in FEET of the SOIL, in each SUB-DISTRICT of LONDON.

NOTE.—In each Sub-district the LARGE FIGURES denote the Mortality from Cholera, and are placed *above the Small Figures* expressing the Elevation : thus, in Hampstead, Cholera 8 deaths in 10000; Elevation 350 feet. The name of each DISTRICT only is given, excepting in the case of the most distant and extensive Sub-districts. Some disturbance in the results of this Table are caused by the Union Workhouses, which were sometimes used as Cholera Hospitals. The lines showing the boundaries of the Districts do not represent their exact shape, but indicate as nearly as possible their relative position.

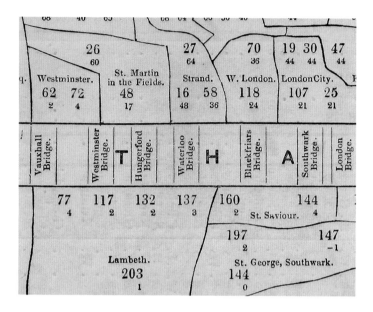

FIGURE 9.7B This detail shows the method by which Farr argued an inverse correlation between altitude (numbers in smaller font) and cholera mortality (numbers in larger font) in the "map-diagram" of London. The shape of the river and the registration districts have been generalized to increase legibility.

embedded registration subdistricts. Points identified the centroid of each registration subdistrict to which Farr added in text both mortality per ten thousand persons and altitude above sea level (smaller numbers). "The relation discovered between the elevation of the soil, and the mortality from cholera, is so important that it was thought right after the above [registration district] calculations were made to submit the principle to another test, by comparing the elevation and the mortality from cholera of *each sub-district*" (Farr 1852a, clxvi–clxix; original emphasis).

From this a general rule emerged: "The amount of organic matter, then, in the atmosphere we breathe, and in the waters, will differ at different elevations; and the law which regulates its distribution will bear some resemblance to the law regulating the mortality from cholera at the various elevations" (Farr 1852b, 163). This did not mean that cholera could not appear at different elevations, only that it would appear more often at lower elevations where air was fouler.

Farr's Law

Like dysentery and other summer diseases, Farr's cholera was lodged in time, its epidemic curve like yellow fever's, beginning in the summer and ending before autumn. It was also a statistical thing instantiated across first London's registration districts and subdistricts and then the nation to create surfaces in which mortality in the population could be considered in relation to geographic and social variables. The cholera landscape thus mapped included a varied terrain of land and water as well as socioeconomic data (housing density, relative income, and the like). Most importantly, perhaps, this was a bureaucratic cholera whose analysis carried the full weight of a system of government data collection and analysis.

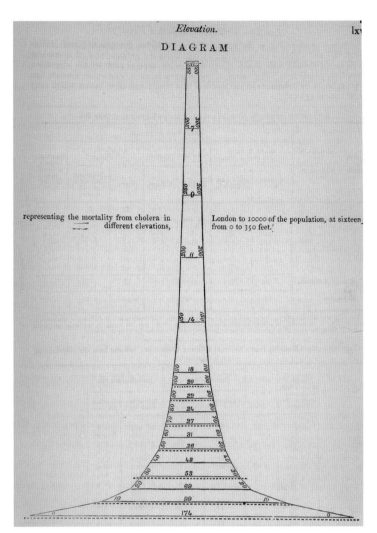

DIAGRAM

representing the mortality from cholera in different elevations,   London to 10000 of the population, at sixteen from 0 to 350 feet.

FIGURE 9.8 Farr's graph demonstrates the inverse relationship between elevation and cholera-related mortality reported in his study.

From these Farr crafted a lawful cholera whose pattern of diffusion was sufficiently regular it could be expressed algebraically, sufficiently accurate that it could be described in a diagram (fig. 9.8). It presented a simple inverse relationship between altitude above sea level (Trinity high-water mark) and cholera incidence. With it one could return to Farr's table-map and map-diagram and *see* that in individual jurisdictions cholera incidence decreased as altitude increased. In his graphic Farr presented both the data and his conclusion based upon it (you can see the result) without sacrificing the real complexity of the disease as an environmental, social, and temporal thing.

Farr's cholera was a meta-disease with something for everyone. It followed protocols of temporal profiling that Sydenham would have applauded. It was an imported disease like Arietta's plague (see fig. 4.6). But it was also as socially grounded as Chadwick's cholera and respiratory diseases were in Leeds (fig. 7.9A). Yes, cholera was epidemic, if not so epidemic as plague had been. Yes, it was a disease of the summer airs; the coxcomb made this clear. And yes, the maps argued, it was as

Hamett had insisted a disease of the urban poor, most virulent in densely settled districts whose sanitary infrastructure left much to be desired (fig. 6.3*A*). For the geographically inclined it was also a disease of the lowland dock and riverside areas of lowland ports where the odiferous wastes of the city accumulated. "It is established by observation that cholera is most fatal in low towns, and in the low parts of London; where from various causes, the greatest quantity of organic matter is in a state of cholera action" (Farr 1852b, 163).

The defining characteristic of Farr's cholera, however, was the inverse relationship between elevation and cholera mortality (fig. 9.8). it was this essential correlation that was linked secondarily to other attributes (for example, income). Demonstrable at both registration district and subdistrict levels, the correlation insisted that cholera diminished as one moved from the lowland riverbank (and poor registration districts) to the higher ground (and wealthier registration districts). And since maps of cholera in Exeter (fig. 8.6*A*), Hamburg (Fig. 7.8) and Rouen (Fig. 7.7*A*) had all shown a riverside concentration, the London case was not presented as exceptional, simply exceptionally argued.

The conclusion for Farr and many of his readers was that cholera was created in a fermentation-like process propelled by the evaporation of contaminated lowland waters into the air. That water-air cholera was the heart of the matter. The interpersonal transmission that Snow advanced, while an important insight, was of secondary importance. The idea of a waterborne disease, suggested by the increased mortality in riverside districts, was explained away in a theory that added to but did not discard the prevailing theory of miasmatic disease.

What made Farr's cholera extraordinary was not simply the wealth of data he brought to bear on his subject but the method of its construction. The conclusion was inherently experimental. the result of a system in which a range of variables of potential significance were tested across an officially formulated data set of incidence. What Farr's cholera did not have, and his approach could not offer, was a definitive statement on the disease agent. Without that, the problem that was cholera remained unresolved. As a *Lancet* review put it, all Farr had done was to "show the conditions under which cholera gains strength and deals destruction" while leaving unsolved the precise nature of the disease. "Concerning this, the fundamental point, all is darkness and confusion, vague theory, and a vain speculation. Is it a fungus, an insect, a miasm, an electrical disturbance, a deficiency of ozone, a morbid offscouring from the intestinal canal? We know nothing; we are at sea, in a whirlpool of conjecture" (*Lancet* 1853, 393).

JOHN SNOW'S SOUTH LONDON CHOLERA

In November 1853 Farr published a supplement to the registrar general's weekly mortality report for London. In it he restated his conclusion that altitude was inversely related to cholera in a table whose subject was not the registration districts or subdistricts but the water companies that supplied them. Noting a continuing debate over the relation between cholera and water quality, Farr reported that the Lambeth Waterworks Company, a principal supplier of South London's registration districts, "changed their source of supply nearly two years ago from Lambeth to Thames Ditton, and from a Table subjoined it will be seen that the results of the present epidemic in the districts supplied by the company, as compared with some others, are rather more satisfactory than they were in 1849" (Farr 1853, 405). In other words, the reported incidence of cholera diminished in Lambeth Waterworks' service area after the company moved its water intakes westward from the polluted central Thames to the less polluted Thames Ditton area.

John Snow saw this as an opportunity to prove his theory. From the start his method had been the intensive study of cholera in local outbreaks with a general attention to patterns of disease incidence in the city at large. Yet another epidemic in South London was the crucible in which he might use GRO mortality reports and available maps to make his case. *If* cholera was water- rather than airborne, *then* the rate of cholera among Lambeth Water customers in South London, roughly equal to that of Southwark-Vauxhall customers in 1849, should now be less, given a cleaner water supply site. Here would be the *experimenta cruces* demanded by Parkes, albeit one Farr had said could not be carried out because of the complexity of potentially confounding social variables and complicating registration district geographies in areas served by the two companies (Farr 1853, 401).

Snow saw no problem here, however, only the opportunity for a natural experiment on the grandest scale. "No fewer than three hundred thousand people of both sexes, of every age and occupation, and of every rank and station, from gentlefolks down to the very poor, were divided into two groups without their choice, and in most cases, without their knowledge; one group being supplied with water containing the sewage of London, and amongst it, whatever might have come from the cholera patients, the other group having water quite free from such impurity" (Snow 1855a, 75).

"To turn this grand experiment to account," Snow wrote, "all that was required was to learn the supply of water for each individual house where a fatal attack of cholera might occur" (Snow 1855a, 75). That would prove easier said than done. Snow faced two problems. First, he had to create service areas for the two principal South London water suppliers: Lambeth Waterworks Company, which had moved its supply point, and its principal competitor, Southwark-Vauxhall Water Company, which had not. Second, he had to locate reported cholera incidence within those supply areas in a manner that would permit a comparison of relative mortality—cholera as a percentage of population—to be accurately assessed. Finally, Snow had to do these in a manner that assured potentially confounding attributes (income, housing density, etc.) would be discounted.

In an attempt to accomplish the first task Snow had drawn a map of South London's two principal water company supply areas. These are distinguished by color, as is a third area in which the two companies competed street by street, their waters so commingled that a separate, conjoined service area was needed (fig. 9.9*A*). Snow would have had to calculate cholera incidence and mortality for all service catchments using mortality data supplied by the GRO. In the map the Lambeth Waterworks service area is colored red, Southwark-Vauxall's blue, and a conjoined area served by the two companies together is colored purple.

Also embedded in the map are boundaries of registration districts (solid red lines) and subdistricts (broken red lines). In relatively unpopulated registration districts Snow added to the map small black bars similar to those used at least since Hamett's Dantzick map (fig. 6.3) to indicate the homes of cholera decedents. These are not explained in the text but appear to correspond with the address of cholera deaths that occurred in the first four weeks of the epidemic and summarized by Snow in an appendix to his text.

These 334 one-line case summaries represented weeks of work—"shoe-leather epidemiology"—by Snow and the "medical man, Mr. John Joseph Whiting" (Snow 1855a, 77). They visited the homes of cholera victims in a sampling of South London registration districts in an attempt to document the source of water used by the deceased. Each summary included the age and death date of the decedent, the street on which he or she lived, the home water company supplier, and the official cause of death. These included Asiatic cholera, cholera, cholera Asiatica, cholera biliosa, cholera maligna, cholera morbus, choleraic Asiatica, choleraic diarrhea, diarrhoea, diarrhoea choleric, epidemic cholera, English cholera, malignant cholera, and premonitory diarrhoea. In some cases, patients were reported to have had first diarrhea and then cholera, or cholera and some other physical complaint. For Snow, unlike Farr, if it was diarrheic it *was* cholera; no further division of cases was required.

The *only* thing of import for Snow was the preferred source of water in the decedent's household. "In two hundred and eighty-six cases the houses where the fatal attack of cholera took place was supplied with water by the Southwark and Vauxhall Company, and in only fourteen cases was the house supplied with the Lambeth Company's water" (Snow 1855a, 79). As Parkes would point out in his review of Snow's study, the comparison at this rather coarse scale of service catchments failed in part because it took no account of potentially crucial geographic and socioeconomic variables. "The Lambeth Company supplies, to a considerable extent, a good neighborhood on elevated ground (including the healthy districts of Streatham, Forest-Hill, and Sydenham); while the Southwark and Vauxhall Company supplies the greatest part of the poorest, lowest, and marshiest districts in London (Parkes 1855a, 461). Farr had stressed the problem of confounding variables like income and geography. Snow had ignored them; and Parkes criticized Snow for that omission.

FIGURE 9.9A (*opposite page*) Snow attempted to create distinct water supply areas to be used in determining relative mortality on the basis of water supply in South London registration districts and subdistricts.

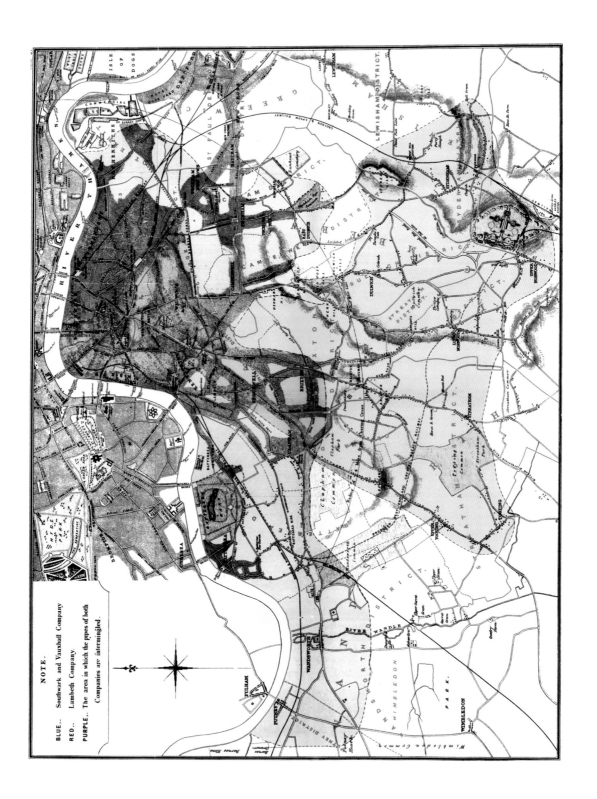

To make his case across the area of the South London outbreak Snow would have had to post *all* reported cholera cases in the more finely grained jurisdictions inscribed in the map of service catchments. And while Snow himself promised, in the second edition of *On the Mode of Transmission of Cholera*, a street-by-street analysis, he could not do this for at least three reasons. First, the map was difficult to read. It was

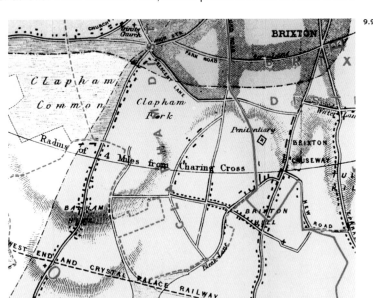

FIGURE 9.9B Snow's map of water supply areas in South London carried includes small dots similar to those used by others to inscribe the location of cholera deaths. Broken and solid red lines describe the boundaries of registration districts and subdistricts.

FIGURE 9.9C Snow's map of water supply areas includes registration district and registration subdistrict boundaries on a transportation map of South London.

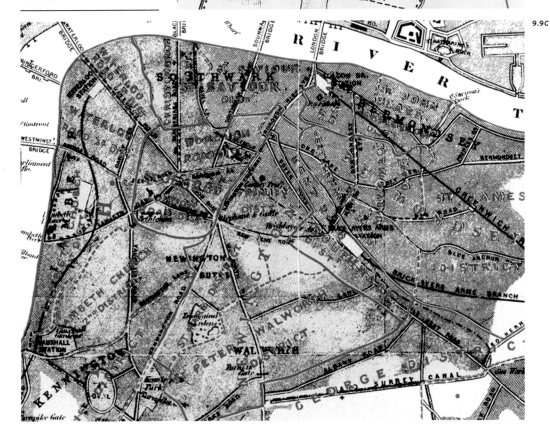

cluttered with irrelevant data, including roads and train lines as well as a circular radius of distance from Charring Cross (fig. 9.9*B*). All were almost certainly artifacts of an older map made for another client made by Cheffins and continued for Snow at the doctor's direction. Equally difficult to read are the interlocking and intersecting boundaries of South London's registration districts and subdistricts (such as St. Saviour and St. Olave), which compete for space with the other attributes of the map (fig. 9.9*C*).

Second, Snow's map made no argument about cholera—presented no clear correlation between disease intensity and water supply—but instead solely proposed the reality of broadly defined water service catchments into which cholera mortality might be first located and then considered. Creating accurate service catchments was not a simple affair, however. Along the river, many residents would simply dip a bucket and pull up water to carry home. Some businesses had their own wells. Residents would not infrequently import water from elsewhere in the city, from wells and pumps they thought especially tasty. Simply put, the catchments leaked.

In an attempt to distinguish between the waters of the two principal suppliers Snow tested samples of their water at their respective intakes. Snow found that a gallon of Lambeth Company water precipitated only 2.28 grains of silver chloride, compared to 91 grains in a gallon of Southwark-Vauxhall water. This seemed to permit the two waters to be distinguished but, as historian John M. Eyler writes, "Only later did he realize that the salient context of the river water varied widely over time, in the recent past by a factor of 20 for the Southwark and Vauxhall Company" (Eyler 2001, 227). In other words, the difference reflected in Snow's test did not adequately serve the study. Even had this test been definitive, however, Snow would have had to test water quality in all the houses in the conjoined service area, and perhaps the entire study region, to make his water service catchments credible. He had neither the time nor the manpower to do this. Thus his testing, like his individual case reports, while interesting, were less than convincing.

Third, at the time Snow was writing his book the GRO records for all deaths were not yet available. As Snow would admit in 1856, "I was unable at the time to show the relation between the supply of houses in which fatal attacks took place, and the entire supply of each district and subdistrict, on account of the latter circumstance not being known" (Snow 1856, 7). Another researcher would have waited a few months for the data to be available but Snow, writing in late 1854, chose to press forward. From a parliamentary report Snow derived an estimate of the total number of houses in all South London served by each of the two water companies (fig. 9.10). Snow used this to calculate a general mortality ratio per 10,000 houses based on a general estimate of cholera incidence in the service areas (Snow 1855a, 86).

The result was unconvincing to his contemporaries, in part because the study could not attend to the differences in mortality within the conjoined service area, the nexus of Snow's grand experiment. Only at the local scale were environmental, geographic, and social variables likely to be irrelevant. At the scale of the water supply areas they were

Snow Table IX	Number of houses	Deaths from cholera	Deaths in each 10,000 homes
S&V Co.	40,046	1,263	315
Lambeth Co.	26,107	98	37
Rest of London	256,423	1,422	59

FIGURE 9.10 In 1855, Snow used a coarse measure of houses served by South London water companies to argue the effect of water supply on cholera. He lacked the data to apply these summary figures to the map where the area of conjoined water supplies was a confounding problem.

literally confounding. As importantly, in his study Snow made no serious attempt to consider other potentially relevant characteristics of the areas that were all potentially explanatory: altitude, housing density, income levels, and so forth (Eyler 2001, 227). As Parkes wrote in another section of his review, "Dr. Snow refers only to the water-supply, and neglects all other circumstances" (Parkes 1855a, 362). And really, how could it be otherwise? Snow had not been able to locate his cholera in the map. Without that, he could not locate other potentially crucial variables, even had he wished to.

JOHN SIMON'S SOUTH LONDON CHOLERA

In 1856 the Board of Health's Dr. John Simon completed Snow's stalled grand experiment in a study "presented to both houses of Parliament by Command of her Majesty" and published by Her Majesty's Stationary Office (Simon 1856). Encouraged in his efforts by Farr, Simon sought to answer the question of the effect of impure water on the incidence of cholera. In his report Simon delivered what Snow had only promised, a study of the relationship between cholera incidence and other diarrheic conditions "in houses enumerated in 1854 as receiving their water supply from different sources in South London" (Simon 1856, 19).

Dissatisfied with previous studies, Simon set out to search for a natural law like Farr's, a regularity of occurrence that could be positively determined within affected populations. Only such a law, demonstrably presented, might settle the question of cholera and water once and for all. He sought to determine if that law might be found in the relationship between water supplies and cholera. "Our Committee had thought it of importance to inquire as fully as possible into the sanitary influence of different qualities of water supply," he wrote. "Copious details of information had been consequently collected; but these could not be brought into an available form against the time when our report was made, and we were therefore reluctantly obliged to construct it [the report without them]" (Simon 1856, 3).

FIGURE 9.11 In 1856 Dr. John Simon completed Snow's grand experiment in a study that carefully compared the effect of South London water source on cholera.

Carefully, Simon marshaled comparative data on both the 1848–1849 occurrence of cholera in South London and the 1853–1854 South London epidemic. Like Snow, he used mortality data made available by the GRO. Unlike Snow, Simon built his analysis at three very different resolutions. First, he considered aggregated mortality occurring in all thirty-one registration subdistricts of South London, second in twenty subdistricts "almost equally supplied by both companies" and finally in a fine-grained analysis of cholera mortality on forty-five individual streets where water supply was totally commingled. This was the street-by-street analysis Snow had promised but not delivered in his revised *On the Mode of Transmission of Cholera*. Only at this last scale would confounding variables (demographic, environmental, and social factors) presumably be wholly eliminated. Only at the scale of the house and street might an apparent regularity produced in coarsely granulated studies be confirmed or denied.

The ratio of mortality across the thirty-one registration subdistricts appeared to confirm Snow's thesis. And there was indeed a major decline in cholera by population in 1854 among Lambeth customers when compared to those served by the Southwark-Vauxhall Company: *"The population drinking dirty water according appears to have suffered 3.5 times as much mortality as the population drinking other water"* (Simon 1856, 6; original emphasis). Further, in 1849, before Lambeth Water Company's intake had been moved, levels among company water subscribers had been roughly equal. Simon compared mortality between 1849 and 1854 and found that the Lambeth customers suffered "*not a third as much as at the time of its unre-formed water-supply.*"

And yet, for Simon, this was suggestive but not definitive. "The recent contribution therefore aims only at giving a more exact knowledge of one cause, not at gainsaying the existence of other causes" (Simon 1855a, 14). In other words, at the coarse scale of the water company supply jurisdictions (and registration subdistricts), confounding variables could easily influence the result. To eliminate that possibility Farr then moved to first the registration subdistrict resolution and then that of the street-level comparison where no confounding variables were likely to be involved. "The real significance of these [coarse] totals is best shown by an examination of the details embodied in them," Simon wrote (Simon 1856, 6). Here, however, was a problem: at the fine-grained resolution of individual streets the apparent certainty of the coarse analytic appeared to fall apart. Of the forty-five streets in his close analysis, eighteen whose water was supplied by Southwark-Vauxhall had no cholera deaths; seven supplied by Lambeth similarly had no deaths.

While there was at the coarse, general scale an apparently strong correlation between water supplier and cholera mortality, this relationship did not hold at the scale of street-by-street analysis. Some houses supplied in 1853–1854 by Lambeth had no cholera deaths at all. Because Simon could not proclaim consistent mortality ratios based solely on water source in his street-level analysis, he concluded he could not prove or disprove the role of water in the generation and diffusion of cholera. The best he could prove was that it did appear to be at least a contributing factor at some scales of analysis. Absent an understanding of the specific "exciting cause of the epidemic manifestation," Simon wrote, cholera appeared "to take effect only amidst congenial circumstances, and that the stuff out of which it brews poison must be air or water abounding with organic impurity . . . fecalised drinking-water and fecalised air equally may breed and convey the poison" (Simon 1856, 15).

We can see the problem Simon confronted—perhaps intuited is a better word—in a recently constructed scatter plot based on GRO data at the fine-grained, house-by-house resolution (fig. 9.13). Scatter plots were first developed in 1833 by J. F. W. Herschel (Friendly and Denis 2005) but were not commonly employed in mid-nineteenth-century disease studies.[8] They are used here to make visual what Simon saw in the data but could not resolve. The scatter plot presents an equation whose numerator was the number of deaths per street and the denominator was the number of houses times the population per household in the pertinent registration subdistrict. The numbers, therefore, are very small. The horizontal axis is mortality for Southwark-Vauxhall customers; the horizontal axis is that of Lambeth customers.

Based on the strong 3:1 correlation in the registration district data, Simon expected to see a similar correlation when the correlation was tested at the fine-grained resolution. But the scatter plot shows no clear correlation with each point on the chart one of the streets whose deaths he analyzed in relation to water company provider.

I.—SYNOPSIS OF RESULTS.

Death-Rates per 1,000 of living Population in Two Epidemic Periods.		In Houses enumerated in 1854 as receiving their Water-supply—	
		from the LAMBETH Company.	from the SOUTHWARK and VAUXHALL Company.
CHOLERA	1848–9 ..	12.5	11.8
	1853–4 ..	3.7	13.0
DIARRHŒA	1848–9 ..	2.9	2.7
	1853–4 ..	2.1	3.3

FIGURE 9.12 Simon's study compared the incidence of cholera and other diarrheic diseases based on water supplier both historically and in the 1854 outbreak.

FIGURE 9.13 A scatter plot of Simon's finely granulated street data shows no obvious relationship between cholera mortality and water supply in South London. The problem was to reconcile this uncertainty with the apparent regularity of mortality across the South London subdistricts.

N.B—*Between the two Epidemic Periods, the Lambeth Water Company had changed its source of supply.*

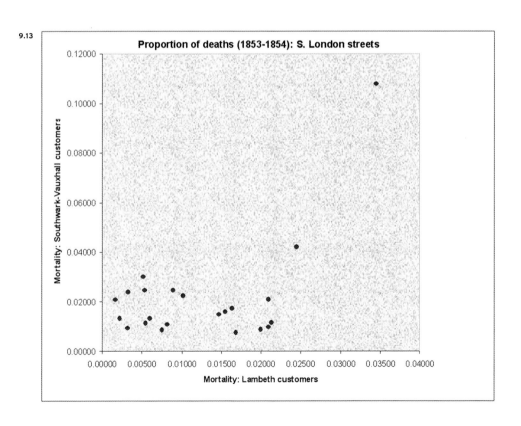

Analyzed with modern statistical tools, linear regression analysis ($r^2=.287$) suggests less than a third of the total mortality can be explained in relation to the water sources. The standard error (±0.18) indicates a likelihood of no overall association whatsoever. It would be the 1960s at least before statistical approaches (Bayesian quadratic regression, for example) were developed in a manner that could return order to the chaos of this scatter plot data. Simon saw the problem and could not therefore show, at the scale of the "natural experiment," a consistent proof of water as the sole source of the epidemic. "We can't say, not yet" he concluded, because he could not match in his study Farr's apparently definitive algebraic relationship between air and water at every scale. What he could demonstrate was that the apparently strong correspondence at the coarse resolution did not necessarily hold (was not obviously demonstrable) at the level of the street-by-street analysis.

JOHN SNOW REDUX: 1856

Snow immediately crafted a rebuttal in which he criticized elements of Simon's study design. There is in Snow's writing style a sense of furious loss: this was *his* study and someone else took it over. Worse, Simon's agnostic conclusion—we can't say definitively on the basis of this data—was not what Snow believed the data argued. One feels Snow's frustration in the lack of Victorian niceties referring to another's work and in his curt dismissal of Simon's analysis, if not his data. Snow would use this same data, a product of Farr's office at the GRO, in an imaginative attempt to model cholera. *If* cholera was solely a result of water quality, Snow reasoned, *then* mortality based on water supply should be identical to the records of mortality supplied by the GRO. The fine-grained analytic, and Simon's caution, got in the way of what to Snow was an obvious truth.

Snow used the Farr-Simon data to construct a table of cholera in the ten South London registration districts (fig. 9.14). With census population figures and number of houses, and thus the estimated population, Snow included the principal water supplier for each district. With figures on cholera mortality for each registration district, now calculated by Simon at the level of the street, Snow calculated mortality per ten thousand persons for each principal water supplier in each registration district. Snow then applied these results to the scale of embedded registration subdistricts. As he had hoped, "the calculated mortality bears a very close relation to the real mortality in each subdistrict" reported by the GRO (Snow 1856, 10).

Snow applied the ratios returned in this study at the district level to an analysis of the forty-five registration subdistricts contained within the registration districts, ignoring the fine-grained problem Simon saw at the individual street level. Recent studies have shown the congruence between Snow's predicted mortality by water supply areas and reported GRO data was less than accurate (Koch and Denike 2006). A series of problems run through Snow's modeling, from simple errors in arithme-

Registration Districts	Number of inhabited houses in 1851	Population in 1851	Estimated constant population per house	"Number of houses and estimated number of persons supplied in 1854 with water as under"				Water supply of houses in which fatal attacks cholera took place				Deaths from cholera in the epidemic of 1854	Mortality per 10,000 supplied with water	
				By the Southwark and Vauxhall Co.		By the Lambeth Company		Southwark and Vauxhall Co.	Lambeth Co.	Pumps, wells, & other sources	Supply not ascertained		Southwark and Vauxhall Co.	Lambeth Co.
				No. of houses	Estimated population	No. of houses	Estimated population							
St. Saviour, Southwark..................	4,000	35,731	7.8	2,631	19,617	1,689	14,201	406	72	10	3	491	207	50
St. Olave, Southwark.....................	2,960	19,735	8.2	2,193	18,638	0	0	277	0	8	28	313	148	-
Bermondsey..............................	7,007	48,128	6.9	8,402	57,884	268	1,785	821	0	25	0	846	142	-
St. George, Southwark..................	6,992	51,824	7.4	3,419	25,039	3,183	23,712	388	99	0	56	543	155	41
Newington...............................	10,458	64,816	6.2	5,224	31,940	5,473	33,531	458	58	2	176	694	143	17
Lambeth..................................	20,447	139,325	6.8	8,077	54,982	11,763	83,786	525	138	24	240	927	96	16
Wandsworth..............................	8,276	50,764	6.1	3,028	18,390	618	3,870	268	7	106	40	421	145	18
Camberwell...............................	9,412	54,607	5.8	4,005	23,472	1,835	10,478	352	33	115	49	549	150	31
Rotherhithe.............................	2,792	17,805	6.4	2,336	14,951	0	0	207	0	46	30	283	138	-
Greenwich & sub-dis. Sydenham...	-	-	-	-	-	-	-	4	4	2	1	11	-	-
Houses not identified..................	-	-	6.6	411	2712	25	165	-	-	-	-	-	-	-
Totals.................................	72,344	482,435	0.7	39,726	267,625	24,854	171,528	37,706	411	338	623	5,078	138	23
Non-ascertained cases distributed in proportion to others.................	-	-	-	-	-	-	-	561	62	-	-	-	-	-
Population (Registrar-General)......	-	-	-	-	266,516	-	173,748	4,267	473	338	-	5,078	160	27

FIGURE 9.14 In 1856 Snow attempted to populate his water service areas with data based on GRO records compiled at first the registration district and then registration subdistrict levels (modified by author).

tic (Carvalho, Lima, and Kreibel 2004) to his handling of non-ascertained cases at the registration district level and their later assignment to the registration subdistrict analysis. These problems were not discussed in Snow's day, however. Instead, his paper, published in the October 1856 issue of the *Journal of Public Health*, seems to have been ignored. It carried neither the authority of a book-length treatment nor the credence of a report to Parliament published by Her Majesty's printers. There was in it none of the caution Simon's report presented or Farr concluded with in his report. And by 1856 Snow's views on cholera were discounted by many as obsessive; his science perceived as a matter of magisterial assertion rather than careful, factual evidentiary argument (Parkes 1855a, 635).

Snow began with the idea of waterborne cholera and then developed the water service catchments in an attempt to demonstrate the existence of solely waterborne cholera in a two-dimensional model, permitting no other variables to be considered. Snow had difficulty locating mortality in the catchments and further problems in the analysis of his top-down model. Simon considered mortality in a one-dimensional model that built street by street from household mortality to mortality at first the registration subdistrict level and then at the district level. From there he was able to argue a multifactoral cholera in a manner that was conservative, judicious, and wholly in accord with existing theory. The fineness of Simon's data had an authority that Snow's top-down modeling of water service areas lacked. In the end, Simon's method of analysis, and his judicious endorsement of Farr's air- and waterborne cholera in the absence of conclusive knowledge of disease agency, seemed the more complete, more logical, and, for most of their contemporaries, better science.

CHAPTER 10

CHOLERIC BROAD STREET: THE NEIGHBORHOOD DISEASE

"The most terrible outbreak of cholera which ever occurred in this kingdom," Snow wrote in the second edition of *On the Mode of Communication of Cholera*, "is probably that which took place in Broad Street, Golden Square, and the adjoining streets." "Within two hundred and fifty yards of the spot where Cambridge Street joins Broad Street, there were upwards of five hundred fatal attacks of cholera in ten days" (Snow 1855a, 38). While struggling to complete his South London study, and with a side trip to Deptford to investigate a local outbreak reported there, in the late summer and early fall of 1854 Snow would simultaneously seek to use this outbreak to prove his theory.

As a young apothecary Snow lived on Firth Street, a few brief minutes by foot from the intersection of Broad and Cambridge streets. It was there he cared for patients while studying for his medical degree, there that he wrote his first professional papers. It was on Firth Street in the 1840s that Snow carried out his seminal work on anesthesia. In the early 1850s, as his income rose with his fame as an anesthesiologist, Snow moved to more fashionable quarters on Stanwick Street at the southern end of Regent Street near Piccadilly. That a cholera outbreak of such fierceness would occur in Snow's old neighborhood must have seemed to him a personal challenge, if not a cosmic insult.

ST. JAMES, WESTMINSTER

Once one of London's fashionable neighborhoods, St. James was a haven for those who abjured the busy world of Piccadilly Circus. In the mid-eighteenth century, it had been home of the family of young William Blake, the future poet, who returned there in his late twenties to open a printing shop (Johnson 2006, 16–17). Old money disappeared in the early nineteenth century as the stately townhouses were broken up into apartments to accommodate the rising population of immigrant workers from Ireland and from the British countryside. By the time of Charles Dickens, Golden Square itself was sufficiently diminished to serve as the shabby setting in *Nicholas Nickleby*.

The beginnings of this transformation can be seen in figures 10.1*A* and 10.1*B*, maps of St. James in 1720 and 1750 from the British Library's Crace Collection. In the 1720 map are the "pest-house fields," lands acquired in 1665 by the Earl of Craven to provide a place where those suspected of plague, or in whom symptoms were manifest, could be housed. On the land were "thirty-six small Houses, for the Reception of poor and miserable Objects . . . afflicted with a direful Pestilence) (Porter 1999, 129). The creation of pest-houses was seen by Lord Craven and more generally the Privy Council as one way to isolate and thus contain the disease. Because of the approximately 30 percent mortality there was also a burying ground for plague victims whose interment was often barred from otherwise appropriate cemeteries.

By 1750 the land on which the pest-houses stood had been reclaimed by merchants who plied their wares at the Carnaby Market. In the nineteenth century the market lands were transformed into housing, and the Poland Street workhouse was erected on the burying place of the 1665 plague epidemic victims.

As population grew, income levels fell until, by 1851, St. James, once a refuge of the upper class, had become a working class neighborhood with 432 people per acre, the most densely populated of all the 135 registration subdistricts in Greater London (Johnson 2006, 18). While in Kensington there were an average of 2 houses per acre, in the heart of St. James the density was 30 houses per acre with, according to the 1851 census, an average of 10 persons per house (fig. 9.6*A*). To accommodate this social transformation the area was physically reformed as well. Regent Street was constructed at the district's western border to separate the streets of the working population from the upscale homes of Mayfair. New buildings, commercial and residential, populated the old pest-field. It was in this area that the 1854 cholera outbreak began.

"I requested permission, therefore, to take a list, at the General Register Office, of the deaths from cholera, registered during the week ending 2nd September, in the subdistricts of Golden Square, Berwick Street, and St. Ann's Soho, which was kindly granted . . . Eighty-nine deaths from cholera were registered, during the week, in the three subdistricts . . . on proceeding to the spot I found that nearly all the deaths had taken place within a short distance of the pump" (Snow 1855a, 39). "The pump," one of more than fifteen in the study area[1], was located at the intersection of Broad and Poland streets. Here, Snow would argue, was the origin of the outbreak and a proof that his theory of cholera as solely waterborne. The pump was a block from the southeast corner of Lord Craven's pest-fields.

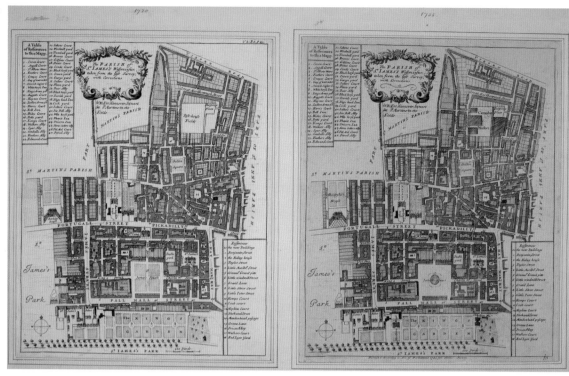

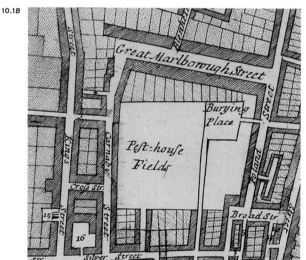

FIGURE 10.1A This 1720 engraving by cartographer Richard Blome presents St. James as a well-to-do parish. In the 1800s its population would grow with a vastly increased emigrant population.

FIGURE 10.1B In 1720 the seventeenth-century plague burial site remained undeveloped land in St. James, Westminster, a block from Broad Street. By the mid-nineteenth century the former pest-fields would be overbuilt with housing.

REVEREND HENRY WHITEHEAD

Snow was not alone in investigating this local outbreak. The Board of Health carried out its own investigation. So, too, did the Metropolitan Commission of Sewers in hopes of countering a popular belief that bad airs traveling through new sewer lines caused the outbreak. The first published study, however, was by Reverend Henry Whitehead, a young curate assigned to St. Luke's Church in the heart of the Broad Street district. Today we might call Whitehead an outreach worker. He came "to live the life of human friendship in this world of friendlessness, want and woe" and stayed at St. Luke's for five years, serving three successive vicars (Rawnsley 1898, 29). A graduate first of Chatman and then of Lincoln College, Oxford, Whitehead's primary mission was in the streets. When the outbreak began it consumed his days as he traveled to give comfort to the dying, and later, to the family members who survived.

In 1854 Whitehead published his seventeen-page pamphlet "The Cholera in Berwick Street" in part to correct the popular impression that the neighborhood was wholly devastated by the outbreak. "It was very mortifying to persons interested in the welfare of the neighborhood, to see the papers teeming with letters describing whole streets as having hundreds lying dead in them, at a time when the deaths in each street were really no more than one or two each day; and equally unsatisfactory was it to hear of employers refusing work to the inhabitants, long after the disease had disappeared—as if, too, a coat or a pair of boots would carry it [cholera] into a shop" (Whitehead 1854, 17).

Over the course of several months Whitehead interviewed every family resident in his parish, sometimes returning to one house four or five times until he could find a respondent (Chave 1958, 95). As Whitehead put it, writing about himself in the third person: "The writer does not choose to rest this statement on mere loose assertion. His previous acquaintance with the people and their houses, added to personal observation, and the observation of his colleagues, of the progress of the pestilence, has enabled him to ascertain—what probably, for obvious reasons, no one else could or can ascertain—the name of each deceased person, and the room in which he or she died, or in the case of removal or departure, the room hitherto occupied by the deceased" (Whitehead 1854, 8–9).

In this first study Whitehead focused on what he knew best: "It is the writer's intention to confine his observations to the district of St. Luke's Berwick Street, with the houses and people of which he has long been acquainted" (Whitehead 1854, 1). For those who did not know the parish, Whitehead included a map that served first to define the area of his authority and second as an index to the locations he described in his report (fig. 10.2). Where were the inhabitants of the "model lodging houses now building," a site of sanitary pride in the parish, so devastated by cholera if mere cleanliness and sanitation were causative factors? How far was this model lodging from the Poland Street workhouse whose poor and dispossessed inhabitants were almost untouched by the outbreak? The map presented the geography of Whitehead's parish in which cholera's occurrence was to be seen, its pattern of incidence questioned.

Whitehead's short pamphlet was an inventory and analysis of 373 cholera deaths, "nearly all of which took place in the first fortnight, and 189, at least, in the first four days" (Whitehead 1854, 3). He listed these August and early-September deaths street by street, providing for each street the number of houses, the total population, and the total number of deaths. Whitehead filled his short report with a series

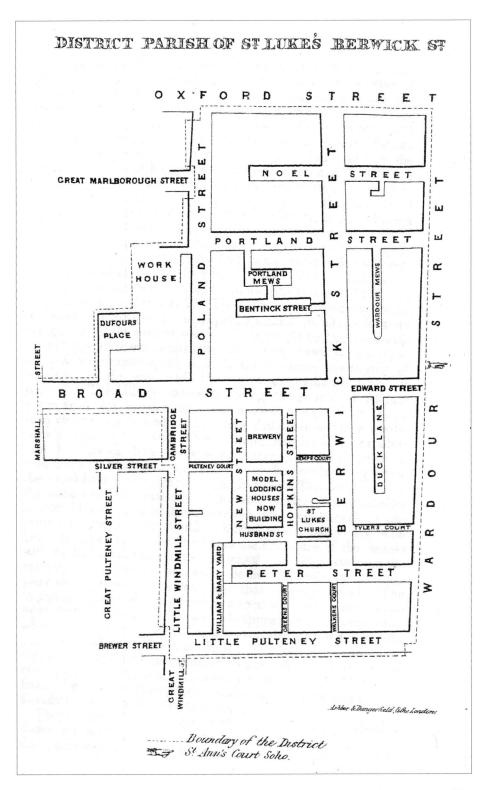

FIGURE 10.2 The Reverend Henry Whitehead included this map of his parish in his 1854 report, "The Cholera in Berwick Street," where St. Luke's Church was located.

of tables breaking down mortality by house population, age, and deaths over time. It was Whitehead who first noted that "there were only four deaths among the regular inmates of the workhouse" in a district where hundreds of deaths were elsewhere occurring (Whitehead 1854, 6). Because newspapers had reported that, "the vast majority of the deaths occurred in the upper rooms" of multiple family homes, presumably because bad airs wafted upward, Whitehead created a table of mortality by house floor and found the theory mistaken (Whitehead 1854, 8–9).

At least at the beginning, the problem cholera presented for Whitehead was as much theological as scientific. "At an average distance of 15 yards from St. Luke's Church stand four houses, which collectively lost 32 persons" (Whitehead 1854, 6). The intensity of the outbreak in his (and thus God's) domain was almost insulting to the young curate. In his studies he found "the very old and the very poor have not supplied nearly so many victims as might have been anticipated," populations in which medicine and common sense said deaths should be especially severe. Might it be, he wondered, the regular church attendance of these seniors—his daily congregants— that saved them? Might cholera be an act of God, a testing of faith, rather than simply the effect of some unseen animalcule or miasmatic force?

While Whitehead's investigations focused on his parish he was well aware that "the streets and parts of streets throughout which the disease may be said to have literally performed a *house-to-house visitation*" extended beyond the parish boundaries (Whitehead 1854, 2; original emphasis). He described the precise boundaries of this greater visitation in which his parish was embedded: "Take a point on the east side of Poland Street, half-way between Portland Street and the level of Great Marlborough Street; draw from thence two straight lines, one to the north-west corner of King Street, Regent Street, and the other to the east end of St. Anne's Court." Joined they formed a four-sided figure "enclosing with singular exactness the area within which only a few houses escaped, and outside of which comparatively few suffered" (Whitehead 1854, 2). While he did not add this description to a map—he assumed the reader's familiarity with the area—he did draw the polygon shape in his report to aide in visualization. In figure 10.3, Whitehead's polygon of the intense area of cholera occurrence is overlaid on the mapped boundaries of the parish described in figure 10.2.

In this manner, Whitehead created two cholera zones: the first was that of the parish to which he was assigned, in which 218 cholera deaths occurred; and the second was that of the greater area of intense mortality in which an additional 165 deaths were recorded. Those zones have been mapped on a simplified version of John Snow's cholera map (fig. 10.3). Absent from Whitehead's report was any mention of the Broad Street pump that would later figure so prominently in Snow's study. In the late summer and early autumn of 1854 Whitehead did not consider cholera a waterborne disease. Rather, he assumed, with most of his contemporaries (ecclesiastic, medical, and secular), that cholera was miasmatic in its diffusion and natural in origin: "It may be, as one writer has philosophically

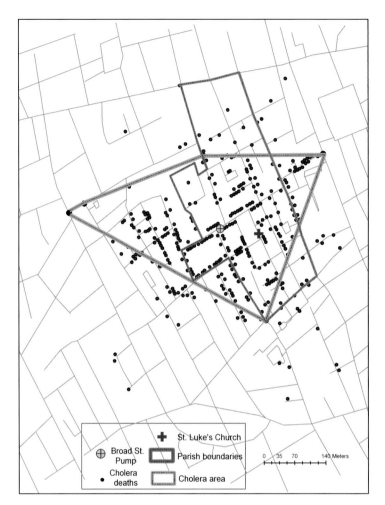

FIGURE 10.3 In 1854 Whitehead described cholera mortality in his parish (the red rectangular area) within the general area of greatest cholera activity. This map was made by the author based on Whitehead's textual description of cholera in his parish and its surrounding area.

and reverently suggested, that great and universal atmospheric changes periodically occur, fraught with ultimate benefit to the whole human race, compared to which the premature death of thousands, nay millions, is but as a grain of dust in the balance" (Whitehead 1854, 14).

THE BROAD STREET PUMP

When the outbreak "commenced in the night between 31st August and the 1st September," Snow later wrote, "I suspected some contamination of the water of the much-frequented street-pump in Broad Street, near the end of Cambridge Street" (Snow 1855a, 38–39). The pump was the closest public water source to the homes of the neighborhood's first cholera victims and therefore, Snow reasoned, the likely origin of the outbreak. "On proceeding to the spot, I found that nearly all the deaths had taken place within a short distance of the pump," Snow continued. "There were only ten deaths in houses situated decidedly nearer to another street pump. In view of these cases the families of the deceased persons informed me that they always sent to the pump in Broad Street, as they preferred the water to that of the pump which was nearer" (Snow 1855a, 39–40).

After requesting pertinent mortality reports from the GRO. Snow almost immediately suspended his investigation. Still a practicing anesthesiologist, Snow was also engaged in his study of the South London epidemic. Were that not enough, in early September he also traveled to Deptford to investigate an "equally violent irruption" where he suspected "a leakage had taken place into the pipes supplying the places where the outbreak occurred" (Snow 1855a, 55). Snow later acknowledged the effect on the Broad Street study of these other initiatives: "I should have been glad to inquire respecting the use of water from Broad Street pump in all these instances [of deaths] but was engaged at the time in an inquiry in the south districts of London . . . and when I began to make fresh inquiries in the neighborhood of Golden Square, after two or three weeks had elapsed, I found that there had been such a distribution of the remaining population that it would be impossible to arrive at a complete account of the circumstances" (Snow 1855a, 41).

For Snow, no longer a resident of St. Luke Parish, the outbreak was one he sometimes described as located near Golden Square. But for those like Whitehead, who were more intimately involved, it was the Broad Street or Berwick Street outbreak. Snow did what might be called an institutional survey in the area nearest the Broad Street pump. He interviewed "the keeper of a coffee-shop in the neighborhood" who on September 6 told him she knew of nine customers who drank from the local pump and were already dead of cholera. Snow learned the Poland Street Workhouse, which Whitehead had noted was surprisingly free of cholera, had its own well. Similarly, Snow was informed that no deaths occurred among workers at the brewery located on Whitehead's map where employees either drank their own product or drew water directly from the brewery's private well (Snow 1855a, 42).

For the type of circumstantial, case-based evidence that Snow's inferential approach required, he needed the help of others closer to the community and its patients. Those who shared their case notes with Snow included the Greek Street surgeon, Mr. Marshall, who also was a member of a Board of Health inquiry; a Dr. Fraser of Oakley Square; and of course Henry Whitehead. Their support gave Snow the luxury of being able to focus his attention, and very limited time, on apparently anomalous outriders, deaths occurring at a distance from the epicenter of the outbreak. "In some of the instances, where the deaths are scattered a little further from the rest of the map, the malady was probably contracted at a nearer point to the pump," Snow hypothesized. As proof, he identified children who lived on Angel Court, Noel Street, Ham Yard, and Naylor's Yard who walked by the pump on their way to school; a cabinet-maker who died in Middlesex hospital but lived at Phillip's Court, Noel Street. All, Snow or his collegial informants were told, regularly drank water from the Broad Street pump (Snow 1855a, 41–44).

While suggestive, these cases could not be considered definitive. Others used extensive case histories to argue wholly different conclusions. Seaman had used brief case reports in his study of yellow fever in New York to argue its miasmatic nature. In 1855, Dr. George Johnson, an assistant physician at King's College Hospital in London, published a 294-page treatise on cholera based on 54 case histories. All were of hospital patients whose autopsy reports he used to demonstrate that cholera must be pulmonary and therefore inhaled (Johnson 1855). An anonymous reviewer praised the study for the thoroughness of its case presentation if not for its treatment protocols promoting the "eliminative plan of treatment": castor oil (Anonymous 1855, 148). In many ways, one saw what one was looking for in both the dissection theater and in the map. To be definitive, Snow needed something more.

In the second edition of *On the Mode of Communication of Cholera*, Snow's strongest evidence implicating the Broad Street well and pump as the source of the outbreak was presented in "a diagram of the topography of the outbreak." Made for Snow by C. F. Cheffins, the prominent London engraver and mapmaker who drew the South London map, this was a street map of the study area in which he embedded, first, public wells and pumps and second, the street location of 596 cholera deaths reported to the GRO.[2] The map was centered on the Broad Street pump, "the spot where Cambridge Street joins Broad Street," the epicenter of the outbreak as Snow defined it (Snow 1855a, 38). Snow cast a wide evidentiary net for his study. It's official boundaries, marked in the map by a dot-and-dash line "surrounds the sub-districts of Golden Square, St. James, and Berwick Street, St. James, together with the adjoining portion of the sub-district of St. Anne, Soho, extending from Wardour Street to Dean Street, and a small part of the sub-district of St. James Square enclosed by Marylebone Street, Tichfield Street, Great Windmill Street, and Brewer Street" (Snow 1855a, 46). Outside that area were water sources and cases to the west of Regent Street and north of Oxford Street.

Snow's mapped argument was similar structurally to that of Seaman in his maps of yellow fever in New York City (figs. 5.6–5.7). First, Snow proposed a study area composed of city streets and important structures like the Poland Street Workhouse. He defined this area through reference to registration subdistrict boundaries that by then were generally accepted administrative regions. Second, Snow brought forth a set of possible sources, local wells and pumps. Third, Snow added to the map a set of choleric deaths reported to the GRO, those for which he had street addresses. Whether the diagnosis was English cholera or premonitory diarrhea, mapping the cases together asserted all were part of a single range of occurrence whose members were to be treated as equal. Snow then made this connection: *if* cholera is waterborne in nature, *then* its incidence should be closest to a complicit water source. And so it appeared: "It will be observed," Snow wrote, "that the deaths either very much diminished, or ceased altogether, at every point where it becomes decidedly nearer to send to another pump than to the one in Broad Street. It may also be no-

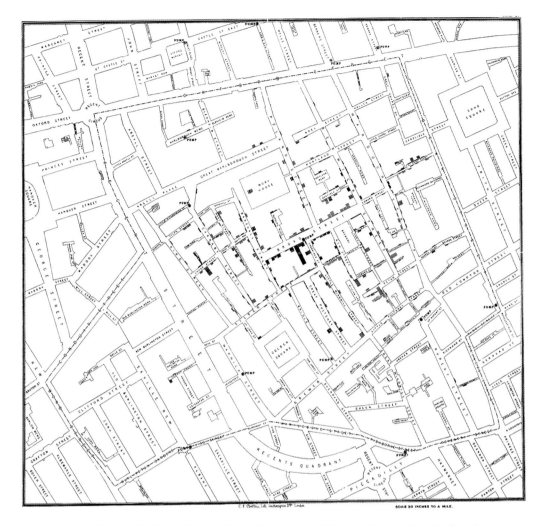

FIGURE 10.4 In this map of the Broad Street outbreak, Snow attempted to show the centrality of the Broad Street pump to the incidence of disease.

ticed that the deaths are most numerous near to the pump where the water could be more readily obtained" (Snow 1855a, 47).

Some today dismiss Snow's now iconic map as nothing but an illustration without intellectual or scientific force: "Snow's map of the epidemic area was simply the visual representation of a deduction from a theory of transmission developed earlier, which in turn was grounded in a theory of the pathology of cholera" (Brody et al. 2000, 66). Mark Monmonier, for his part, dismisses entirely the idea that Snow's map "or those of his rivals were of any value in generating insightful hypotheses. Snow's was pure propaganda . . . and copycat propaganda at that" (Monmonier 2002, 155).

These and similar critics miss the point. Snow developed a spatial theory that was tested in the map. This was not propaganda but an attempt at science. The map was the embodiment of Snow's proposition that *if* cholera was waterborne *then* its source had to be water, in this case, the Broad Street pump at the epicenter of the outbreak. The map was perhaps the critical statement of Snow's argument based on the available data. In it the cases of Snow's (and Whitehead's) circumstantial reportage became members of a class of cholera cases whose location to the maps was transformed into truthful conclusion. "It may be noticed . . ." was all the proof Snow offered (1855a, 47; 1855b, 109). The argument *was* the map. *If* one agreed that proximity implied causality, *then* the map's evidentiary value would be overwhelming, assuming no other confounding sources resided at the epicenter of the outbreak.

In a second version of the Broad Street map (fig. 10.5) included with Snow's report to a St. James Parish committee inquiry into the outbreak, Snow added an additional seventeen deaths for which addresses had earlier been unavailable; he also rectified minor topographic errors.[3] To this second map Snow added a wandering, dotted line creating an irregular area nearer to the Broad Street pump than to other pumps in the study area (Snow 1855b). Snow did not describe how this area was constructed but it is almost certainly based on walking time between local pumps, a kind of "Manhattan metric" based on pedestrian rather than shortest distance pathways. What today would be called a nearest neighbor boundary[4] bolstered Snow's argument. Because more deaths appeared nearer to the Broad Street pump than any other, logically, for Snow, the pump was necessarily the origin of the outbreak.

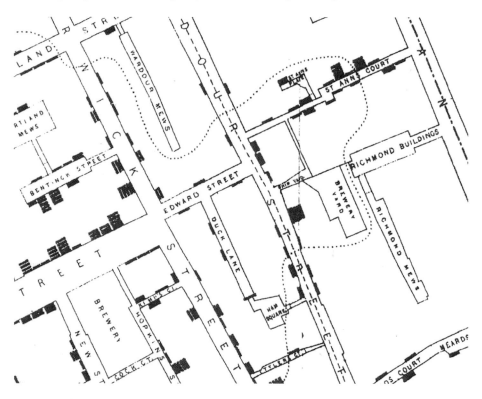

FIGURE 10.5 In a second map of the Broad Street outbreak, Snow included a dotted line to create an irregular polygon enclosing the area closer to the Broad Street pump than to other pumps.

Perhaps the most interesting thing about Snow's mapping was its omissions. He did not calculate the number of deaths in his walking area compared to the rest of the study area. He did not create a similar area for, say, the Rupert Street pump in a manner permitting him to say that two-thirds of all deaths were in the Broad Street pump vicinity and less than 12 percent in the second pump's vicinity. Nor did Snow attempt to calculate mortality ratios based on the population of the areas he constructed. He assumed others would see in the map what he saw and that they would draw the same conclusions. But without even basic quantification of the mapped incidence Snow could not translate the obvious concentration of deaths near the Broad Street pump into a meaningfully rigorous argument. Nor in his mapping did Snow attempt to consider with any rigor other possible sources of contagion, for example the old, seventeenth-century plague burial pit (Fig. 10.1*B*) many—laypersons and professionals alike—believed to be a likely source of disease generation. On September 7, for example, "An Old Subscriber" wrote to *Bell's New Weekly Messenger*, quoting Harrison's *History of London* on the location of the old pest-field from the 1665 plague epidemic.[5] To some, the old plague burial site over which new homes had been built and under which new sewer lines had been laid (all agreed the smell from these sewer lines was noxious, and therefore potentially disease generating) was as obviously complicit as the Broad Street pump was to Snow. On September 24, *Bell's* ran a story based on the report of a Dr. J. Rogers, the medical officer of health on Dean Street at the edge of the Broad Street area.[6] Rogers, who visited many of the ill, condemned the "sickening and nauseous odours" emanating from the Berwick Street sewers. "Life destroyed by exhalations from sewers," the headline insisted, condemning the "infamous gully-holes in the street."[7]

In his updated monograph Snow dismissed this and other alternate theories in passing, making no serious attempt to investigate their involvement himself. "The situation of the supposed [plague] pit is, however, said to be Little Marlborough Street, just out of the area in which the chief mortality occurred. With regard to effluvia from the sewers passing into the streets and houses, that is a fault common to most parts of London and other towns. There is nothing peculiar in the sewers or drainage of the limited spot in which this outbreak occurred" (Snow 1855a, 55). As proof he cited a report by the sewer commissioners that had exonerated its sewer lines laid in the 1850s. A more complete treatment of these other theories was needed, however, if Snow was to convince others not simply that the local pump was a source of the Broad Street outbreak but the only possible source.

CAUSATION

Others with an interest in the cholera outbreak in St. James were able to draw different conclusions based on the same data Snow drew upon, mortality reported by the GRO. In September 1854 Farr's office published an interim report on cholera and diarrhea deaths occurring between August 19 and September 9 in the districts of Berwick Street, Golden Square, St. James; St Anne, Soho; and All Souls, Marylebone, "to assist in the investigation of the epidemic outbreak." That report was republished September 13 in both the *Daily News* and the weekly *Bell's Life in London*, which five days earlier had reported that, "owing to the favourable change in the weather, the pestilence which has raged with such frightful severity in this district has abated, and it may be hoped that the inhabitants have seen the worst of the visitation."[8] The temperatures had dropped, in other words, at the same time that deaths were diminishing. To others, the relationship of declining temperatures and declining mortality seemed obvious proof of cholera's climatic origins, and thus of a miasmatic distribution. Where Farr had been at pains to note climatic variables, including temperature, rain and wind, incorporating them in his argument, Snow simply ignored them.

If the Broad Street pump was complicit, what was the source of *its* contamination? In neither of his reports on the Broad Street outbreak did Snow identify the index case from which the contamination spread in the manner of Seaman, who in his study of yellow fever in New York City had identified the sailor from the ship *Polly*. Without an understanding of the local source of the outbreak—it would be Whitehead who identified the index case—Snow's analysis was open to interpretation. "This certainly looks more like the effect of an atmospheric cause than any other," Parkes wrote of the Broad Street study in his review of Snow's second edition. "If it were owing to the water, why should not the cholera have prevailed equally everywhere where the water was drunk? Dr Snow anticipates this by supposing that those nearest the pump made most use of it; but persons who lived at a greater distance, though they came farther for the water, would still take as much of it . . . There are, indeed, so many pumps in this district, that wherever the outbreak had taken place, it would most probably have had one pump or another in its vicinity" (Parkes 1855a, 458).

REVEREND WHITEHEAD'S SECOND MAP

Some of these concerns were answered in an extraordinary map (fig. 10.7*A*) included in the parish inquiry committee report made in collaboration with the Board of Health. "Unlike Snow's better known version, this map included all streets and news, updated the number of cholera deaths in the area by tracing people who left the neighborhood for hospitalization, and tallied deaths of nonresidents who worked in or visited Golden Square" (Paneth, Vinten-Johansen, and Brody 1998, 1547). To the map Whitehead added, for his parish inquiry committee report, a large circle centered on Broad Street to show the area of cholera activity in his parish and its environs.[9] This circle served as a visual analytic, drawing attention to the epicenter of the outbreak in St. Luke Parish.

Produced by the lithographers Day and Son, the map simultaneously borrowed from and improved upon a map by Edmund Cooper, an engineer for the Metropolitan Commission of Sewers (fig. 10.6). In 1854 Cooper was ordered to investigate the then

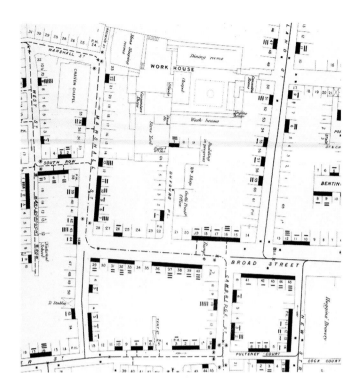

FIGURE 10.6 Edmund Cooper, engineer for the Metropolitan Commission of Sewers, mapped a partial set of cholera deaths (black bars) and sewer lines in the Broad Street area. The location of the former plague burial area is incorrectly located and sized.

common suspicion that new sewer lines might be the source of the Broad Street outbreak. Cooper mapped 351 cholera deaths reported to the GRO in the first fortnight of the outbreak against an inventory of sewer lines, noting the year in which they were laid. Included as well were the locations of sewer grates from which foul sewer airs might be expelled. A thick black bar symbolized homes of cholera victims; thinner bars were stacked under those symbols to show the number of deaths occurring in each house. Half the homes surveyed were not connected to the new sewer lines, Cooper reported; instead they used cesspools to store household waste. Looking at the map— he used no other analytic—Cooper reported no obvious correspondence between the cholera cases he mapped and the location of sewer lines in the study area.

Cooper mapped an oval to symbolize the old Craven Hill plague pit that he shrank and incorrectly located on Little Marlborough Street, northwest of the epicenter of the outbreak (fig. 10.1). This was one apparent source of Snow's easy dismissal of the old plague site on the basis of its distance from the outbreak's epicenter. While "non-medical people" thought the old pest-field more central, Snow wrote, "The situation of the supposed pit is, however, said to be Little Marlborough Street, just out of the area in which the chief mortality occurred" (Snow 1855a, 54). He had this on the authority of a sewer commissioner, who had it on the authority of Cooper's map. Unfortunately, Cooper was wrong.

Whitehead included in his map Cooper's oval as well as the correct size and location of the former plague burial site on which houses had been built. The boundary of that area, far greater than Cooper's mapped oval, was a block from the Broad

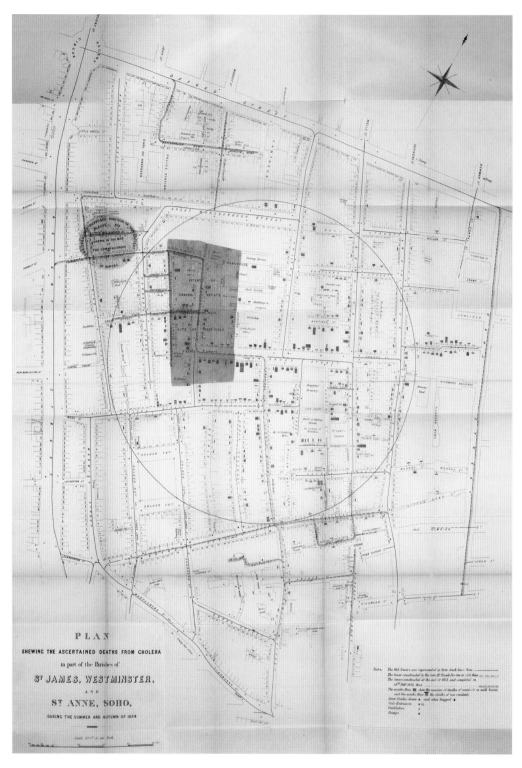

FIGURE 10.7A Reverend Henry Whitehead mapped 684 deaths that occurred during the Broad Street outbreak and considered their location in relation to a range of potential environmental contaminants, including water sources.

Street pump. In his map Whitehead had included the sewer lines and grates from Cooper's study, as well as fifteen public pumps and wells in his study area. To all this Whitehead mapped 684 deaths, symbolized by the now-familiar black bars, reported by the GRO from August through October. As on Cooper's map, these deaths could be located both by street and house number (and unlike Snow's map by Cheffins, which did not include house numbers on the individual streets).

The map that resulted presented a broadly ecumenical meta-argument: *if* cholera is influenced by environmental attributes, *then* the pattern of disease occurrence should reveal a correlation with one or more of those attributes (water, sewer lines, plague burial site, or the like). Each of these might be individually tested in the map. *If* the airs from the old plague pit were the source of the outbreak, *then* deaths would be clustered near that site. *If* the source of cholera was in sewer lines built since 1850, *then* cholera cases would be seen to cluster along those lines in the map. But *if* cholera resulted from contamination of water, *then* perhaps one or more of the fifteen water pumps inscribed in the map would be shown to be complicit in the outbreak.

Each separate proposition assumed proximity of clusters of deaths to a suspected source might argue one or another association. Whitehead's principal analytic was a circle, encompassing almost all the reported cases of cholera, whose center was almost precisely at the location of the Broad Street pump. In effect, by using his dividers to create a cholera field (today we would call the circle a "buffer"), Whitehead located the epicenter of the outbreak at the Broad Street pump. This, and his close association with Snow during the months of their mutual investigations, convinced Whitehead of the pump's complicity in the outbreak. As he would write in 1865 in a *MacMillan's* magazine article, Whitehead became convinced, "slowly and I may add reluctantly that the use of this water was connected with the commencement and continuance of the outburst." The old ladies who seemed immune to cholera contracted it less frequently because their homes did not have children to run to the pump and bring them water. They thus drank from it less frequently (Chave 1958, 95). It was not God's presence but the stair climbing limits of age that had protected Whitehead's elderly parishioners.

In almost every way, Whitehead's map—and the report it supported—was more complete than either Cooper's or Snow's. The total numbers of deaths and water sources were greater and the location of deaths more precisely mapped. And while, like Cooper's and Snow's, Whitehead's data set was based on GRO mortality reports, his local knowledge gave his report an unparalleled ethnographic depth. "The ordinary course of my duties taking me almost daily in the street, I was under no necessity to be either hasty or intrusive, but asked my questions just when and where opportunity occurred, making a point of letting scarcely a day pass without acquiring some information and not caring how often I had to verify it" (quoted in Chave, 1958, 96). What Snow had believed impossible—"I found that there had been such a distribution of the remaining population that it would be impossible to arrive at a complete account of the circumstances" (Snow 1855a, 41)—Whitehead accomplished.

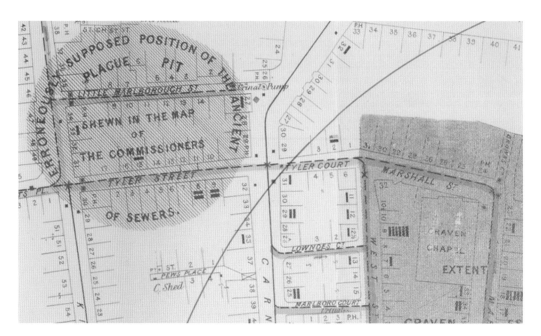

FIGURE 10.7B Whitehead correctly located the Craven Estate plague pit, incorrectly located in Coooper's map, that many believed complicit in the Broad Street outbreak. Dotted and solid lines symbolize different sewer lines.

It was not simply that Snow did not have the time and that Whitehead did. Snow, never the most garrulous of men, was less suited to the business of visits to the home of the bereaved local worker than was the young curate. Snow was no longer the young apothecary with a local practice on Firth Street. By the 1850s Snow was a famous anesthesiologist, a specialist whose practice was not bound by the community in which he lived. Whitehead's mission was to serve the troubled of the neighborhood: they knew him and he knew them in a way that Snow could not.

Not surprisingly, therefore, it was Whitehead who identified the index case of the outbreak and the method by which the Broad Street pump was contaminated. London constable Thomas Lewis and his wife had moved to 40 Broad Street in the 1840s, living with their growing family through the 1850s in a former single-family building that had been cut into flats. Their daughter Sarah was born in March of 1854 and in August first became ill with "promontory diarrhea." She died on September 2. As the GRO weekly mortality reports put it: "At 40, Broad Street, 2d September, a daughter, aged five months: exhaustion, after an attack of diarrhoea four days previous to death" (Johnson 2006, 178).

Mrs. Lewis told Reverend Whitehead the diarrhea had begun on August 28, five days before Sarah's demise. Mrs. Lewis soaked her daughter's soiled diapers in pails of water and then tossed the wastewater into the cesspool at the front of the house. While lined with bricks, the cesspool, less than a yard from the edge of the Broad Street well, later was found to leak when examined by local engineer Jeremiah York for the parish inquiry committee (York 1855). The water line of the cesspool was higher than that of the well, whose walls were surrounded by soil contaminated by the cesspool.

Here was both the index case *and* a likely source of the well's contamination. Because only the Lewis household had easy access to the cesspool at their house the chain of infection stopped when Sarah Lewis died even though other residents of 40 Broad Street would later become ill (Chave 1958, 96). Not surprisingly, perhaps, Snow was enormously pleased with Whitehead's argument, and with engineer York's physical support of it.[10] Unfortunately, that supporting material was discovered only after Snow finished writing both *On the Mode of Communication of Cholera* and his report for the parish inquiry.

THE LIMITS OF BROAD STREET

Famously, Snow had petitioned to have the handle of the pump removed to prevent local citizens from drinking its water. "I had an interview with the Board of Guardians of St. James's parish, on the evening of Thursday, 7th September . . . In consequence of what I said, the handle of the pump was removed on the following day" (Snow 1855a, 40). But by then, Snow acknowledged, "the attacks had so far diminished before the use of the water was stopped, that it is impossible to decide whether the well still contained the cholera poison in an active state, or whether, from some cause, the water had become free from it" (Snow 1855a, 51–52).

When the second edition of *On the Mode of Communication of Cholera* went to the printers Snow had not yet received the news of Whitehead's discovery of the index case near the pump. Nor did he have in hand York's report on the physical state of the Broad Street well. All Snow knew was that a cursory external examination showed "no hole or crevice in the brickwork of the well, by which any impurity might enter; consequently in this respect the contamination of the water is not made out by the kind of physical evidence detailed in some of the instances previously stated" (Snow 1855a, 52).

To this negative evidence was added the acknowledged quality, and broadly assumed purity, of the Broad Street well water. England's first great authority on microscopy, Arthur Hill Hassall, who examined the water from the Broad Street pump for the Board of Health inquiry, declared it "relatively bereft of microscopic animal life" that might be identified as a contaminant (Paneth, Vinten-Johansen, and Brody 1998, 1547). Snow admitted that the absence of observable contaminants—he carried out his own microscopic examination—forced him to "hesitate to come to a conclusion" on the pump water's complicity in the outbreak (Snow 1855a, 39). He could demonstrate the centrality of the well and pump but only assert, deductively, its pollution: "Whether the impurities of the water were derived from the sewers, the drains, or the cesspools, of which latter there are a number in the neighborhood, I cannot tell" (Snow 1855a, 53).

Snow therefore could argue, on the basis of his mapping, the centrality of the Broad Street well and its apparently pure waters; he could also argue inductively its complicity in the outbreak. That, however, was not the definitive proof he had hoped to achieve and, apparently, believed he had produced.

Nor, of course, could Snow prove beyond any reasonable doubt his assertion that "each epidemic of cholera in London has borne a strict relation to the nature of the water-supply of its different districts, being modified only by poverty, and the crowding and want of cleanliness which always attend it" (Snow 1855a, 56–57). In its separate report, the Board of Health's Cholera Inquiry Committee came to a very different conclusion: "We do not find it established," they wrote, "that the water was contaminated in the manner alleged, nor is there before us any sufficient evidence to show whether inhabitants of that district, drinking from the well, suffered in proportion more than other inhabitants of the district who drank from other wells" (General Board of Health 1855, 52).

It was not that evidence was lacking but that the evidence was insufficient to prove that the Broad Street pump was *solely* responsible for that neighborhood outbreak. There was room for doubt in this local case, and therefore skepticism toward Snow's broader theory in which cholera exemplified a class of waterborne disease. The Board of Health committee members wondered about those who drank from the well and did not get sick, and those who appeared to have become sick even though they lived far from the Broad Street pump and were not all known to prefer its waters.

Snow mapped the problem, but that mapping did not guarantee a solution. Too many potentially confounding variables had yet to be ruled out. As John Simon said of another map in his report on the 1854 epidemic: "When the 211 deaths are mapped upon a house plan of the city (as may conveniently be done by stamping a black ink mark at each place where one of these occurred) the broad features of the epidemic will be visible at a glance . . . [T]heir distribution may be noticed especially in two directions: many, dotted about in confined and crowded courts, where domestic cleanliness is rare, and atmospheric purity impossible; many, on the southern slope of the city, where it is a habitual complaint that stenches arose from the sewer" (Simon 1856, 27).[11] In the map the epidemic was "visible at a glance," but in the mapped environment others saw a range of potentially explanatory conditions, each related to a different theory of disease. Simon saw in the map the close quarters of the poor and the unsanitary conditions that resulted as disease inducing. In South London he saw areas of "atmospheric impurity" that on the basis of Farr's 1852 study and the medicine handed down from Hippocrates through Sydenham to the mid-nineteenth century seemed to suggest a miasmatic correlation. Snow only saw water, and in his refusal to consider the other possibilities in his map, as Whitehead had in his, the jury of his contemporaries remained unconvinced.

Equally limiting was the scale of the study itself. Neighborhood-level studies did not easily translate into definitive general proofs of the nature of the disease. There was nothing obvious about the Broad Street pump and its water. "Bad as was the produce of the Broad Street well,—containing the results of organic decomposition filtered through but scanty thickness of surrounding soil," Dr. Simon wrote, "this quality of water was not peculiar to it." This was the general state of the wells of London: "Everywhere filtering from a dangerous proximity to cesspools and sewers; everywhere loaded with nitrates or ammonia; everywhere containing evidence that they represent the drainage of a great manure-bed; and everywhere liable at any moment to contain excremental matter only imperfectly oxidized" (Simon 1856, 12). At the scale of the particular one could agree with Lea or Snow arguing the likelihood of a local outbreak's source but it was impossible, Simon argued, to generalize from the eccentrically local to the universal.

It did not help that in the second edition of *On the Mode of Communication of Cholera* Snow proposed waterborne disease as a general category that included with cholera other diseases, including plague, typhoid fever, and yellow fever. "I have become more and more convinced that many other diseases are propagated in the same way" (Snow 1855a, 125). The evidence here was at best anecdotal and speculative: "The natives of Gurhwal, a province in the north-west of British India, in which the plague has been present for the last thirty years, believe that it may be transmitted from one person to another in articles of diet, such as a jar of ghee" Snow noted. And yellow fever, he continued, resembled cholera and the plague by flourishing in "low alluvial soil, and also in spreading greatly where there is a want of personal cleanliness" (Snow 1855a, 127). From there to the fact that communicable diseases might generally share a solely waterborne source was a simple step only for Snow.

Columbia, Pennsylvania: 1854

Other researchers using similar techniques worked with equal diligence and at least equally complete mortality reports in other countries. Most of these concluded, in the 1850s, that the data supported a cholera that was principally miasmatic and airborne. Unlike Snow, most considered the subject of the nature and origin of the disease an open question and not one that easily could be definitively settled. In 1855, for example, Dr. T. Heber Jackson published an extraordinarily thorough study of an 1854 cholera outbreak in Columbia, Pennsylvania. Jackson's goal was to consider "all such conditions as might reasonably be supposed to have exerted an influence upon the development and propagation of the epidemic" (Jackson 1855, 123–31).[12] This included the local geography, which meant the waters of the Susquehanna River and the terminus of two canals. Was it the water, impurities transferred to the local reservoirs? Some thought so: "It was well known to all the physicians that very many who confined themselves, from habit, to the use of pump-water, were attacked with fatal effect" (Jackson [1855] 1958, 126). This ruled out the reservoirs because as Jackson had noted, pump waters drew from local wells.

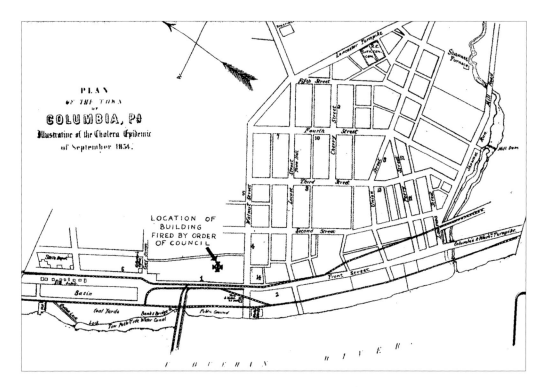

FIGURE 10.8 In this 1855 map of the 1854 cholera outbreak in Columbia, Pennsylvania, the location of individual deaths, listed to the left, did not argue a single source or origin. For Dr. T. Heber Jackson, the conclusion was that cholera was environmental, and likely miasmatic.

But was it the well water that was the origin of the disease? The path of contagion was not clear and the number of deaths seemed to argue a different origin and spread. For the origin, many looked at local immigrants, who were among the first to be attacked in the epidemic. It spread . . . from them, many believed. "Contagious disease[s] do not seize upon great numbers at once," Jackson explained, "but progress from case to case." But why did it take hold, there, among the immigrants and others? "It would appear that the cause of the appearance and spread of malignant cholera in Columbia, was manifestly connected with the air and locality; that it was endemico-epidemic" (Jackson [1855] 1958, 128).

The nice thing about Jackson's report is the ecumenical manner in which he carefully considered each theory—popular and scientific—and attempted to work his way through the data using this or that thesis—air, immigrants, environment, or water (river or well)—to come to his conclusion. The progression of ideas stated, considered, and tentatively rejected made of his article a review of the contending ideas that for him argued an environmental, and probably miasmatic disease. With his study Jackson included a simple schematic of Columbia, the bare geography instantiating the city with its streets, rail lines, and the river itself. Embedded in the map were the locations of 27 of the 127 deaths that occurred across the epidemic. For 13 cases there was no locational data and presumably he saw no reason to map the other 87 deaths.

Like Jackson's argument, the map placed the theories in the space of the city, suggested and rejected alternatives, and supported in the end a tentative rather than definitive conclusion: it's in the environment, somehow. Unsatisfactory, perhaps, from Snow's perspective and that of other strident advocates of this or that theory, but honest in its assessments and knowledgeable about the limits with which mid-nineteenth-century science could argue definitively the etiology of an invisible disease whose symptoms were glaringly evident in the patient.

CONCLUSION

Neither the Broad Street nor the South London study settled the cholera question. Snow's arguments did not, as he hoped they might, rewrite the idea of disease itself. "Intellectually and rationally" theories of airborne and waterborne disease, each with its evidenced geographies, were too evenly balanced for a determination between them (Ackerknecht 1948b, 566). Balancing Snow's arguments were other studies, in England and elsewhere, which argued on the basis of reams of data that cholera was a miasmatic or at least a multifactorial disease. Notable in England was Dr. H. W. Acland's *Memoir on the Cholera at Oxford*, an extraordinarily careful and detailed study of cholera in that city. The conclusion was, à la Farr, that altitude above water and therefore air quality was the strongest correlation with cholera deaths. For each

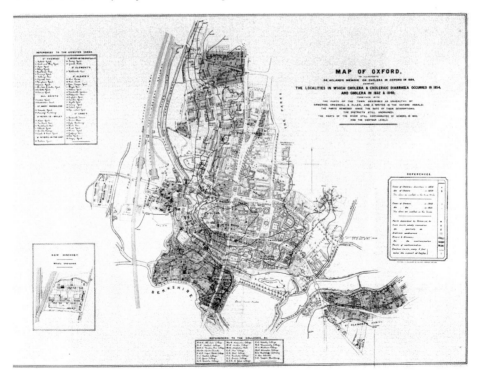

FIGURE 10.9 Acland's map of cholera in Oxford argued the airborne nature of the disease, using contour lines to show altitude as a correlate to disease intensity in the city.

of the 290 cases occurring in 1854 (128 of them fatal), Acland published tables of the age, date of onset, diagnosis, outcome (death or recovery), location, occupation, residence by street, and the sex of the patient. Beautifully mapped, Acland included contour lines to construct an argument that altitude was inversely correlated to cholera mortality.

Even if Snow did not convince his contemporaries, they did listen. In 1852 Farr acknowledged the likelihood of the fecal-oral route of interpersonal transmission. In the 1850s critics like Simon rejected Snow's thesis as definitive but accepted water as a possible source and vehicle of transmission. When cholera returned to South London in the 1860s, Whitehead wrote popular articles on the 1854 Broad Street outbreak for *MacMillan's*, as well as a more technical article in 1867 for the *Transactions of the Epidemiological Society of London*. In these retellings Snow was very much the hero.

Farr, still at the GRO, was "prepared in 1865 to closely scrutinize the water supply" (Morris 1976, 210). It was not simply the accumulated weight of Snow's arguments, which he had always been willing to entertain. It had been Farr who had first introduced the idea of the South London registration district water supply as a field of study. In the 1865 outbreak Farr traced the likely source to an East London Water Company emergency reservoir where open ponds of water appeared to be tainted by sewage. The company objected, arguing that Farr's analysis was based on obsolete maps whose accuracy it questioned. The *Medical Times and Gazette* defended Farr, his methodology, and his focus on the East London Water Company reservoir, calling the company's objections "puerile," and insisting that its culpability was proven (East London Water Company 1866, 254).

It would not be until 1883, when bacteriologist Robert Koch definitively identified *Vibrio cholerae* as the previously invisible, waterborne agent, that the nature of cholera would be settled. Others had previously seen the bacterium, but it was Koch's rigorous methodology that made his study conclusive. Still, as late as 1874, members of the International Sanitary Conference in Vienna "unanimously affirmed" the concept of cholera as a portable disease whose source might be "excremental pollution, excremental sodden earth, excrement-reeking air, or excrement-tainted water" (Woodworth 1875, 47). Any or all of them might be complicit. Without a method of *seeing* the disease agent the multifactoral, the idea of environmental cholera continued to dominate. With the fact secured and the bacterium identifiable, Snow's work was at least temporarily forgotten, one more relic of an older, science replaced by a new methodology.

214

In England some would argue that while Snow proposed the thesis that cholera was a waterborne disease it was Whitehead who truly made the case in his study of the Broad Street outbreak. "This doctrine, now fully accepted in medicine, was originally advanced by the late Dr. Snow, but to Mr. Whitehead unquestionably belongs the honour of having first shown with anything approaching to conclusiveness

FIGURE 10.10 A young priest at the time of the Broad Street outbreak, this photograph shows Whitehead at the age of fifty-eight years.

the high degree of probability attaching to it," wrote one of Whitehead's biographers (Rawnsley 1898, 40). After all, it was Whitehead who discovered the index case and in his tireless, repeated visits to the homes of the victims in his parish was the master of the "shoe leather epidemiology" which Snow used to eloquently and passionately argue his thesis.

Quoting an 1871 valedictory of Whitehead, Rawnsley wrote, "In the Broad Street outbreak of cholera not only did Mr. Whitehead faithfully discharge the duties of a parish priest, but by a subsequent inquiry, unique in character and extending over four months, during which time he sat up night after night till 4 a.m. arranging the evidence he had steadily collected during the day in the course of his laborious duties in that densely populated parish, he laid the first solid groundwork of the doctrine that cholera may be propagated through the medium of drinking water, polluted with the intestinal discharges of persons suffering from that disease" (Rawnsley 1898, 40). For those few still concerned with cholera's history, or medical history generally, it was the "Snow-Whitehead" theory, one in which the disciplinary matrix of local investigation, spatial mapping, and temporal analysis of public health data came together.

CHAPTER 11

CHOLERA, THE EXEMPLAR

John Snow was rescued from historical obscurity by an early-twentieth-century textbook written by William Thompson Sedgwick, a professor of biology and a lecturer in sanitary science and public health at MIT (Sedgwick 1911). In edition after edition (the first was 1901, the last in the 1920s) Sedgwick's text instructed students in, the subtitle said, "the causation and prevention of diseases" through the application of principles of sanitary science designed to advance public health as both a discipline and as a medical specialty. In his textbook Sedgwick presented infectious disease first as a failure of the sanitary infrastructure of the city and second as a failure of personal hygiene. Two diseases with a similar etiology dominated the case studies in Sedgwick's text: typhoid fever and cholera. For him, Snow's Broad Street study exemplified the method by which the environmental determinants of infectious diseases like these might be taught.

"As a monument of sanitary research, of medical and engineering interest and of penetrating inductive reasoning, it deserves the most careful study. No apology, therefore, need be made for giving of it here a somewhat extended account" (Sedgwick 1911, 170). Sedgwick's ten-page retelling in close, small type of the Broad Street outbreak revived the fable that Snow's demand that parish officials remove the handle to the Broad Street pump stopped the outbreak in its tracks and proved cholera was solely waterborne. Snow's friend, Benjamin Ward Richardson, first proposed the story of the pump handle in a collection of Snow's papers (Chave 1958, 103; original emphasis): "The Vestry was incredulous but had the good sense to carry out the advice," wrote Richardson. "The pump handle was removed and *the plague was stayed*."

Sedgwick used Richardson's story—denied by Snow (1855a, 51)[1] and later by Whitehead (1865)—to make Snow the far-seeing, solitary hero who single-handedly stopped cholera, proving to the world its waterborne nature. Because he was promoting a methodology and an idea of disease, Sedgwick was able to ignore the failure of Snow's work to convince contemporaries. Because his was a teaching tale, designed to inspire as well as instruct, Sedgwick did not need to mention the hundreds of researchers in scores of countries whose work created the intellectual foundation on which Snow built his argument, citing their findings in support of his idea. Nor did Sedgwick mention the labors of the many GRO district and subdistrict registrars who collected from local physicians the causes of deaths that became the records used by Cooper, Farr, Simon, Sutherland, Whitehead, and Snow. Of them all, Sedgwick mentioned only Reverend Henry Whitehead who in the mythologizing of the teaching text was transformed into Snow's helpmate, a Tonto to Snow's Lone Ranger.

Sedgwick emphasized Snow's inductive logic to argue that, if only one spatially locates both the deaths and possible sources of an outbreak, the logical relation between them will be revealed in the local sanitary infrastructure. The message of Sedgwick's map (fig. 11.1) was clear: this is how it is done, and Snow was the creator of the method (Koch 2004). In his map "after the original version by Dr. John Snow,"

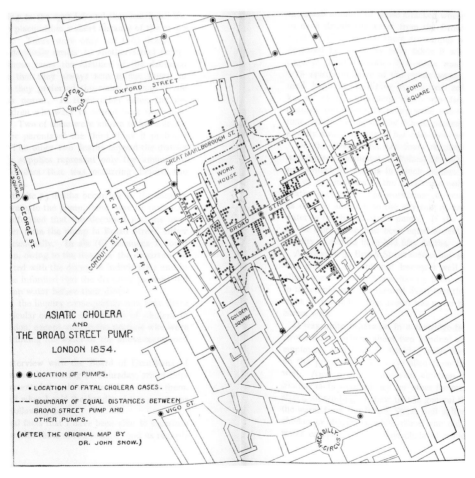

FIGURE 11.1 William Sedgwick recreated the Broad Street study as a simple tale of a single researcher who solved the problem of cholera. Similarly simplified was Snow's Broad Street map, which Sedgwick modified as a teaching example.

Sedgwick widened the streets and, presumably for purposes of graphic clarity, transposed the symbol of deaths from bars to dots. He also enlarged the pump symbols to make them stand out. Deaths on Heddon and Leicester streets west of Regent Street, obvious in Snow's original map (see fig. 10.4), disappeared behind the legend in Sedgwick's version. With the walking range of Snow's map for the parish committee the centrality of the pump to the outbreak was . . . obvious.

Sedgwick's teaching map made graphic what by 1901 was bacterially understood: *Vibrio cholerae* is waterborne. Reality was defined by microscopy and the narrative of the map was refashioned in its image. Mapping was a tool of disease studies and disease was a bacterial something whose source of local contamination could be identified by mapping the landscape.[2] Snow was the principal exemplar whose methods were the basis of the sanitary science of public health Sedgwick taught. Just as Horatio Alger's books promised every poor lad a chance to make good in society, to become respected and financially secure, Sedgwick's textbook promised every sanitarian a chance to identify a disease and discover its source through the diligent and self-reliant, inductive approach to disease Snow exemplified. Those who today criticize the simplifications of Snow's map and cholera studies by medical geographers (Brody et al. 2000; Vinten-Johansen et al. 2003, 396–99) fail to recognize that the real simplification was by Sedgwick and his successors. Medical cartographers simply followed their lead.

W. H. FROST

Sedgwick inspired a generation of professionals whose focus was the physical environment in which a class of bacterial, epidemic diseases whose agents could be seen through the microscope had been shown to prosper. Perhaps the most famous in the United States[3] was W. H. Frost, a public health physician who in 1919 was seconded to Johns Hopkins University in helping create the School of Hygiene and Public Health. Frost was the son of a Virginia country physician; as a student he read Sedgwick's text and after getting his medical degree took and passed the examination for the U.S. Public Health and Marine Hospital Service (PHMHS). In 1905 he was sent to New Orleans where he and other public health officers investigated a yellow fever outbreak after local attempts to contain the disease were unsuccessful. Members of the PHMHS were assigned to each ward of the city both to collect data and to enforce screening and fumigation programs, then new approaches, in an attempt to control the mosquito population. "This was one of the earliest and most dramatic demonstrations in the United States of the power of medical science, working through community organization, to effect a rapid and complete eradication of an insect-born disease" (Maxcy 1941, 5).

Later posted to the service's "Hygienic Laboratory"—commissioned by Congress to investigate the source of disease—Frost was "primarily assigned to routine examinations of Potomac River water in relation to the typhoid problem in Washington" (Maxcy 1941, 7). In 1910 Frost published a study of a typhoid fever outbreak in Williamson, West Virginia, modeled upon the approach promoted by Sedgwick in his textbook (and exemplified by Snow). "Here Frost was dealing with a very simple but characteristic outbreak, one whose solution was clear from the start. Unlike yellow fever, typhoid fever *was* caused by contaminated water. In investigating an outbreak, one looked for the point where sewage contaminated the water supply. It required no great imagination, no painstaking collection of facts and of guarded inference. Yet Frost made a conscientious statement of the facts and circumstances so there might be no doubt about the matter. The criminal neglect of current sanitary knowledge was obvious, yet the publication contained no words of reproof or censure" (Maxcy 1941, 23–24).

As he had been trained to do—as Sedgwick's Snow had done—Frost gathered data from local health officials and then charted a timeline for the outbreak.[4] This temporal profile had its origins in the methodologies of yellow fever in the eighteenth century and, of course, cholera in the nineteenth century (for example, see fig. 9.2). In this case, Frost charted not disease incidence as a function of local temperature but instead the percentage of cases over time based on the possibility that interpersonal contact contributed to the transmission of the disease (fig. 11.2). The tables of cases that permitted him to produce his graph were inscribed in a map.

Because typhoid fever was known to result from a bacterial contamination of the water supply there was no question of the outbreak's agent, only its local source.

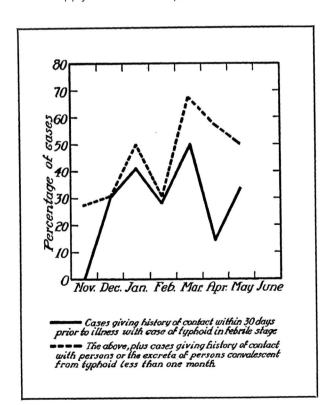

FIGURE 11.2 In addition to his map and tables, Frost contributed a chart showing the typhoid fever cases, month by month, adding to it cases that appeared to involve interpersonal transmission.

CHOLERA, THE EXEMPLAR

CHAPTER 11

There were 152 cases of diarrheic symptoms assumed to be typhoid fever in this outbreak. Microscopic examination of a few stool samples collected from that population "showed clear colonies resembling the typhoid or paratyphoid bacilli, almost as numerous as the red colon colonies. Several of the clear colonies were identified and found to be culturally identical with *B. paratyphosus, B. enteritidis Gartner*, and others of the so called 'Enteritidis' group bacilli case" (Frost 1941, 61). Those few samples were used to assert the presence of a common class of typhoid fever sufferers in Williamson, West Virginia.

Frost then mapped the epidemic using different symbols for typhoid fever cases month to month, May to December, on a simplified map of Williamson and East Williamson (fig. 11.3). The map thus created a class of disease events not simply *in* time but *over* time. Central to the map was the principal local water source, the Tug River, a fork of the Big Sandy River, to which an arrow indicating current was attached. Also in the map was a water intake location near the powerhouses. Unfortunately for Frost, there was no clear, single source of contamination in this case, no cluster showing where the outbreak originated like the one Snow identified in Broad Street.

As Frost admitted in his report, "It cannot be said that any marked regional [locational] prevalence of typhoid fever has been demonstrated . . . In general, the distribution of cases seems to have been roughly proportionate to the distribution of population, with notable exceptions in several well-marked foci" (Frost 1941, 46).

Why include the map with the study if it offered no clear insights? Because that was the method Sedgwick promoted and Frost had been taught. It was the way such diseases were studied. In the map, publicly reported tables of incidence were located

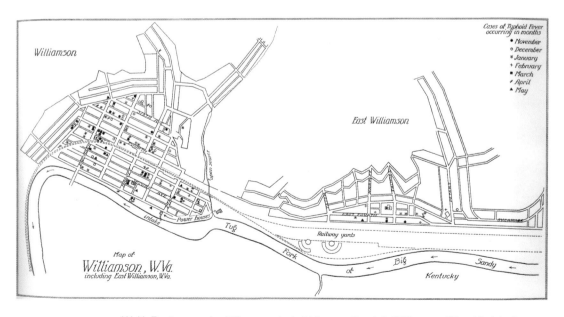

FIGURE 11.3 W. H. Frost mapped a 152-person typhoid fever outbreak in Williamson, West Virginia, in which a contaminated water supply was proposed as the source of the outbreak.

and clusters identified. The river and the city's water intakes were assumed sources of disease. Why? Because it was typhoid fever and typhoid fever *was* waterborne. That one could not see in the map a single an obvious cluster of disease with a sole environmental source at its center was unimportant.

The map signaled that Frost had employed the appropriate methodology and the work therefore was professionally done. Unlike Snow's map of the Broad Street outbreak the purpose of the Williamson map—and the study it anchored—was neither to develop nor to test a theory of disease but instead to apply what was known, reinforcing at once the profession of public health officials and the efficacy of their methodological toolset with yet another example. The work also served to assert federal action in the care of its citizens through the dispatch of federal health officers to the site of a local outbreak. The whole was a testimony to the benefits of the evolving public health bureaucracy: "your tax dollars at work."

Science is built from a repetition of the known with the piling on of one example after another until everyone agrees on a set of facts, and of equal import, a method for first their definition and then their consideration. Whether it was cholera or malaria or typhoid fever, it *was* important to investigate and document the outbreaks that occurred. Frost's study emphasized the methodology of public health within the political and social movement that demanded a sanitary infrastructure that would help to diminish the likelihood of future outbreaks. By mapping a known condition, even where the mapping was inconclusive, Frost reinforced the utility of the accepted, methodological matrix already known to work (after all, it worked for Snow!). The problem, as Frost and his contemporaries discovered, was that not all diseases were to be understood in this way. The methodologies by which cholera and typhoid fever had been studied in the past would not serve to explain a whole class of conditions that were neither necessarily bacterial nor obviously waterborne.

CHICAGO: TYPHOID FEVER

In fact, typhoid fever didn't even explain . . . typhoid fever. An outbreak in Chicago, Illinois, in 1902 was especially severe in several inner-city wards where Dr. Alice Hamilton was staying at Jane Addams's famous Hull House (Hull House 1895; Schultz 2007). The center of a socially active urban sociology movement, Hull House members extensively mapped their community, the Nineteenth Ward, and attempted to draw social and economic conclusions based on that work. Hamilton took on the typhoid fever epidemic whose local severity could not be explained, she wrote, by the quality of local water alone. She demonstrated this in a map of Chicago's political wards in which she embedded the home locations of typhoid fever cases. The dense clustering of cases in the Nineteenth Ward (fig. 11.4), whose water was drawn from the general urban supply, argued to Hamilton a failure of sanitary infrastructure. "To those who studied the distribution of cases of typhoid fever it soon became evi-

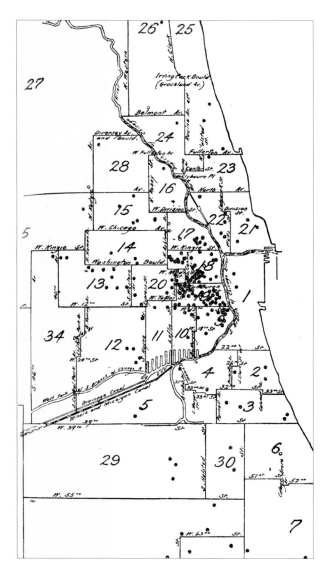

dent that the number was greatest in those streets whose removal of sewage is most imperfect (Hamilton 1903, 577).

House-by-house, Hamilton built five distinct classes of sewage disposal and to this inventory added the reported incidence of typhoid fever (fig. 11.5). *If* the incidence of typhoid fever was related to the lack of sanitation, the map argued, *then* one would expect the greatest incidence in the houses with the worst sanitation. The mapped results seemed, at best, ambiguous. "The cases of typhoid fever were not more numerous, on the whole, in houses where the disposal of fecal matter is most defective. Houses with good plumbing which are close to those with badly-drained cesspools are just as apt to have case of typhoid as are the badly drained ones" (Hamilton 1903, 258).

In the map Hamilton observed what appeared to be a cumulative, neighborhood effect irrespective of the waste disposal system of any one house. Typhoid fever was most intense on streets where at least some houses lacked good sanita-

FIGURE 11.4 Dr. Alice Hamilton mapped the incidence of typhoid fever in Chicago in 1903 to demonstrate the waterborne disease was especially severe in neighborhoods with a poor sanitary structure.

tion. To Hamilton, this suggested an intervening vector, the housefly. Others had written on flies as possible disease vectors, contaminating food and persons with fecal waste found in the household privy. The failure of local sanitation at individual houses clearly created breeding grounds in Chicago's Nineteenth Ward where flies proliferated. Here, she proposed, was the unseen vector, one secondary to but dependent on large-scale elements of district sanitation. Hamilton then carried out a series of tests to demonstrate that houseflies were able to carry the typhoid fever bacillus. Proving that flies could be infected suggested the likelihood they might serve as transmission vectors. No other explanation served as well, Hamilton concluded, in explaining the pattern of disease occurrence in her ward.

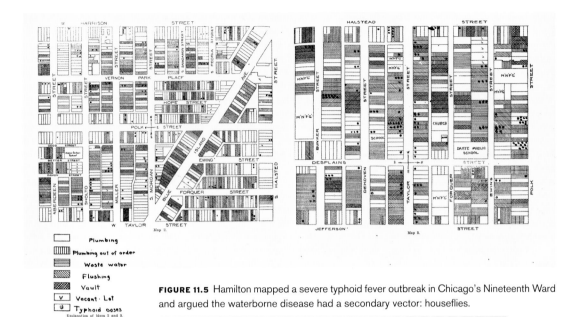

Plumbing
Plumbing out of order
Waste water
Flushing
Vault
Vacant · Lot
Typhoid cases
Explanation of Maps 2 and 3.

FIGURE 11.5 Hamilton mapped a severe typhoid fever outbreak in Chicago's Nineteenth Ward and argued the waterborne disease had a secondary vector: houseflies.

Here were Seaman's mosquitoes, found in New York during the eighteenth-century yellow fever epidemic, prevalent in the hot summer months when Sydenham's classificatory system predicted such disease outbreaks. To Snow's waterborne disease was added a second stage of diffusion, one resulting from a lack of sanitary infrastructure and the resulting human waste sites from which disease could be transmitted by flies. The conclusion wasn't as simple as washing one's hands. Hamilton's waterborne typhoid fever promoted entomology with a social face that led back to Chadwick's concern over income inequality and Farr's focus on general sanitary infrastructures. If typhoid fever was still in a process of study and definition, the situation was far more complex, with emerging diseases whose agency remained as uncertain in the early twentieth century as cholera's had been in the 1840s.

Polio: 1910

Named in 1840 by Jakob Heine, a German orthopedist, poliomyelitis outbreaks were first reported in North America in the early decades of the twentieth century. While established as a virus by 1909, in those first years of virology "bacteria and viruses were only vaguely differentiated" (Rogers 1996, 14). This emergent disease was assumed reflexively to be a consequence of the dirt and disease that pervaded the city, and especially densely packed immigrant areas. This had been the first assumption of physicians facing plague in London in the seventeenth century, yellow fever researchers in the eighteenth century, and cholera researchers like Hamett in the nineteenth century. A lack of personal and communal hygiene created an environment in which disease flourished . . . why should this polio thing be any different?

Between 1910 and 1912 W. H. Frost was sent to investigate separate polio out-breaks in Iowa (1910), Ohio (1911), and finally Buffalo, New York (1912). Unlike cholera or typhoid fever, this paralytic illness tended to strike young, Caucasian, middle-class children, rather than elderly residents in densely settled immigrant wards and racial ghettos. In Mason City, Iowa, the first case occurred on April 11, 1910, when a kindergarten student was struck "in one of the best residential sections of the northwestern section of the city" (Frost 1941, 172). The disease spread among other students at the school, and then somehow to students at other schools. On-site investigation of the "sanitary condition of premises," the homes of the afflicted children, yielded a mass of data on home cleanliness, ventilation, local sewage and garbage disposal, and so forth. It yielded no clues, however, about why polio struck this house rather than that house, in this neighborhood but not another.

Frost mapped the locations of cases of polio based on local health department reports to build a spatial portrait of the outbreak in hopes of seeing a concentration in one or another area that might give a hint as to its origin (fig. 11.6). The map's argument was familiar: first, a study area in which the paralyzing thing was resident was posted. Then a class of polio cases, distinguished by the month of onset, was embedded in a street set that created the habitable study area. To this were added major public buildings, schools, and churches where children could be expected to congregate as possible origin sites. The result argued that *if* cholera had a local environmental source, *then* it would be visible near where cases were most concentrated. While mild clustering could be seen in the map, neither its origin nor source location was suggested. What the map really said was that whatever this polio might be, its secrets were not going to be divulged as easily as Sedgwick's map (fig. 11.1) had implicated the Broad Street pump.

Frost then shifted scale, mapping the homes of patients at a single school in an area of active polio activity, the eight-block area around public school number 8, where polio's index case had occurred. In the map Frost included the school, the church, and the homes of the students who went to the school and presumably attended the church a block to the east. These homes were differently symbolized depending on the status of each diagnosis. In some cases the patient was paralyzed, and in some cases that paralysis passed—"aborted," as Frost would say. In other cases symptoms were polio-like but clinical eccentricities made the diagnosis uncertain. Cases were *probably* polio but could equally have been something else, for example, meningitis.

In proposing a class of disease events in his mapping, Frost made all these mapped polios—paralytic, aborted, suspected, and unknown—into a single event class whose members could be considered separately or jointly in the map (fig. 11.7). Frost added to each location a date of onset, embedding temporality, and finally, a line joining the home of a polio patient to the school, and for some, the local parish church.

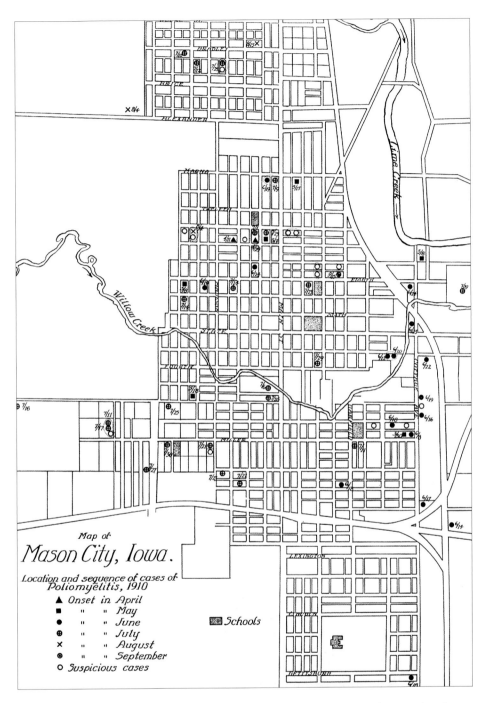

FIGURE 11.6 W. H. Frost mapped a poliomyelitis epidemic in 1910. All that was clear was that clusters occurred around schools where victims, primarily young people, studied.

School number 8 and a local church *were* central to this subset of patients in the Mason City outbreak. The map brought forth a relationship between the cases and the locations but what this meant, if anything, was unclear. In the early twentieth century, before the growth of suburbia, people tended to live a spatially insular life.

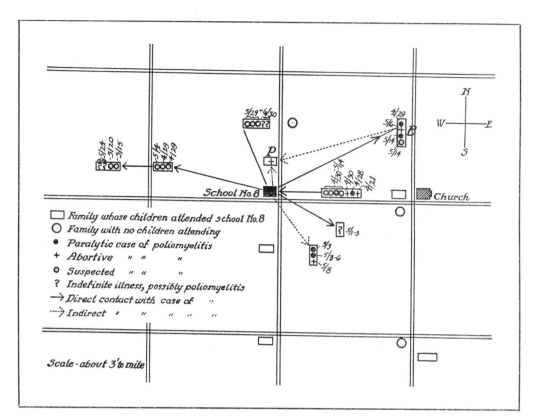

FIGURE 11.7 Frost mapped polio cases that appeared to originate at a local public school. Attempts to link cases suggested an interpersonal transmission but revealed no etiology or single contaminated disease source.

Children walked to nearby schools and families to nearby churches. These institutions were central—literally and figuratively—to the neighborhoods in which people lived. All that could be said on the basis of the mapped data was that public school number 8 was centrally located in a neighborhood where students were visited by these different polios. The map revealed nothing about either the origin or the source of the disease or its mode of transmission.

Finally, Frost attempted to get a handle on polio's pattern and rate of diffusion across the state (fig. 11.8). Using county-level public health data, Frost mapped reported cases by month, using different symbols to inscribe temporality. The map thus argued a state jurisdictional reality whose resolution was the individual county. "The disease is seen to have been prevalent throughout the state, but if the reports received may be relied upon at all it is evident that the distribution in various sections was not uniform. The prevalence was considerably greater in the northern than in the southern half of the state. The disease would seem to have been especially prevalent in Cerro Gordo County, where the first epidemic was reported, in the eight adjacent counties, and in Grundy County, one county away from Cerro Gordo" (Frost 1941, 221).

If the data was reliable, *if* paralytic, aborted, suspected, and unknown polios were all correctly diagnosed and accurately reported variants of a single condition, *then* a pattern of diffusion across the state emerged. The disease was introduced to the state in one county, where intensity was greatest over time, and spread contiguously from there to other communities, diminishing in intensity over time with distance from Cerro Gordo County. This spoke to the origin of the outbreak in Iowa but not to its local source or its method of diffusion. Nor did this speak to the relationship between the Cerro Gordo County polio outbreak and those occurring elsewhere in the United States. Just as Arrieta's plague campaign was focused solely on his province and Seaman's yellow fever was assumed to be a creature of New York City, so too was Frost's polio a spatially limited creature.

Others noted similar patterns elsewhere and were similarly incapable of explaining them. Frost worked not alone but within a vast international coterie of medical and public health researchers writing about their attempts to understand this newly pandemic, paralytic disease whose principal victims were children. "Rather striking instances of similar rapid radial spread are given by Wickman in his account of the epidemic of 1905 in Sweden and by Ling in his description of the Swedish epidemic of 1911. Such rapid spread over a large territory practically disproves the hypothesis that epidemic outbreaks are due to place infection independent of and not transferred by ordinary traffic, it being quite improbably that place infection of numerous locali-

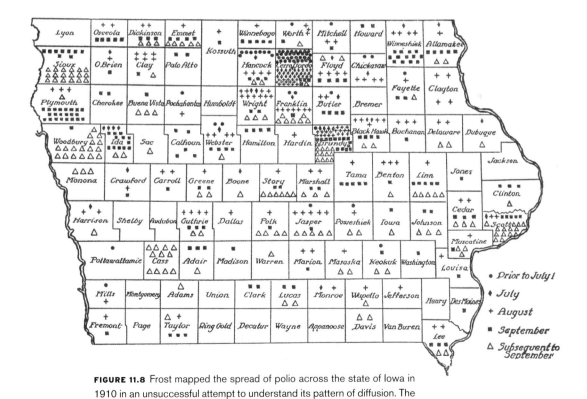

FIGURE 11.8 Frost mapped the spread of polio across the state of Iowa in 1910 in an unsuccessful attempt to understand its pattern of diffusion. The result does show a decreasing intensity as one moves from the epicenter of the outbreak to toward the state border.

ties within the same general territory should develop independently bout the same time. The rapidity of spread has, too, in some instances been such as to suggest that man, the most rapidly and widely traveling animal, is the carrier of infection" (Frost 1941, 252–53). Like Arrieta's plague, this polio was a spatially extensive, mysterious creature somehow linked to human commerce and traffic, anthropogenic in a way that could not be easily explained.

In a similar vein, Frost and others of his generation were at a similar loss to explain the virulence of the 1918 influenza pandemic. There are few maps of the 1918 epidemic in the public health literature of the era, and for good reason: the disease spread so rapidly, and its intensity in a population was so great, that mapping was pointless. It would require the mainframe computers of the 1970s before an accurate mode of pandemic influenza was possible. A more rigorous virology and a more robust modeling would be required before the precise nature of the disease and its progression could be practically described (Pyle 1986; Hunter and Young 1971).

Frost's disease maps emphasized the power of the state to collect and analyze disease-related data. They also served, although not for Frost, as testimony to the limits of Snow's inductive epidemiology. From Virginia through Iowa, in study after study, mapping served more or less for diseases like typhoid fever and cholera but not for emerging diseases like polio and killer influenza. The real lesson of the maps, and the disease theories they sought to apply, was that a Broad Street–like pattern of simple diffusion from a single point of seemingly obvious contamination occurred only rarely.

FROST: PUBLIC HEALTH

In 1919, while still an active federal public health officer, Frost was seconded to Johns Hopkins University where he became an instructor at its then-new School of Hygiene and Public Health. He taught as he had been taught, emphasizing case method analyses and the basic descriptive algebra of mortality ratios that was Farr's legacy. In 1927 Frost published his masterwork, *Epidemiology: Public Health Prevention Medicine*. The subject had progressed markedly from the days of Sedgwick . . . but nowhere near far enough. With influenza, polio, and other emergent viral disease states, Frost was left, as Snow had been, with disastrous disease outbreaks whose agents were unclear and whose sources anything but obvious.

Why did Frost promote John Snow as the hero of cholera and his work a "near perfect model" (Frost 1836, ix) for public health studies if his methodology did not answer the questions the public health officer struggled with? Perhaps because John Snow was the name Sedgwick gave to the disciplinary matrix whose application to disease studies defined a discipline and its methods. John Snow had become the shorthand by which an approach that had worked for simple, waterborne, bacterial diseases had come to be known. The idea that this methodology might equally serve the analysis of emergent diseases of unknown etiology must have been . . . comforting.

The way in which the broad limiting conditions in the distribution of a disease are fitted in with the details of local distribution so that each set of observations contributes to the interpretation of the other, is excellently illustrated in John Snow's masterly analysis of the epidemiology of cholera. As his methods of collecting, combining, and presenting the data used are unusually simple and direct, his argument will be summarized in substantially the same sequence that is followed in the original. (Frost 1927, 532)

In his retelling Frost referenced Sedgwick as a source for his knowledge about Snow's Broad Street study. To that Frost added a précis of Snow's study of the South London outbreak, broadening the tale to bolster his promotion of Snow as *the* pioneer of epidemiological methodology. In 1941 Frost published a collection of Snow's writings, including the entirety of the second edition of *On the Mode of Communication of Cholera*, as well as Richardson's elegiac memories of Snow for good measure. The reasons for skepticism by Snow's contemporaries were ignored as Snow became for Frost, as he had been for Sedgwick, the nineteenth-century hero who through sheer brilliance created the methodology by which cholera was conquered and modern disease studies were to be pursued.

THE HERO TALE

The transformation of John Snow is curious because Snow's cholera studies were not a success. Snow's critics asked questions he could or would not answer and Snow thus did not convince his contemporaries of what he fervently believed to be true. Cholera was eventually revealed as water- rather than airborne, but science is not about being proven right *someday*. Science is about convincing a jury of one's peers of the rightness of a set of evidentiary propositions tested with a generally accepted methodology. That, John Snow did not do. So . . . why make John Snow a hero? Indeed, why nominate any single hero in a discipline that is ineluctably interdependent, the labor of many?

The question is important because the myth is not harmless. For more than a hundred years students in disease studies have been given Snow as an example to follow and that is a problem with consequences, as University of London's Roger Cooter has argued: "By failing to inquire seriously into the contexts for, and meanings of, discovery and innovation, it [the Snow myth] supports a wholly individualist impression of how modern biomedicine was installed and operates . . . When the focus is on the researcher rather than his or her subject and its messy context then false science and false breakthroughs are encouraged. As a contemporary example there is the work of Hwang Woo-Suk, the Korean scientist who falsely announced a breakthrough in cloning to promote not the science but himself and his country, only to eventually have his work discredited" (Cooter 2006, 1647–48).

That is it, exactly. Damage is done when the emphasis is on the hero and not the subject of his or her quest. When that happens the science is lost. Richardson's ele-giac—if not always accurate—memories helped shape this hero tale that Sedgwick modified to meet his instructional needs. The Snow that Frost would promote in the 1930s was fashioned out of this backcloth, and perhaps, Frost's need to believe that the disciplinary matrix of public health that Snow had come to personify would serve the struggle to understand emerging diseases like polio whose etiology was unclear and against which medicine was powerless.

In addition, this Snow and his cholera reinforced a very American preoccupation with ideals of autonomous action and independent thought over communal effort. Schooled in this same tradition are the generations of physicians and medical re-searchers inspired by the tales of medical heroism in Paul de Kruif's *The Microbe Hunters* (1939), a collection of biographies celebrating researchers whose solitary brilliance improved the lives of us all. "Such novels often celebrate the quietly heroic aspect of medicine, fashioning physicians in the mold of Joseph Campbell's arche-typal hero" (Verghese 2005, 1846).[5] The form demands a bold investigator as indi-vidualistic as Dashiell Hammett's solitary private eye, Sam Spade, a microbe rather than man hunter equally brilliant and ready to risk all (Roueche 1991).

Who else would fit the bill? Whitehead the curate might have been fashioned into a kind of Sherlock Holmes, an amateur scientist of admirable excellence. But public health needed a medical professional, not a priest, to argue its legitimacy. And in an age uncertain about the status of religion the curate, no matter how gifted, could only serve the hero tale as a helpmate: Watson to Snow's Sherlock Holmes. Farr's origins were as humble as Snow's and his contributions to public health more extensive. Americans do not make heroes out of bureaucrats, however, no matter how brilliant their work. Those who labor in statistics are similarly disqualified.

And of course, the great thing about this fiction is that Snow was right. "Snow had a *good idea*—a causal theory about how the disease spread—that guided the gathering and assessment of evidence" (Tufte 1997, 29; original emphasis). Indeed, Snow had a *great* idea he pursued with fervor. Forget that his assessment was un-convincing and the narrative comes into focus. Here, for example, is Sandra Hempel's description of John Snow, quoted by Christopher Hamlin in a review of her novelistic treatment of Snow, *The Medical Detective* (2006): "What distinguished Snow . . . was his tireless determination to pursue every scientific investigation relentlessly to its logical conclusion. There were no short cuts, no leaps of faith and no unquestion-ing acceptance of untested traditional wisdom, but organized procedure, sound ex-periment and careful observation. This step-by-step approach, which seems to have come naturally to Snow, was in fact a precursor of the modern method that is now the basis of all medical research" (Hamlin 2007).

Indeed, Snow was determined to the point of obsession. Time after time he rushed to print only to scramble afterward to publish addendums as data the original study lacked became available. Snow did this with the publication of his first edition of *On the Mode of Communication of Cholera* in 1849, as we have seen, and again after the second edition in 1855. Snow was so convinced he was right he did not seriously consider alternate theories, even to exclude them definitively, as Farr did in his 1852 study. In this rush Snow skipped lots of steps, pronouncing too often as proven what he only assumed to be true. It was this habit that most angered Snow's contemporaries.

In June 1855, for example, Snow testified as an expert witness before a parliamentary committee on proposed public health and nuisances removal bills. There he stated as fact that while the smells of tanners, soap boilers, tallow-chandlers, and other such trades were certainly obnoxious, they were not disease producing. Yes, the foul stench might cause vomiting and "when very concentrated, will produce sudden death; but when the person is not killed, if he recovers, he has no fever or illness" (Parkes 1855b, 634). But the real cause of illness was in the water, Snow insisted, not ever the air. "Is this evidence scientific?" E.A. Parkes asked in *The Lancet*. "Is it consistent with itself? Is it in accordance with the experience of men who have studied the question without being blinded by theories?" (Parkes 1855b).

Snow was right that the foul odors of the tanners and tallow-chandlers did not cause diseases like cholera. Parkes was correct in saying Snow made assertions that he could not prove and that seemed not only partisan—he testified at the request of the noxious trades association—but also nonsensical. At issue was the perceived nature of disease itself. If the foul odors made people sick, and sometimes deathly ill, as Snow admitted, then, at least as Parkes defined disease, they were disease *producing.* But for Snow disease had become by definition waterborne and the noxious airs of the industrial companies, while perhaps unpleasant (causing vomiting and fainting), were therefore not generative. Snow might have said, "The odors won't cause typhoid fever or cholera but it will make folk ghastly ill and that is enough to ban these businesses from the city." He didn't. But then, he was there testifying at the request of the attorney for the nuisance trades, a "hired gun" whose prestige as an independent scientist was compromised, perhaps, by his status as a champion for interested parties (Lilienfeld 2000).

Here is the deeper danger to the myth of Snow. In its promotion is lost the complexity of the definitional problem that cholera presented and diseases generally present. Are illnesses to be defined symptomatically (Parkes) or by suspected source (Snow)? Today we ask are they genetically inherent, environmentally promoted, socially grounded, or the result of individual choice?[6] When the focus is on the expert and not the evidence, such questions are not asked, especially when the expert is a partisan. Lost as well is the importance of the questions advanced by Snow's contemporaries. In promoting Snow we deny the voice of his jury and its concerns, and thus the nature of Robert Boyle's juried science itself. We return instead to a Hobbesian insistence on logic's revelatory power in a natural world. Snow sat astride these two sciences but in his argument relied more on inductive logic than the increasingly distinct science of his peers, abjuring the voices of his jury.

Snow was caught in a period of transition from one to another kind of data, from one kind of evidence to something very different. He relied on carefully constructed, typically inferential arguments based on the close observation of small data sets. Snow employed to great effect the anecdotal and circumstantial case histories he and other physicians collected. Time and again his writings assert the evidentiary value of this kind of data and on its basis the necessity of his conclusions. If he erred it was in asserting the necessity rather than merely the likelihood of his conclusions. The inferential offers probability but not the certainty Snow asserted. It left room for doubt.

BROAD STREET REDUX

Snow's approach built upon a methodology long in service by 1849. Alas, for Snow, what would have served a generation earlier no longer was convincing in the mid-1850s. The science being birthed in that era was more statistical and general, a matter of seeing disease in data sets and ratios rather than the accumulation of case histories. They were, and remain, important, but in Snow's day were already serving as tools for hypothesis building more than as proofs of theories. New approaches were becoming the accepted standard that Snow did not meet. The dot map was being replaced by more statistically oriented surface maps like Chadwick's map of Leeds (fig. 7.9A) and Grainger's map of cholera (fig. 7.10). Mortality ratios embedded in those surfaces were increasing employed as public data sets burgeoned beyond the capacity of a simple mapped display, or easy hand calculation.

With this in mind, return to Snow's iconic Broad Street maps (figs. 10.4–10.5) and their argument. "The brilliance of Snow's map lay," writes Harvard historian of science Steve Shapin, "in the way that it layered knowledge of different scales—from a bird's-eye view of the structure of the Soho neighborhood to the aggregated mortality statistics printed in the Weekly Return to the location of neighborhood water supplies—all framed by particular understandings of how people tended to move about in the neighborhood, of the physical proximity of particular cesspools to particular wells, and of the likely behavior of specific, still invisible, and still unnamed pathogens. A city is a concentration of knowledge as much as it is a concentration of people, buildings, thoroughfares, pipes, and bacteria. Maps like Snow's allowed the modern city to remake itself and to understand itself in a new way" (Shapin 2006).

All maps layer knowledge (Wood and Fels 2008; Wood 1992), structuring the world in a way that permits encoded data to be assessed. There was nothing about Snow's map that was special in this regard. Cities had been seeing themselves through the various layers of myriad maps at least since the first edition of Braun's and Hogenberg's *Civitates Orbis Terrarum*. Yes, Snow's map was a vehicle for the encoding of data on disease. Again, however, Snow was no innovator. He employed a long accepted, indeed almost a requisite methodology in his study. And in this case Snow's mapped topography argued not his own data, as did Lea's simple map, but the boon of a bureaucratic revolution that had transformed data on mortality and morbidity into a public resource. It was not just about Snow, not at all.

Quantifying Broad Street

If Snow had chosen to work, as Hempel says he did, in a methodological, step-by-step manner, how would his study have been different?[7] Within the science of the day, might Snow have been more convincing? There were two things that would have served. For both, data was available in the maps of the day and accessible to the science of Snow's age. First, he would have quantified what was, in fact, a not very rigorous visual argument. "It will be observed that the deaths either very much diminished, or ceases altogether, at every point where it becomes decidedly nearer to send to another pump than to the one in Broad Street," Snow wrote (Snow 1855a, 46). That was the heart of Snow's evidentiary argument. Observing this cluster the logical conclusion was, to him, inevitable: the pump was the source of the outbreak.

What, however, did the map prove? Snow could not count the number of deaths in the observed cluster in the map prepared for the second edition of *On the Mode of Communication of Cholera* (fig. 10.4), because its boundaries were unclear. Did Carnaby, King, or Marshall Street define its western boundary, for example? Where did the cluster begin and end to the south? The visual argument said too much and too little at once. This problem was solved in Snow's second map, included with his report for the parish inquiry, in which the irregular polygon based on walking distance created a subset of all cholera deaths based on proximity to the Broad Street pump (fig. 10.5). Surprisingly, Snow never counted the number of deaths within this Broad Street pump service area. Had he done so he would have been able to state that almost two-thirds of the 596 mapped deaths (63.9%) occurred within the Broad Street water service area. This would have been a stronger if not in itself a conclusive statement.

To prove his thesis to his contemporaries Snow would have had to show not simply that the majority of deaths were closer to the Broad Street pump but that mortality *by population*, a ratio Snow calculated across his other cholera studies from 1849 to his death in 1858, was greatest nearest the pump. After all, if the population of Broad Street were ten times that of, say, Carnaby Street, then one would expect, all other things being equal, Broad Street to have ten times the cholera. Without a population denominator the cluster Snow mapped around the Broad Street pump, while visually suggestive, proved nothing. Even Lea of Cincinnati knew this (fig. 7.12), including coarse mortality ratios in his study.

The material for the calculation of a population denominator was available. Whitehead's and Cooper's maps (figs. 10.7A and 10.6) both gave street addresses for every house in the study area bounded by Oxford Street to the north and Regent Street to the west. This permitted the population for each street segment in Snow's bounded study area to be calculated. In his 1852 table-map (fig. 9.6) Farr had given, on the basis of 1851 census enumerations, a population of ten persons per house for the St. James registration district. There were, therefore, two denominators readily available: for one ratio the number of cholera cases would be divided by the number of houses per street segment and for another cholera mortality first would be divided by the number of houses per street multiplied by ten persons per house.

It took me about an hour to count and then recount the number of houses on each street segment for Snow's Broad Street service area (fig. 10.5) using the data in Cooper's and Whitehead's maps as references In the Broad Street service area were 381 cholera deaths in 223 houses with a population of 2,550. From this a population mortality of 149.41 per 1,000 persons was easily calculated.

In itself, however, the ratio was meaningless because the real question was the *relative* mortality of the Broad Street pump service area compared to others in the study. Snow constructed his single service area based on local knowledge. He might, with a day or two to spare, have created similar distance-based catchments for other pumps. Because Snow did not, another way had to be found to create these service catchments. The simplest way was to create a set of Thiessen polygons with a ruler and pencil on photocopies of the maps. These are sometimes called Dirichlet tes-sellations after J.P. Lejeune Dirichlet, a German mathematics professor and contem-porary of Snow's (Dirichlet died in 1859) who first described how to create service areas around a specific set of objects in space, in this case the pumps.[8] Maybe Snow knew of his work, maybe not.

It took me an hour or two to manually create with pencil and ruler a set of pump catchments on a copy of Snow's map. The number of deaths within the Thiessen polygon was almost exactly the same as those counted in Snow's irregular polygon. It then took perhaps three days of off-and-on labor to count street segment populations and cholera mortality in the handmade catchments for the entire study area. Street by street this data was penciled into photocopies of Snow's map based on the ma-terial in Whitehead's and Cooper's maps so that the number of deaths in relation to the number of houses and their population could be calculated across the map. The manual counting was prone to error and it therefore took four tries for each section before consistent numbers were returned.

I later redid the work in a computer-based mapping program, ArcGIS 9.2. This required first spatially adjusting the maps of Cooper, Whitehead, and Snow in the software so they were precisely overlaid. It then took the better part of four days first to catalog in a single database the deaths, streets, sewers, and water pumps embed-ded in each of the three maps. Because of the similarity between Whitehead's and Cooper's maps, in figure 11.9 only Snow's and Whitehead's are shown. Rather than "stacking" deaths, the practice in the mid-nineteenth century, deaths at each house were aggregated in the GIS in a manner facilitating analysis. Once completed, creat-ing the Thiessen polygons centered on water sources was a trivial procedure, easily completed.

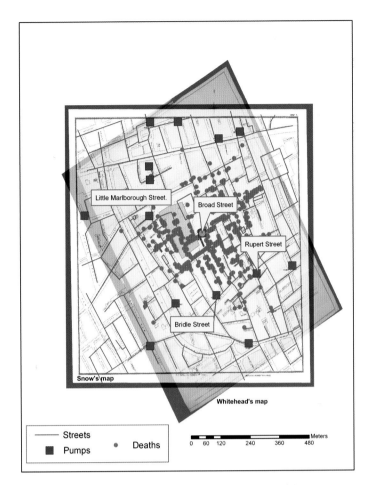

FIGURE 11.9   To study Snow's methodology and that of his contemporaries, their maps were layered in a geographic information systems program and "spatially adjusted" so all maps would overlap. In this environment it is possible to compare the maps and develop a mortality denominator in different pump catchments.

The result—by hand or by computer—repaid the effort. For demonstration purposes, only the principal water service areas adjacent to the Broad Street polygon are included in figure 11.10. This attenuated study area included almost 94 percent of all deaths mapped by Snow. Focusing on these service areas avoided certain technical problems, for example edge boundary concerns, and differences in the total study area of Cooper's and Whitehead's maps which were, unlike Snow's, bounded by Regent and Oxford streets. The results strongly supported Snow's thesis, with 28.71 deaths per 1,000 in the Rupert Street pump service area, 25.88 deaths per 1,000 in the Little Marlborough Street pump service area, and 12.34 deaths per 1,000 persons in the Warwick Street pump service area. Similarly, the number of deaths per house, a simple nineteenth-century measure of intensity, dropped as one moved outward from the Broad Street service area.

Similarly, intensity of deaths per house was higher in the Broad Street service area than in the other service areas. That is, the number of deaths per house was higher in the Broad Street water service area than in any other service area. Figure 11.10 provides a summary of the comparison of the pump service catchments in a manner that would certainly have been strongly suggestive to Snow's contemporaries.

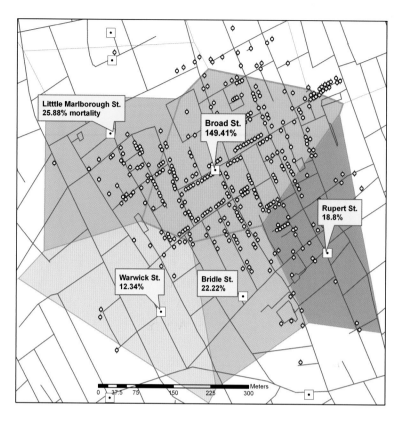

Litttle Marlborough St.
25.88% mortality

Broad St.
149.41%

Rupert St.
18.8%

Warwick St.
12.34%

Bridle St.
22.22%

Meters
0   37.5   75   150   225   300

Cholera mortality per 1,000 persons for central pump catchments.

Pump catchments

Bridle St.   Broad St.   Rupert St.

Little Marlborough St.   Warwick St.

· Pumps   o Cholera deaths

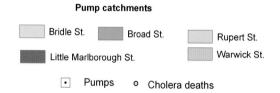

FIGURE 11.10 Mortality per 1,000 persons was calculated for the central water service catchments by taking the number of deaths per street segment in public water catchments, and then dividing in each by the number of houses multiplied by an estimated 10 persons per house.

This type of analysis, using basic addition and division, well within the realm of mid-nineteenth-century expertise, would have transformed Snow's observation ("it will be observed that the deaths . . ." [Snow 1855a, 47])—into acceptable, perhaps even convincing, evidence.[9]

This did not prove the source of the outbreak was the Broad Street pump, however, only that epicenter of the outbreak was at or near that damn pump. A general criticism of Snow's work was that "Snow had not eliminated other explanations or the role of coincidence" (Eyler 2001, 226). To claim he had proven the outbreak originated at the Broad Street pump would have meant disproving, with equal rigor, the likely engagement of the other proximate environmental sources others suspected. A quick look at Whitehead's map (Fig. 10.7A) shows the problem. The edge of the old plague burial site some thought complicit in the outbreak, breeding bad airs, was a short block from the Broad Street pump. As importantly, the new sewer lines built after 1850 (included in Cooper's, the Board of Health's and Whitehead's maps) that many believed to be a source of miasmatic transmission—the air was famously foul—ran under both Broad Street and the old plague burial site.

Snow dismissed both the old burial site and the sewer lines as possible sources of contagion, citing the Metropolitan Commission of Sewers study in which engineer Edmund Cooper inventoried the existing sewer lines and gratings in a map of 315 cholera deaths occurring between late August and mid-September (Cooper 1854). In his map Cooper symbolized the seventeenth-century plague burial site as a relatively small oval area distant from the outbreak's epicenter. Unfortunately for Snow, the former plague burial site, on which nineteenth-century housing had been built, was far larger than Cooper had asserted. Maps 10.1A and B confirm that the Board of Health's and Whitehead's maps correctly located the old burial site's borders within a block of the Broad Street pump.

Cholera deaths were strewn across housing built on the old burial site and along its streets that were served by the newer sewer lines that also ran under much of Broad Street itself. This much had been evident at least since Whitehead's first publication with its map of the parish and his description of the range of cholera in the area surrounding it. Again, an impartial look at any of the maps of this period make this clear (fig. 11.11A–B).

Within the constraints of the science of his day, Snow could have used the data mapped by Cooper and Whitehead to discount the likelihood these other sites of potential contagion were the source of the outbreak, however. To demonstrate this I first drew a ten-meter buffer around the burial site area mapped by Whitehead to allow for winds that some believed carried miasmatic odors, mentioned in popular reports, into neighboring streets.[10] To create a numerator I counted the number of deaths in the buffer around the old pest-field. For a denominator I counted the number of houses within the buffer area centered on the old pest-field. Multiplying that by a thousand gave a mortality ratio per a thousand persons.

For modern researchers the question of the sewer lines and the air spreading from them provides a more complex problem. For Snow and his contemporaries, however, it would have been sufficient to first identify the number of houses on streets with sewer lines built after 1850 and the number of deaths occurring in them. Again, deaths served as numerators and the number of houses was used to calculate a denominator returning a simple mortality ratio. This would have been feasible in Snow's era and indeed, well into the twentieth century.

Calculating the combined effect of both the old plague burial site *and* the post-1850 sewer lines together (if the pest-field airs were complicit and if they sewer pipes carried them, then both should be considered as a single source and vector) presents a series of technical problems in the analysis of overlapping but spatially noncommensurate areas beyond the science of Snow's day. For example, because all sewer lines, irrespective of age, were joined in an integrated disposal system,

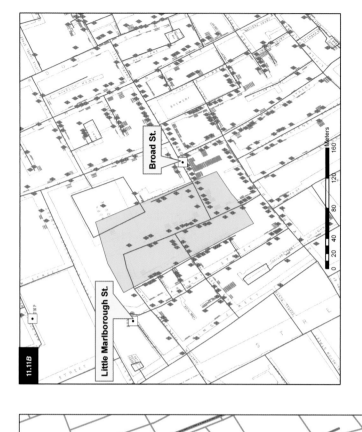

11.11A

Broad St. pump

Sewer lines by age

———	1830
———	1851
▪▪▪▪▪	1853
━━━	1854

◆ Water pumps

● Cholera deaths

☐ Old plague pit

▨ Buffer of old pest-field

11.11B

Little Marlborough St.

Broad St.

Meters
0 20 40   80   120   160

Cholera houses in the area of the old pest-field are mapped, with wells, on top of Snow's famous map. The proximity of the old pest-field, correctly located, to the Broad Street pump is made clear in this manner.

☐ Cemetery, correct location

⊡ Pumps and wells

◆ Houses with cholera

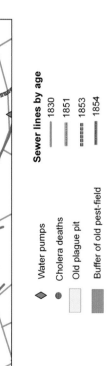

analyzing the effect of only post-1850 lines—or only those within Snow's irregular polygon—would have been inadequate. And because the then new sewer lines ran under Broad Street, the old plague burial site, and elsewhere in the greater study area, correctly analyzing their effect would have required a form of analysis not available to mid-nineteenth-century researchers. Simply averaging the cholera deaths per thousand persons, a likely recourse in that age, would not have returned an accurate assessment of their effect.

TABLE 11.1 Calculating mortality in Broad Street water service areas

Location	Cholera			Service area		Mortality per 1,000 persons (relative risk ratio [RR])	
	Number of deaths[*]	Number of houses[*]	Number of cases per house	Total number of houses	Population		
Broad St.	381	223	1.71	255	2,550	149.41	7.99
Rupert St.	38	60	0.63	209	2,090	18.18	0.33
Little Marlborough St.	59	43	1.37	228	2,280	25.88	0.49
Bridle St.	30	28	1.07	135	1,350	22.22	0.43
Warwick St.	19	16	1.19	154	1,540	12.34	0.23
Marlborough Mews	7	5	1.40	48	480	14.58	0.30
Berniers St.	13	14	0.93	115	1,150	11.3	0.22
Newman St.	21	14	1.50	N/A	N/A		
Total cholera deaths in principal water service areas	547						
Total deaths[*]	596			1,922	19,220	31.01	
Cemetery buffer	113	57	1.98	108	1,080	104.63	2.50
Sewer buffer (all lines built after 1850)	184	118	1.60	541	5,410	34.01	0.56
Post-1850 sewer lines in Broad St. service area only	139	73	1.90	219	2,190	63.47	1.51

[*]According to Snow's findings.

FIGURE 11.11A (*opposite page*) In this map, the proximity not only of the old pest-field but also of foul smelling post-1850 sewer lines to the epicenter of the outbreak can clearly be seen. It was this that convinced some of Snow's contemporaries the sewers were a possible source of disease origin, if not also diffusion.

FIGURE 11.11B (*opposite page*) In this detail of the working maps the proximity of the old pest-field to the Broad Street well and pump is clearly expressed. Given the then-dominant theory of airborne disease, ignoring this proximity in the analysis was a serious error.

Even had Snow and his contemporaries been aware of these problems, the simple mortality ratios returned from the separate analysis of relative mortality in the pump catchments, the old pest-field buffer, and streets with new sewer lines would have been sufficient. Table 11.1 fairly definitively proves, within the context of mid-nineteenth-century science, the overwhelming likelihood that the origin of the Broad Street outbreak was in the service area centered on the Broad Street pump. Mortality per one thousand persons was far greater in this service area than in any other in the study area. It was also far greater than mortality per one thousand persons calculated for either the post-1850 sewer lines or the old plague burial site. While lower than that of the Broad Street service area, mortality on the old plague burial site was high.

Modern researchers would add to this analysis a set of risk ratios to better assess the relative risk of both the Broad Street pump in relation to others in the study area as well as to other possible sources. These ratios are calculated in a two-by-two contingency table in which the numerators are cholera deaths in a pump service area (Broad Street) and the study area's cholera deaths minus those in the specific water service area. Denominators are the population in the local pump area and in the total study area, minus the local population in the other denominator. It is easier to see than describe:

$$\frac{\text{Cholera deaths in Broad St. pump area}}{\text{Broad Street pump area population}} \qquad \frac{\text{Other study area deaths (non–Broad St.)}}{\text{General study area population} - \text{Broad Street's}}$$

The same type of argument was used to calculate relative risk for the old plague burial site with its buffer, comparing mortality in a specific population to the general mortality from cholera (minus the plague burial site). This was not a methodology or language familiar to Snow or his contemporaries. For them, if not for us, however, the result using simple mortality ratios would have been far more convincing than the approach Snow used. It would have transformed Snow's visual assertion into a statistical argument, based in the map, which promoted the Broad Street pump to the most likely source of the outbreak. Critics might have quibbled about the size of his buffers, or other technical matters, but the difference in mortality was so marked that even without modern statistical tools the results likely would have been (again, within the science of the day) as near definitive as was possible.

The question becomes: why didn't Snow take an extra few days to make his argument in this manner? Why not at least *count* the cases and say, "Look! Two-thirds of all cases reported to the GRO were nearest to the Broad Street Pump." The answer is . . . we don't know. One may read Snow's failure to quantify his Broad Street findings as arrogance, a refusal to speak in the language of the contemporaries he knew were the real judges of his theory (Snow 1849a, 31). A more generous interpretation is that Snow simply didn't believe the additional work was necessary. The map was clear to him, its logic so compelling he believed he needed do no more. Snow *knew* he was right and, not perceiving the necessity, he was too busy to do the extra work.

In addition, Snow was compelled, perhaps, by a sense of urgency to publish as soon as possible. People were dying and it was imperative to nail down the source of outbreaks to prevent the transmission of this disease. Logic told Snow that cholera had to be waterborne and if the data was incomplete . . . so be it. Let others do the detail work; he was obliged to give his theory. Even if understandable, the result was unfortunate. Snow rushed and in his hurry chose not to carefully consider the concerns of other researchers. He knew he was right and was . . . impatient. This virtually assured his theory would not be accepted by contemporaries, who were properly cautious in adopting a new disease theory based on limited case studies without even basic mortality ratios. And really . . . carefully considering Snow's data in the absence of our modern knowledge we, too, would have been unconvinced. Who among today's researchers would accept as proof of a new, eccentric theory of disease a neighborhood study in which not even mortality ratios were quantified?

In the end, everyone was right. William Farr demonstrated there was an inverse relationship between cholera mortality and altitude. In 1961 a bacteriologist rescued Farr's study from obscurity, confirming the numerical relationship Farr observed and lauding Farr's study as the first epidemiological model of air quality (Langmuir 1961). And, of course, Snow was right. Cholera was a waterborne disease. But John Simon was correct as well. Without precise knowledge of the disease agent even the best statistical analysis could not definitively assign cholera to air or water exclusively.

And also, in the end, everyone was wrong: Farr's to us fanciful cholera, created through the agency of local water dispersed in the air, was just fantasy despite the evidence amassed to prove it. Snow was dead wrong in his assertion that cholera exemplified a broad class of conditions that included plague and yellow fever. Simon's general rejection of Snow's waterborne cholera was, as he would admit decades later, based on traditional disease theories that were in the process of dying themselves (Simon 1890).

The exemplar in the final analysis was not Snow but cholera and the methods of study that developed around its pandemic visitations. Cholera exemplified disease as a thing of social and political as well as clinical concern. Its investigation was to be promoted at every scale through the evolving matrix of resources—civil, medical, and social—that into the twentieth century came to define disease studies. At every scale that matrix linked the bureaucracies of nations responsible for the collection of local, metropolitan, regional, and national health data, promoting disease studies as a communal activity and responsibility. At the same time, the mass of available data demanded its specific subjects be argued not as a collection of individual cases on a large scale map but as a statistical thing embedded in the map. Whatever its nature and symptoms, the disease thing was to be understood experimentally as well as inferentially. The elements of Snow's method of analysis did not disappear any more than cholera or the dot map disappeared (CDC 2002); they were simply incorporated into the matrix by which old diseases had come to be understood and new ones investigated.

PART III

THE LEGACY
AND ITS FUTURE

CHAPTER 12

CANCER AS CHOLERA

The story was never really about cholera, or about John Snow. They were just coordinates anchoring ideas of health and disease seen as part of a history we give to students and tell ourselves. We want it to be simple—a straightforward something we can see and therefore know, and in the knowing, control. Disease is rarely simple, however. Almost always it conflates the environmental, the genetic, and the social in its evolution and subsequent diffusion. The story's central character could have been malaria, plague, typhoid fever, yellow fever, smallpox—the only disease modern science has eliminated (Henderson 1999)—or any other condition whose virulence has long tasked our mortality. Whatever the subject, in the telling we create from symptoms we see an object, "x is . . . this," understood through a methodology of knowing across our scales of experience (personal, communal, national, and international) that is generally applicable, outbreak by outbreak, study by study, into the future. What we learn is grounded in what we think we know, the simplified past offering a method of assembling the uncertain present in the hope its tomorrow will be somehow manageable. It was the memory of plague that powered the approach of the researchers who first encountered yellow fever. The methodologies of yellow fever's knowing informed the cholera studies that came after. All this, and new laboratory techniques, created the potential for Alice Hutchison's typhoid fever, and more generally, the early-twentieth-century work of W. H. Frost.

From the work of Arrieta in the seventeenth century through the current search for understanding of diseases today, the pathway of knowing has been the same. This or that congress of symptoms observed in the patient, named and placed within a classificatory system, is understood within a general theory of disease whose limits are tested in specific outbreaks. Studies of this or that disease employ a matrix of materials (physical and intellectual) assembled by researchers to provide insight into the new health threat while deepening understandings of disease in general. The pat-

245

terns of physical change and of worldly occurrence may be coarsely constructed (the buboes of plague, the simple number of cholera decedents on Broad Street), or more finely conceived (the level of CD4 and T4 cells in an HIV-positive patient's blood, a study of West Nile virus diffusion in age-stratified counties of the United States). Either way it is seen, recorded, and analyzed.

This pattern of knowledge assemblage and construction is everywhere in disease studies. Consider, for example, the history of cancer studies over the last 150 years.[1] From the 1860s to the present cancer was transformed from an unfortunate individual occurrence into a public epidemic against which we "war" mightily (Davis 2007). The story of cancer as a public health concern repeats the steps familiarly found across this book's history, from plague to yellow fever to cholera and typhoid. It is the same tune in another key, one whose ending is yet to be written.

ALFRED HAVILAND

In Graunt's 1662 *Table of Casualties* (fig 4.8) cancer was grouped among several miscellaneous causes of death (including gangrene and fistulas) that shared no common symptomology. Its transformation into a public health concern began on November 30, 1868, with a lecture by Dr. Alfred Haviland to the Medical Society of London. His study of increasing cancer rates, based upon General Register Office mortality data, was reported in the *Standard*, making cancer rates a public as well as professional subject of concern. To make his case Haviland had to argue the specificity of cancer as a deadly thing seen in maps that showed pockets of high cancer mortality rates in some areas of Great Britain and not in others (Haviland 1869, 9). In this first presentation, cancer as a public disease shared equal billing with heart disease, another object of Haviland's research. Benjamin Richardson, John Snow's old friend, described Haviland's work in a self-produced pamphlet.

> Having given a general view of the climatological relations in England and Wales of Heart Disease, the author directed the attention of the Society to the large map of Cancer. He said, on looking attentively at this chart, the first great fact that drew attention was, that nearly the whole of Wales and the north-west portion of England was colored red, which indicated a low amount of mortality from cancer over this larger area; coincided with this was the fact that these parts of England and Wales belong, geologically, to the oldest formations, namely the Silurian and carboniferous, and physically included the highest and best drained mountainous districts in the country. (Haviland 1869, 10)

Several years later, Haviland collected his notes and additional research to produce a book on cancer as a general health problem.

Like the yellow fever pioneers (Pascalis and Seaman, for example) and the cholera researchers (Farr and, of course, Snow), Haviland first made his argument to a professional association and followed it with a privately published pamphlet. Like them,

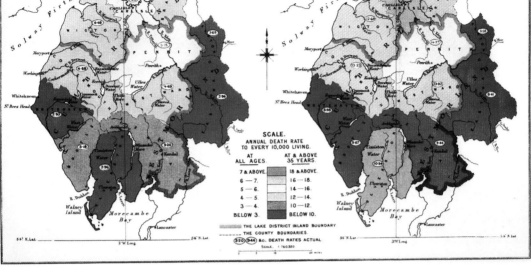

MAPS OF THE GEOGRAPHICAL DISTRIBUTION OF CANCER (FEMALES).
IN THE ENGLISH LAKE DISTRICT,
CUMBERLAND AND WESTMORLAND,
1851———1870.
BY ALFRED HAVILAND, M.R.C.S.E., &c.

FIGURE 12.1 Alfred Haviland mapped the geography of cancer in an attempt to first locate the areas with a high cancer rate and then prepare an argument based on environmental determinants of the disease.

he used official mortality data to create a public health crisis. And again, like Farr and Snow, Haviland then published his findings in a book in which mapping was a critical analytic. The idea was not simply to post cancer in the map but to map cancer incidence in relation to environmental conditions "perchance some light might be thrown upon the etiology of that fatal class of malignant diseases, registered as causes of death under the term cancer, if the available statistics, as published by Dr. William Farr, C.B., were to be treated on the same geographical principals as had been demonstrated in the cases of phthisis and heart disease" (Haviland 1875, 100).

Just as Farr's table-map of cholera in metropolitan London had proposed an inverse relationship between cholera incidence and altitude above sea level (fig. 9.8), so too did Haviland's map argue an inverse relationship between altitude and cancer incidence. "What may be called the cancer fields of England," Richardson wrote in his summary of Haviland's mapped argument, were in the low-lying river districts of industrial areas like London the Thames. The result began the transition of cancer from a clinical diagnosis into a statistical conclusion based on mortality ratios: "If we take the extremes of mortality indicated in the divisional map, we shall be enabled to see what physical, geological, climatic, or social characters are coincident with the high

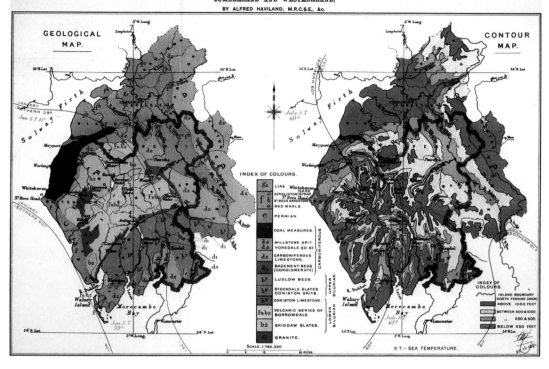

THE GEOGRAPHICAL DISTRIBUTION OF DISEASES:
MAPS OF THE GEOLOGY AND CONFIGURATION OF THE ENGLISH LAKE DISTRICT,
CUMBERLAND AND WESTMORLAND,
BY ALFRED HAVILAND, M.R.C.S.E., &c.

FIGURE 12.2 General mortality in Cumberland and Westmorland registration districts could be investigated in terms of specific diseases and local geographies that created an environment promoting specific diseases.

and low mortality in eastern and western [population] groups" (Haviland 1875, 69).

Haviland's work sparked a number of studies that made cancer a national and then international object of medical concern. In the second Thomas Morton Lecture on cancer and cancerous disease in 1888, Sir Spenser Wells used GRO records to demonstrate that, in England, Ireland, and Wales, "[t]he increase in the number of deaths from cancer is now, and has been for many years past, greater than the proportional increase in population" (Wells 1888, 1201). It soon became clear that the incidence of cancer was increasing not only in England, however, but also elsewhere in the Commonwealth. As a result, cancer became a new, demographic reality that was pandemic. As Dr. Gordon McDonald wrote in a review of New Zealand's cancer statistics, for example, "The mortality from cancer exactly doubled itself in ten years, although the population had only increased by about a fifth" (McDonald 1890, 252).

In effect, McDonald argued, the increase in cancer mortality in New Zealand's districts was a problem identical to that of England where, "[t]he cancer coloring [in maps] is dotted from end to end of the British Isles. It does not follow the course of great rivers or the spread of any particular geological formation but is found, more or less, in all localities" (McDonald 1890, 252). In effect, cancer had become a cartography whose demographic reality was identified by its mapped coloration. Just as the *Lancet* authors in 1831 had argued the mapped ubiquity of cholera (fig6.4*A–B*) as an argument against an environmental agent, McDonald argued that cancer was sufficiently prevalent that Haviland's theory of environmental determinism was probably incorrect. Why its incidence was increasing was the question. "The time has arrived when the cause of the disease itself must be thoroughly investigated," Haviland wrote, "and in relation to the soil and the atmosphere ascertained" (Haviland 1892, viii).

Haviland's cancer was miasmatic and localized in a manner physicians from Hippocrates through Sydenham (and Snow, in the end) would have appreciated. There was no biological model that explained either the nature of tumor growth and little attempt at this time to categorize specific cancers (lip, penis, stomach, tongue, vulva, and so forth) as etiologically distinct conditions. It was all just . . . cancer, whatever its biology and wherever on the body it might occur. In 1898 Martinus Beijerinck advanced the idea of a *contagium vivum fluidum*, an unseen kind of bacteria, perhaps, while others argued with an equal lack of evidence that cancers might result from parasitic infections.

In the first decades of the twentieth century, newly rigorous methods of microbiol-

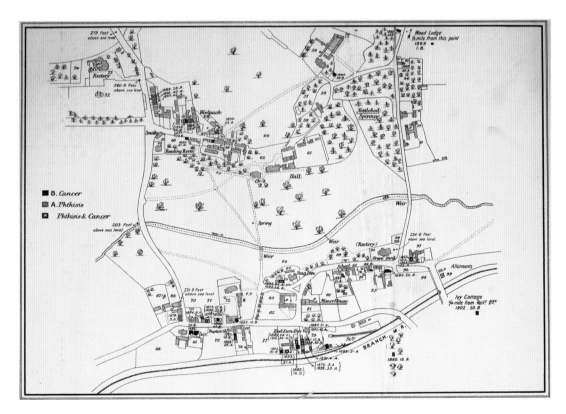

FIGURE 12.3A Dr. D'Arcy Power mapped cancer occurrence for the years 1872–1888 in a British village of 1,036 persons to demonstrate the uneven incidence of cancer cases and to identify the location of "cancer streets" and "cancer houses."

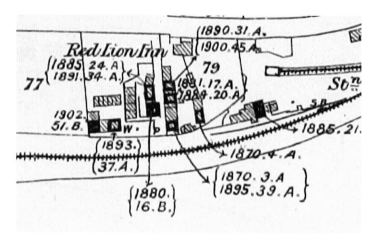

FIGURE 12.3B Power created a cancer topography in which patients were identified by number and the year in which they died. The result argued over decades the existence of cancer clusters on specific streets. The origins of the disease appeared to be bio-geographic rather than an outcome of social class or status.

ogy did not serve to uncover an invisible agent of this disease. That, we now know, would require the advances of twentieth-century electronic microscopy. Research therefore continued in the time-honored method of mapping public health data against suspected environmental causes. As D'Arcy Power put it in a 1902 study, "Like many other medical men I am interested in the increase of cancer which has undoubtedly taken place within the last fifty years, and I have made many attempts to discover the cause. I thought, at first, that the question might be solved in the laboratory, and I spent much time in examining the various appearances which were put forward so confidently as cancer parasites" (Power 1903, 697). Alas, none were found.

For Power and other physicians the map was the workbench on which a cancer *in situ* might be explored in areas of local medical practice. Regional and national cancer maps of mortality data awakened the nation to the idea of cancer as an increasingly serious statistical problem. Studies seeking to determine its precise causes were carried out, however, at the level of the local physician's practice. Just as Seaman had sought to understand yellow fever by mapping the neighborhood of a violent outbreak in New York City in 1793 (figs. 5.6 and fig. 5.7), and Lea had considered cholera on his street in Cincinnati (fig. 7.12), Power and his contemporaries mapped cancer's occurrence street by street and house by house in their practice areas. "Although the actual numbers are not very imposing in this series of cases, a glance at the maps will show the remarkable manner in which the disease is distributed" (Power 1899, 420).

In one map series, not shown here, Power first used 25-inch ordinance survey maps to locate the exact location of cancer cases in a village of 1,036 inhabitants during the years 1872–1888. Finding clusters of recurrence on individual streets and in some homes, he then attempted to determine if those places of high incidence could be distinguished on the basis of patient age, microgeography, through hereditary patterns (did parents or siblings have the disease?) by income (high or low), and other variables. In another map (fig 12.3A) the village itself was presented, transformed from the ordinance maps, as a small collection of buildings on streets that paralleled the railway. Trees symbolized agricultural areas in a similar if more sophisticated version of Arrieta's seventeenth century symbology (figs. 4.4, 4.5A–B). Among the environmental determinants consider was altitude above sea level, as it had been in Farr's (fig. 9.7 A) and Shapter's (fig. 8.6A) cholera maps.

Within Power's community the homes of patients who died of cancer or phthisis (consumption, wasting) were identified. These were his patients', and those of his medical colleagues, cancers that they saw and attempted to treat. And so, in the text, he provides a brief synopsis of the cases just as, a half century before, cholera researchers like Snow included the particulars of the cases they mapped. For each home in which a cancer occurred Power lodged in the map the dates of reported deaths. "Cancer houses," those where deaths from cancer appeared repeatedly, were obvious at a glance, as were the "cancer streets" where the disease appeared to concentrate just as cholera had appeared to cluster in Snow's map of Broad Street (fig. 10.4). Unlike at Broad Street, no simple explanation, no single source of disease, was obvious in the local cancer maps, however. "It seemed to cling to certain spots and groups of buildings irrespective of their size and age; it has attacked the hall and manor-house as well as the cottage of the humblest laborer. It has occurred in new buildings of brick as well as in the old thatched homestead" (Power 1899, 420). Seeing cancer in the maps Power could see it's clustering but no clear rational for its pattern of occurrence.

Failing any clear correlation between heredity (cancer as family history) or socioeconomic variables (cancer as a result of poverty) or a pattern of occurrence suggesting interpersonal

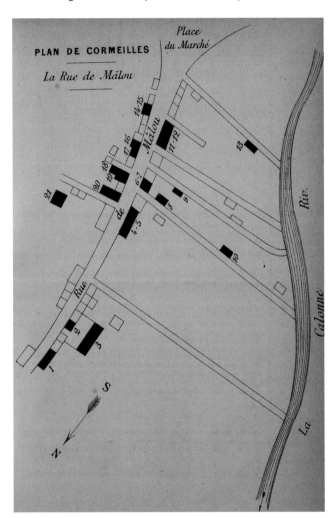

FIGURE 12.4 Tables of mortality showed a higher incidence of cancer in rural towns like Cormeilles than in larger cities like Rouen. Arnaudet mapped a decade of cancer cases in Cormeilles in an attempt to uncover local peculiarities that might explain that greater incidence. Numbers at cancer homes (black squares) were keyed to a table of cases in his paper.

transmission, the logical conclusion for Power was geographic and environmental: "It is impossible, I think, to resist the conclusion that the marshy grounds from which the rivers arise in this district, as well as their wooded banks, have some causal connection with the numerous cases of cancer which have been observed in their immediate vicinity. This part of the country must at one time have been preeminently malarious . . . It is not improbable therefore that the hypothetical cause of cancer might have increased facilities for its work and growth in such surrounding as those which have been described" (Power 1903, 702).

Scores of late-nineteenth- and early-twentieth-century researchers similarly considered cancer rates at the fine scale of the town or village, seeking in most cases an equally localized environmental explanation. They searched for waterborne contagion, airborne pollutants, and some form of insect or parasite that somehow might be the source of cancer in local homes. In Normandy, France, for example, the physician Arnaudet reported in 1890 that cancer rates for the years 1883–1893 were far higher in rural towns like Cormeilles (10.47 percent) than in large towns like Rouen (3.8 percent) or Le Havre (2.4 percent) (Arnaudet 1890).

Given the surprisingly high incidence in the supposedly healthier rural areas, Arnaudet focused on the incidence of cancer in Cormeilles. Numbers attached to each house in his map of cancer referred to a table in which cases were identified by patient age, cancer location on the body, and so forth. The excessive mortality in the area, and in the individual cancer streets themselves, argued for him, as it had in England for Power, a disease whose origins lay in specific geologies, airs, and waters. The high number of gastric cancers noted (*cancer du tube digestif*) suggested a subset of cancers resulting from local diet, he concluded, perhaps the inordinate use of local ciders rather than presumably more stomach-friendly wines. Joseph-Francois Malgaigne had posited a similar explanation for variations in hernias among French military recruits in the 1830s, mapping his argument using military recruitment statistics (Malgaigne 1840).[2]

As a result of this class of studies, cancer became a subject of international attention around which clinical, cartographic, and statistical interests focused at three distinct scales of investigation (and resolution). These included neighborhood and local studies, regional studies, and national studies summarizing local and region work. In a remarkable study bringing together clinical, cartographic, and statistical approaches, Charles Edward Green published a review of "the cancer problem" in 1917. "Are there cancer houses?" Green asked in his paper. "Personally I believe they are in spite of the fact that it has been argued on the theory of probabilities that such houses do not exist" (Green 1917, 5). Here was a problem of statistical scale in demographic studies: what appeared clear in the large-scale, local map was not necessarily statistically (or visually) obvious when incidence was calculated at the scale of the region or nation. There was, of course, the problem of scale that stymied John Simon in his 1856 study of cholera in South London.

FIGURE 12.5 Charles E. Green mapped cancer mortality among British registration district populations to demonstrate statistically significant variation in disease intensity in areas with different environmental properties.

Because the number of cases considered in neighborhood studies was so small, Green argued (as had Simon in the 1850s), maps and studies at this scale could only be suggestive, never certain, in their conclusions. Another problem was that even if residents of a house or street showed a higher incidence of cancer than did others in a community, the problem was not necessarily the house but might be the biogeographic environment in which it was situated. "The question does not lie, to my mind, so much with the house as with its surroundings," wrote Green, by which he meant the general geology of the area's substructure and the nature of the chimneys that poured out smoke inhaled by the citizenry (Green 1917, 6).

Having dispensed with the extremely local, Green built his study by calculating cancer mortality in the registration districts of London, the same field of study worked by William Farr and later by John Simon and John Snow. At this scale there were statistically significant variations: "The amazing fact that cancer accounts for one death in every 5.88 deaths from all causes in St. Marylebone while it only accounts for 1 death in 54.10 from all causes in Stepney. In other words, cancer is 10 times as common in St. Marylebone as it is in Stepney" (Green 1917, 7). In effect, he argued, cancer was to be seen as a statistical reality lodged in regional districts rather than a simple, medical reality embedded simply in the neighborhood physician's practice. What to make of these anomalies was unclear, however. For a broader perspective Green then used GRO data to calculate mortality ratios across England's registration districts in a manner similar to but more mathematically rigorous than the statistics employed in

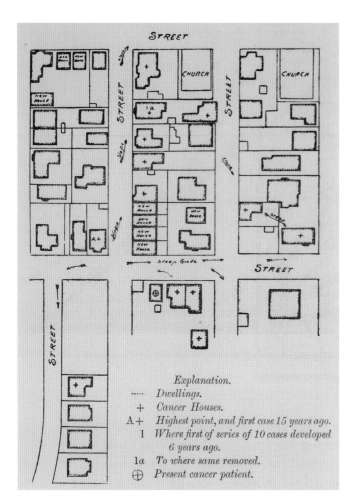

Explanation.
...... Dwellings.
+ Cancer Houses.
A+ Highest point, and first case 15 years ago.
1 Where first of series of 10 cases developed 6 years ago.
1a To where same removed.
⊕ Present cancer patient.

FIGURE 12.6 In mapping a neighborhood in which cancer was prevalent Green's map attempted to show a path of infections that over a decade carried the disease from house to house.

the 1890s by Haviland. Like Haviland (and earlier, the *Lancet* editors surveying cholera literature in 1831), however, Green could find no clear environmental factor that said "cancer" and therefore used that coarse, national scale to identify areas of unusual local and regional incidence where causal factors might be identified.

Just as cancer was statistically far more prevalent in the St. Marylebone registration district than in Stepney in Greater London, the tables of cancer statistics by registration district uncovered other registration districts with higher cancer rates than their neighbor's. Registrars in the town of Masham reported only two cancer deaths in ten years, for example, while in Leyburn fifteen deaths were reported and in Middleham there were in that decade twelve deaths (fig. 12.5). Geographically, Masham was built around a flat square market place and was generally level, while Leyburn and Middleham were built on slopes with various gradients (Green 1917, 19). As Farr himself might have noted, there was something about altitude that appeared to increase the incidence of cancer in this registration district's population, with cancer incidence greater in areas with greater slope.

With cases occurring and reoccurring in specific houses and on specific streets,

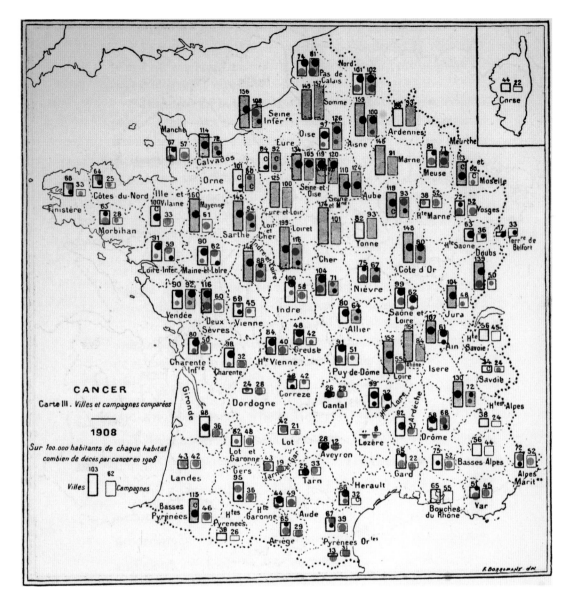

FIGURE 12.7 Green mapped different types of air pollution and relative cancer incidence in French districts using national cancer statistics. Red dots symbolize wood-burning *villes* and *campagnes*; black dots symbolize those that burn coal, and a "c" shows which burn charcoal.

it was hard to deny the likelihood that cancer might somehow be passed from person to person or result from some local contagion. To demonstrate this, Green modified a map reproduced from another's publication to show a "cancer district" at the resolution of the house and street scale (fig. 12.6; Green 1917, 23). The black-and-white map presented a contained neighborhood of five streets and six city blocks populated with houses and other buildings. In this environment those houses in which one or more cases of cancer had been reported over the previous ten to fifteen years were all symbolized by a "+." The home in which the first diagnosed patient lived was distinguished by a "1a" above to the plus sign. The highest geographical point in the neighborhood was indicated by an A+. The first case occurring within the previous decade, the "index case" is distinguished by a "1."

All this created a rather sophisticated set of cancer homes distinguished temporally (fifteen years ago, ten years ago, current), and geographically (by altitude). They were related one to another through a set of arrows attempting to show the relation between cases over time on the basis of proximity and presumed social interaction. The conclusion of was that cancer was a contagious condition that can pass from person to person and place to place over time in a potentially predictable manner.

While the idea of cancer as a contagious or an infectious disease seems preposterous to us that is only because our knowledge of the disease is more complete than was Green's. Lacking the ability to *see* a cancer agent, the tools Green had were cartographic and inferential, those employed by Snow and his predecessors. In the mapping, Green sought apparent clusters, identifying areas of intense occurrence in which he and others might test theories of disease transmission and process (such as contagious, environmentally based, infectious, and hereditary).

Green's theory was that cancer resulted from an airborne contagion whose relative rates of occurrence reflected different atmospheric pollutants. Here was an argument Farr would have appreciated. "If coal soot or some of its ingredients can cause epithelioma in chimney-sweeps, as is proved by the figures of the registrar general, or coal tar bring it on in coal tar workers, as is proved by the Home Office having drafted special regulations for them, is it not reasonable to suggest that the inhabitants of a district where the products of coal combustion are imperfectly removed should be more subject to cancer than dwellers in a district where these are more perfectly removed"" (Green 1917, 26). He attempted to prove this using French cancer mortality data collected at the scale of towns (*villes*) and districts (*campagnes*) in France's larger health *départements.*

This cancer (fig. 12.7), lodged in the map, was pathological at base. Its mode of transmission was atmospheric and therefore active at varying scales of address. To demonstrate this Green correlated cancer incidence in local health districts with the use of one or another cooking and heating fuel: charcoal, coal, or wood. The assumption was that different heating materials created different atmospheric pollutants that might have a different effect on cancer generation and growth in the nation.

Because cancer was everywhere, and everywhere at variable rates, Green sought to identify an airborne environmental agent that was similarly ubiquitous. The result was a complex series of propositions that defined the nation at the scale of the French bureaucratic *départements* in which Green proposed three possible fuel sources in each *ville* and *campagne*. In the map the result were rectangles sized to reflect cancer rates in each jurisdiction. Above each rectangle was the number of cancers per hundred thousand persons for each area. Black dots were used to symbolize coal-burning areas; red dots symbolized wood- and a "c" symbolized charcoal-burning areas.

This work could only occur within the broad view of cancer as a national disease whose incidence was not equal but varied, city to city and reportage district to reportage district. It was within that context that the specific relations between cancer and potential environmental triggers took place. One sees this not only in the map attempting to correlate air pollutants and cancer but more generally, but also in a dot map of cancer in France in the costal region near Nairn. Figure 12.8 shows a detail from a large map included in Green's report, attempting to relate cancer incidence to local environmental contaminants is yet to come. What the map does, however, is significant. It presents cancer as a reality to be detailed spatially within the towns and districts of the Nairn region. In it the roads, fields, and villages of the region are presented within the political jurisdictions of the local area. The sharp cluster of red dots near the city, and their distribution in the region, make clear where the focus of future studies must be.

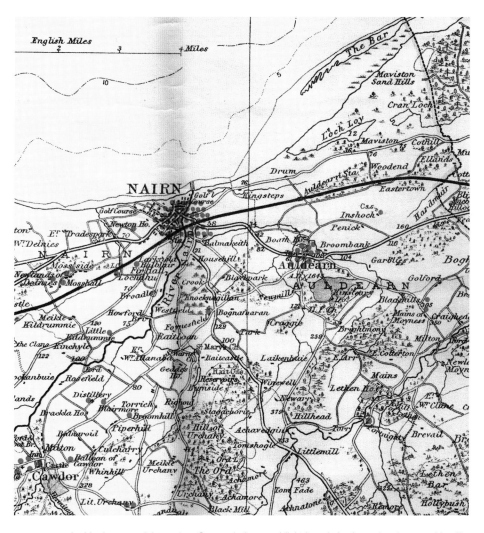

FIGURE 12.8 In this dot map of the region, Green tried to establish the relative intensity of cancer, identifying a "hot spot" in the area of the town of Nairn. This served both to assert the varying intensity of cancer in the population and identify areas of special research interest.

CANCERS AS A STATISTICAL EVENT

Green's study did not reveal cancer as the result of an airborne pathogenic agent. Nor did any of the other then-contemporary studies shed real light on the nature of the epidemic disease. One problem was that "cancer" was too big a term. Nobody knew but many suspected that the cancers on different parts of the body might have different histories and different explanations. There was, for example, the cancer of the chimney sweep and the cancer of the coal worker: Might other cancers be similarly located in a single profession? Was it even appropriate to talk of cancer as a single disease? A related question concerned the populations within which cancer should be considered. Obviously, cancer of the penis was male and cancer of the vulva was female. Their populations were distinct. Might it be necessary to interrogate each cancer in terms of distinct populations characterized by age, occupation, sex, and perhaps socioeconomic status?

In 1915, cancer was transformed from a health into an economic problem, a pandemic with actuarial importance. The Prudential Insurance Company published in that year *The Mortality from Cancer through the World*, a statistical tour de force that culled cancer mortality data from health agencies around the world to create an international statistical portrait of cancer (Hoffman 1915). In this transformation of clinical observation into economic and statistical realities, local cancer studies by physicians like Power and Arnaudet fell into relative disuse. Because the samples of the local medical practices were too small to be statistically significant the information returned at the fine scale of the local was assumed to be suspect. And really, this type of study was not interested with the care of cancer patients, or the nature of cancer itself, only with its incidence and the likely actuarial costs of its presence in society.

And yet, as Dutch researchers of the 1920s noted, clustering of cases at the scale of local medical practices was a statistical reality easily seen in the mapping, albeit a reality that raised questions without answers. "At the moment when we make a causal relationship between living in a particular spot and the incidence of cancer in the homes, that's when the big question mark arises" (Bakker, Van Dam, and Bonne 1926).[3] Similarly, regional variations in cancer rates were in Holland, as elsewhere, a statisitcal reality evident in maps (de Jong 1926). In the absence of a rigorous biological model of cancer, the *only* real method of disease investigation involved mapping individual cases at different resolutions and varying scales to study geographical differences and commonalities within distinct populations (Bakker, Van Dam, and Bonne 1926).

The actuarial and clinical were combined in the 1920s by physician-bureaucrat Dr. Percy Stocks, the head of statistics at the GRO in London. The inheritor of William Farr's old office, Stocks used national mortality data to report upon the diseases that plagued the nation. His principal passion was not cancer, although it was a disease on which his analysis focused, but the methods by which a statistical relationship could be described and then applied to all national health data collected by local registrars in Britain or elsewhere (Stocks 1924).

CANCER OF UTERUS, VAGINA, OVARIES, BREAST, SKIN, LUNG, STOMACH, INTESTINES, RECTUM, BONES and ALL SITES in COUNTIES (including the large towns). Actual deaths of FEMALES in 1921-30 per cent. of the number calculated from the distribution of the female population by age and class of district.

	Uterus		Vagina, Vulva	Ovaries, Fallopian Tubes	Breast Females	Skin Females	Lung Females	Stomach Females	Intestines Females	Rectum Females	Bones Females	All Sites Females	
	25–65	65 up	25 up	25 up	25 up	25 up	25 up	25 up	25 up	25 up	25 up	25–65	65 up
England and Wales ...	100	100	100	100	100	100	100	100	100	100	100	100	100
ENGLISH COUNTIES:—													
Bedford	(94)	(94)	(82)	(98)	109	72	(92)	112	116	(107)	(78)	94	(100)
Berkshire	91	86	79	137	110	66	(114)	83	(96)	123	(96)	91	103
Buckingham ...	78	111	(115)	(95)	106	(91)	167	74	(102)	(101)	(90)	87	(103)
Cambridge	88	131	147	(106)	112	(83)	138	77	112	(110)	(73)	(102)	(102)
Cheshire	89	(99)	122	(94)	(100)	120	75	117	107	93	(105)	103	110
Cornwall	125	111	78	88	94	116	(87)	120	94	(98)	(114)	108	(100)
Cumberland ...	(106)	(100)	(87)	65	88	(110)	67	118	128	(96)	(100)	107	(99)
Derby	(103)	(103)	127	(106)	(100)	77	(98)	(99)	93	(99)	(106)	(100)	97
Devon	(102)	(101)	(102)	116	(101)	(93)	(113)	88	84	87	(108)	97	96
Dorset	91	78	(108)	(108)	(97)	119	(80)	87	(97)	(97)	(110)	94	97
Durham	133	115	(98)	70	77	122	59	129	(102)	86	89	105	(99)
Ely, Isle of ...	141	137	171	67	(96)	(82)	(140)	141	123	(89)	(120)	116	115
Essex	91	(104)	85	109	105	85	118	94	104	105	111	(100)	(101)
Gloucester ...	82	(96)	(107)	125	112	70	(93)	80	96	(103)	80	95	96
Hereford ...	(96)	(104)	(93)	133	111	(111)	(114)	88	84	(98)	(110)	(103)	(97)
Hertford	72	(91)	58	(104)	105	122	78	94	106	(95)	69	90	103
Huntingdon ...	(115)	(91)	(86)	136	(98)	200	25	(98)	140	(86)	(60)	(101)	114
Kent	104	(101)	(103)	117	110	(96)	(104)	80	(98)	108	(91)	98	(101)
Lancashire	97	93	(96)	84	97	124	93	117	(99)	86	(100)	(101)	(100)
Leicester	92	110	131	(96)	110	(98)	(86)	(99)	108	107	(87)	(100)	105
Lincoln, Holland ...	122	135	162	(105)	(100)	(92)	(60)	128	120	(108)	(100)	(100)	120
*Lincoln, Kesteven	(96)	(91)	183	119	(94)	65	(100)	86	(100)	123	56	94	(101)
Lincoln, Lindsey ...	131	120	125	117	(100)	(108)	(104)	86	94	(106)	(87)	107	(101)
‡London	88	93	86	124	104	71	148	91	96	114	(98)	98	99
Middlesex	96	108	(95)	125	114	78	146	90	107	128	115	107	106
Norfolk	110	119	(109)	110	(104)	(95)	(89)	77	(101)	(96)	119	(100)	(100)

FIGURE 12.9 In the 1920s, the inheritor of William Farr's position, Percy Stocks, applied a newly rigorous method of statistical analysis to cancer incidence using British registration district data. Here, "cancer" is a broad rubric for a host of specific diseases which could be separately investigated.

In this work cancer itself metastasized. What had been a single disease, albeit occurring in different parts of the body, became a class of related but distinct diagnoses distinguished by the age and sex of the patient as well as the physical locus of the cancer itself. Stocks statistically distinguished specific cancers—breast, chest, colon, oral, skin, and so forth—within local, regional, and national populations, applying to each table of this or that cancer new, rigorous, standard mortality ratios.

At an international conference on cancer sponsored by the British Empire Cancer Campaign[4] in 1928 Stocks promoted his new statistical approach: "A more accurate, but more laborious method of eliminating the simple arithmetic effect of age distribution would be to standardize the cancer-rates by the indirect method, employing age intervals of five years from 45 to 80 and grouping together ages over 80" (Stocks 1928, 510). The results were used to search for social and occupational differences in cancer incidence, differences based on urbanity or rurality, and the ethnicity of patients. In this statistical turn one saw cancer as a numerical reality whose significance was statistic, not as the source of patient symptoms. Even mortality disappears except as a verifiable datum in the lists whose object is the statistical portrait of a class of diseases whose commonality is known only at third hand through the data retrieved from local death certificates and hospital records.

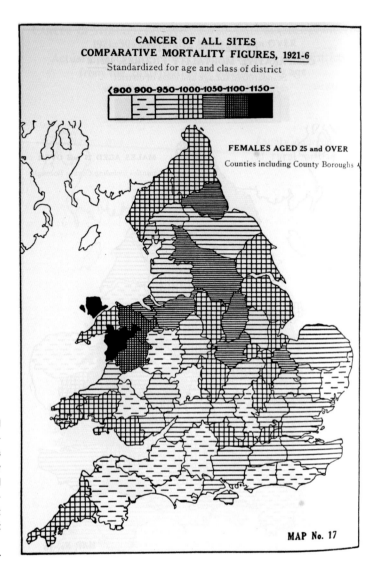

CANCER OF ALL SITES
COMPARATIVE MORTALITY FIGURES, 1921-6
Standardized for age and class of district

<900 900-950-1000-1050-1100-1150-

FEMALES AGED 25 and OVER

Counties including County Boroughs

MAP No. 17

FIGURE 12.10A Using standard mortality ratios and age stratification, Stocks argued cancer as both a condition specific to body sites (lung, stomach, uterus) and as a general public health concern affecting all citizens, albeit at different rates in different populations.

By 1936 Stocks had created a mini-atlas for the *British Empire Cancer Campaign Annual Report*, the maps reporting British cancer mortalities in the years 1921 to 1930 by age, sex, and general cancer location (such as intestines, liver, lung, rectum, or stomach) (Stocks 1936). A map titled "Cancer of All Sites" for males, and another for females, presented variations in cancer rates as statistical conclusions based upon the accumulated evidence of the series maps of specific cancers reported by district registrars. "All sites" were both "all sites where cancer occurred" and "all sites geographically," the districts of England from which the local reports of cancer were reported to the GRO.

Stocks presented his mini-atlas of cancer at the 1936–1937 British Empire Cancer Campaign meetings in 1936. It was the culmination of more than a decade of work in which he first demonstrated the utility of standard mortality ratios, employing confidence indices, and then systematically applied the results to registry data for each cancer type reported in each of London's registration districts. The maps of cancer in British registration districts presented not cancer as experienced by patients or their physicians, concerned with local clusters, but cancer as an outcome of statistical manipulations that defined specific cancers as significant realities, lodged in bureaucratic jurisdictions, for which a computed degree of confidence could be expressed. It was not Stocks's task to investigate the causes of the cancers he mapped, only to assert cancer as a spatial reality in the population of a reportage region.

In this transformation from the experiential to the statistical lung cancer appeared as a significant concern. It was not until the actuarial studies of Hoffman, that lung cancer as a distinct disease became a general health concern. And it was not until Stocks's work in the 1920s that the range of its occurrence became evident. But it was also only in the first part of the twentieth century that constant tobacco use began to become widespread, a social addiction rather than an occasional pleasure. And it would have taken that long for workplace pollutants in various industries to combine with increased tobacco usage among workers to create the field within which this cancer would become itself an epidemic concern.

In many ways, the identification of epidemic lung cancer was an artifact of the creation of statistical indices capable of identifying specific cancers within age-defined populations. That said, however, there were two or three significant problems with this broad transformation of cancer from a village reality into a statistical thing. Not only were the computations for accurate standardized mortality ratios complex and time consuming, they were dependent upon the accurate reportage by local physicians to registrars of the precise nature of a cancer causing death. In an era before a microscopic taxonomy of cancers was possible this was often problematic. Was a woman with multiple cancers of the breast and lung a breast cancer patient, a lung cancer patient, or both? If cancers were observed in the liver, pancreas, and intestine of a deceased patient how could the primary cancer be defined?

Of no less importance was whether local registrars accurately reported cases (and whether physicians accurately reported their cases to the registrars). Registrars in geographically remote, less heavily populated districts might be less likely to both receive good data from local physicians and pathologists, and working with smaller staffs, to report vigilantly the niceties of individual cases. There was as well the problem of mobile populations whose movements made it sometimes difficult to know what geographical location was the one to which the deceased should be assigned. If someone died in London but had lived for thirty years previously in Newcastle-upon-Tyne were did one locate that death? The death certificate would be filed in London but the slow-growing cancer likely had its origins in Newcastle.

These problems were endemic to the field of statistical disease studies. While recognized as important in countries like England (and elsewhere in Europe), they were especially acute in North America where significant elements of a very mobile population existed in rural areas with at best haphazard systems of health data collection.

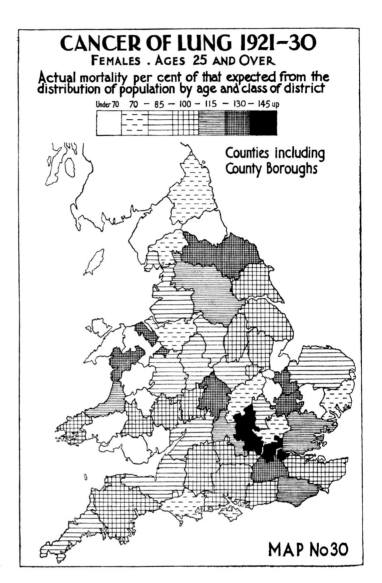

CANCER OF LUNG 1921–30
FEMALES . AGES 25 AND OVER

Actual mortality per cent of that expected from the
distribution of population by age and class of district

Under 70   70 – 85 – 100 – 115 – 130 – 145 up

Counties including
County Boroughs

MAP No 30

FIGURE 12.10B With the general mapping and statistical arguments about "cancer" as a disease family, Stock mapped individual cancers in his work. Here is one of the first maps of lung cancer as a distinct health problem among females over the age of twenty-five years.

It was for this reason that, especially during the years of World War II, the Statistical Abstracts of the United States for years carried a caution warning of its data:

Because of the unprecedented movement of the population during the war, the changes in the birth and death rates for the war years needs to be interpreted with care. Part of the increase or decrease in the rates may be accounted for by the changes in the age, race, and sex composition of the area. Of particular significance in this connection are: (1) the withdrawal of a large number of physically fit males of military age from the civilian population, (2) the concentration of such a population in military establishments in certain areas, and (3) the movement to overseas duty of a large number of men in the armed forces. All of these factors make rates for the war years less comparable with those of previous years (US Bureau of the Census 1946).

It is hard not to read this without a smile, and a wince. This "unprecedented population" movement did not consider the thousands of woman who also joined the military in a variety of capacities. Similarly ignored was race. Statistically, America was fundamentally white. In the mid-1940s began an immense migration of African American men and women from the southern states to industrial northern states. Some of this was the effect of the war, men who went into the armed forces and women whose labor was desperately needed in the manufacturing plants that stoked the military with armaments and supplies. And yet, through the 1970s, data on disease within the African American population was at best scanty and certainly underreported. The reasons, in retrospect, are obvious: African Americans were less likely to have family physicians; the circumstances of their deaths were more haphazardly investigated. In the generally racist social climate of the day the nonwhite population was on the periphery. This marginalization was carried into the otherwise increasingly rigorous pattern of disease reportage that had begun in the United States earlier in the century.

Slowly, the methods of data collection and reportage improved for all citizens. In the 1960s computers were introduced to permit both automatic calculations of large fields of data and the graphic transformation of that data into maps (Koch 2005). The results—in both England and the United States—were new atlases of cancer asserting its reality as a national health problem whose incidence varied statistically across the nation. In England G. M. Howe produced the first modern atlas of cancer in Great Britain in the early 1960s. He seems to have been unaware of Stocks's earlier work, explaining at length in his introduction the idea of standard age-adjusted mortality ratios as if they were a new statistical invention (Howe 1963, 7–10).[5]

In figure 12.11, is evident the aggregation of cancers that while affecting loci in the body were otherwise associated by environmental or other factors. By the time of Howe's first atlas it was well recognized that both tobacco use and some environmental pollutants contributed to specific cancers, in this case those of the bronchus, lung, and trachea. While these might be distinguished by various contaminants (pipe smokers were more liable to throat than lung cancers) the relationship between the three made their shared mapping a clinical statement as well as a statistical convenience.

Like Stocks, Howe's aim was only secondarily to investigate the nature of cancer. Rather, his principal goal was to assert an identifiable, statistical variation in general disease incidence in England's registration districts and then to promote the potential of this nationally mapped, statistical reality as an aide to disease research. Just as Haviland had argued a century before, Howe assumed that areas of high incidence of cancer rates would reflect local conditions that promoted cancer generation. It was in these statistically defined, mapped areas, Howe hoped, local cancer generators might be found. To emphasize the point he quoted Hippocrates on the importance of local climatic conditions as a means of understanding disease variation.

Computerization reformed not simply the mapping and calculation but, by the end of the twentieth century, the collection and distribution of the data on which such

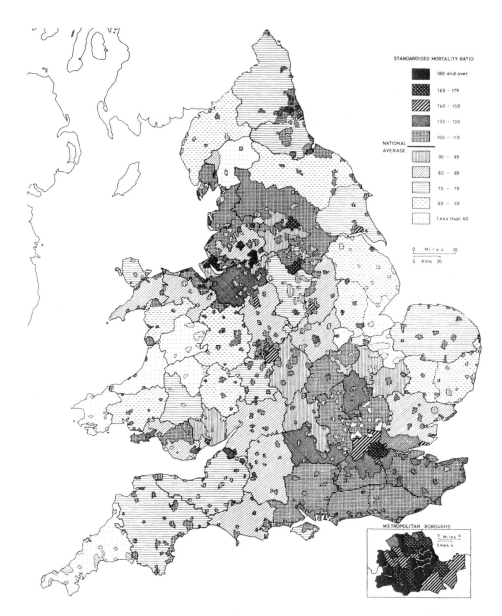

STANDARDISED MORTALITY RATIO

180 and over
160 - 179
140 - 159
120 - 139
100 - 119

NATIONAL
AVERAGE

90 - 99
80 - 89
70 - 79
60 - 69
Less than 60

0   Miles   30
0   Kms 30

METROPOLITAN BOROUGHS
0 Miles 4
0 Kms 4

FIGURE 12.11 G. M. Howe's 1963 atlas of mortality in Great Britain calculated standard mortality ratios for cancers and other diseases to present a statistical portrait of disease incidence in the nation's reportage districts. This map of causally associated cancers of the lung, trachea, and bronchus demonstrates his graphic-statistical approach.

maps were built. Increasingly, death certificates and records of disease incidence were transposed from print into computerized data. The data that resulted gave the appearance of disinterested fact but remained only as good as the reportage that occurred at the local level. In the United States, for example The National Cancer Institute's 1975 *Atlas of Cancer Mortality for the United States* used data from 1950 to 1959 to map different cancers in the United States based on county-level statistics. The cancer that was mapped was, however, only that affecting Caucasian populations (Mason et al. 1975). In 1976 a second atlas covering African American citizens was published, but its data set was incomplete, permitting mapping only at the level of aggregated counties called "Standard Economic Areas," the forerunner of current standard metropolitan areas (Mason et al. 1975). In effect, cancer remained as segregated a reality as were housing and schooling in that era.

These atlases were the direct result of the growing political definition of cancer as a national health problem whose origins lay in either unnatural environments or specific human behaviors. The U.S. surgeon general's report in the 1960s on the relationship of tobacco smoking and lung cancer, for example, identified tobacco use as a statistically significant contributor to the increasing incidence of this cancer. Snow's "If only they would wash their hands they would not get cholera" became, in the late twenty-first century, "if only they did not smoke they would not get cancer." Nor was smoking the only environmental or habitual determinant. By the 1930s a host of carcinogens widely used in industry and across the industrialized world had been identified (Davis 2007).

The result was not a sustained effort to rid the United States (or the world) of cancer-causing agents but instead a massive research campaign to study cancer as a public health threat. In 1971 President Richard Nixon famously funded the first official "war on cancer" with $100 million.[6] The monies from that war chest supported the formation of the National Cancer Institute (NCI), the "war center." Just as Arrieta mapped his military campaign against plague at the end of the seventeenth century ,the NCI created an atlas of US cancers in 1975 that became a blueprint of the "war." Simply, one had to see the enemy to plan its restraint.

The stated purpose of these atlases was to display the disease patterns, primarily to generate etiologic hypotheses for subsequent studies (Pickle 2009).[7] The underlying death certificate data had been available alphabetically by county name (in print and electronically), but organized in that manner it was impossible to see geographic patterns in the data (Mason and McKay 1974). The success of the UK atlas, and of early attempts to identify regional cancer patterns noted in state maps of 106 cancer rates (Burbank 1971) suggested the utility of county maps of cancer rates.

There are several examples of the success of these atlases in uncovering previously undetected but significant clusters of unsuspected cancer incidence. "From the maps, the term 'hot spots' (of intense coloration) arose to denote areas of the country that had the highest rates for specific cancers" (Anderson 1987, 1). Among these hotspots, seen in the map but not previously in the county-by-county statistical ledger, were areas of elevated rates in costal shipyards whose workers were exposed to asbestos during World War II (Blot et al. 1978) and a two-county "hot spot" in Montana, first thought to be random noise on the map, that was found to result from widespread arsenic air pollution from a copper smelter (Lee-Feldstein 1983).

Perhaps nowhere was the utility of statistical mapping better seen than in the investigation of previously undetected clusters of unexpectedly high rates of oral cancer among women in the southeastern United States. What was hidden in the statistical tables arranged by county became glaringly evident when argued in the 1975 atlas's map. Epidemiological field studies, instituted on the basis of the mapped cluster, revealed the cancer as likely occurring because of high levels of snuff use among women in the red areas of the map (Winn et al. 1981).

From the 1970s into the 1990s, women in areas of high incidence changed their tobacco habit from smokeless products, with their higher rates of esophageal and oral cancers to cigarette use. The result was a decrease in oral cancers and an increase in lung cancers, a chance visually evident in the next generation of NCI atlases. In a series of maps produced from county-level data by Dr. Linda Pickle, a senior statistician at the NCI (fig. 12.12), the resulting change in cancer mortality rates is readily seen. The data in these maps is categorized by deaths per 100,000 and mapped in quintiles, with mortality ranging on a color scale from yellow (lower) to dark brown (higher).

Map to map one sees two different cancers, each with a different cause. The upper left-hand map represents the incidence of oral cancer among woman occurring between 1950 and 1954. In the first national cancer atlas the unexpected clusters of oral cancer among women (a disease more often associated with male pipe smoking) required field work to resolve. It was discovered that the majority of cases in these hot spots occurred among women who used "smokeless tobacco" like snuff. But by

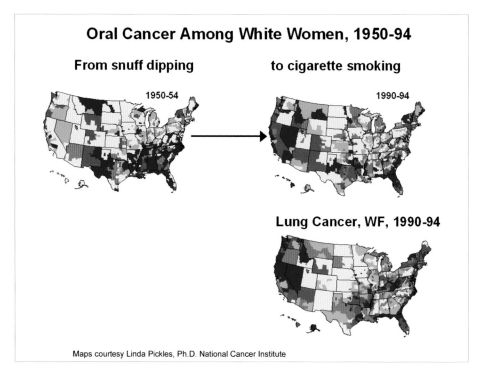

Maps courtesy Linda Pickles, Ph.D. National Cancer Institute

FIGURE 12.12 These three maps of the changing pattern of cancers in U.S. women details the shift in cancer caused by long-term use of smokeless tobacco products in the 1970s (esophageal, trachea) to cigarettes (lung) in the 1990s.

the 1990s the cultural prohibitions on women smoking had disappeared and, as the map in the upper right of the triptych shows, cigarette smoking had become more prevalent.

The result was a decreased incidence of oral cancer among women in counties where cigarette smoking had replaced "smokeless tobacco" as an acceptable addiction. As a result, hot spots in which there seemed to be regional epidemics of oral cancer were diminished. The map at the bottom of the triptych demonstrates this was not an unalloyed health boon. The change from smokeless to smoked tobacco was accompanied over time by an increase in lung cancer rates among U.S. women. Together the maps present two related progressions, the first from one class of cancers to another within a single population over time, and secondly, the assumption this was a result of the shift from smokeless to smoked tobacco use. All three maps present county mortality data based on medical diagnoses to create a surface in which variable intensities of disease incidence are evident. Across the map series the argument develops relating cancer to tobacco use over time.

ENVIRONMENTAL CANCERS

Within the framework of official, statistically rigorous national studies, the use of solely local mapping was devalued in most research circles. Statistically, the populations were too small, the number of potentially complicit variables too great to permit confident statistical statements. It was not that local studies ceased but that they ceased to be taken very seriously just as, as we have seen, John Simon found the local Broad Street study unconvincing in its scale of address. Fine-grained mapping of local phenomena seemed to serve only once a coarse grained, statistical study revealed a disease cluster.

It was only in the mapped statistics of the general that the localized could be credited. This tendency has had unfortunate results. In the 1950s, for example, members of the Navajo nation began to manifest inexplicable wasting illnesses, a higher than usual incidence of various cancers whose origins were dismissed as a "Navajo neuropathy,"[8] perhaps a genetic weakness of the Navajo people. Navajo neuropathy and cancers did not show up as hot spots in the NCI cancer atlases of the 1970s, or later atlases in the 1990s. The population was too small, too rural, and too dispersed for a confident statistical statement of incidence or localized risk. Public health officials insisted repeatedly there were nothing statistically significant about the number of Navajo patients whose cancers and related neuropathies appeared to local physicians as active clusters.

Increasingly, however, the possibility that these insignificant cancers were linked to long-term radiation exposure gained popular if not scientific traction. In the 1940s, the United States convinced members of the Navajo nation that the mining of uranium on Navajo lands was a matter of national necessity that would have no long-term ill effect on the health of local population. The result was a landscape of old uranium mine sites and contaminated water sources serving a dispersed, largely rural population. In 2006, the *Los Angeles Times* published a critical series, "Blight on the Land," using EPA maps to argue the relation between Navajo illnesses and pollution from uranium mining sites (Pasternack 2007). Data on the pollution of Navajo groundwater from old uranium tailing ponds was lodged in the maps of the Environmental Protection Agency.

Nor were Navajos the only ones for whom apparent concentrations of cancer resulting from adverse environmental exposure were proposed. In 1987, Openshaw and his colleagues used a mapped approach to investigate the relationship between cancer incidence and proximity to nuclear power sources in England (Openshaw et al. 1987). In 1988 he tested the approach by interrogating incidence of childhood leukemia in relation to nuclear power sources in Great Britain (Openshaw, Charlton, and Craft 1988). His "geographical analysis machine" used a Monte Carlo simulation technique to assess the importance of cancer incidence across the surface of England in relation to the nuclear power sources brought online in that country.[9]

The cancer data Openshaw analyzed was collected by Chadwick's and Farr's inheritors. The idea of national health statistics collected at the resolution of very local health jurisdiction has become a taken for granted reality across the industrialized world. But they instantiated what had first been argued by Graunt in his analysis of London's Bills of Mortality, created to track both plague and other than prevalent epidemic and endemic conditions (Porter 1999). The results in this case showed significantly higher incidence of some cancers in areas proximate to nuclear power sites. Most significantly, perhaps, while two were expected on the basis of previous studies Openshaw's mapping revealed a third cluster, previously unsuspected cancer cluster near a power plant. For a decade researchers debated Openshaw's cartographic-statistical methodology as opposed to other, non-graphic purely statistical approaches (Fotheringham and Zahn 1996). The conclusion that cancers increase with proximity to power plants seemed lost in the debate over statistical efficacy and efficiencies of calculation.

ELMF CANCERS

Finally, there has been a voluminous but still contentious debate over the relation between the emanations of power lines and transformers and increased cancers within communities living under and near them. The argument was developed in 1979 and made popular in a three-part *New Yorker* magazine series by journalist Paul Brodeur. As he reported in articles, and later, a book (Brodeur 1989), in 1979 Colorado epidemiologist Dr. Nancy Wertheimer reported that children in Denver who lived in homes with high-current wiring were two to three times as likely to die of leukemia, lymphoma, or cancer of the nervous system (Wertheimer et al. 1979). There was no doubt that cancers seemed especially prevalent in these Denver areas and homes.

The question was whether those differences were, in fact, statistically significant and if so were they related to the presence of high-current wiring, transformers, and the like.

Research on the problem has been hampered by the absence of a biological model that might adequately explain the relation between cellular metastasis and the extra-low frequency emissions (ELMF) of power sources. Without such a model the observed effects are uncertain. Much of the current work in this area appears map poor, papers and studies without accompanying maps in the reports. But as Dr. Maria Feychting, a principal researcher on the relation between cancer and electromagnetic power sources explained[10] that maps remain a critical medium in which studies relating cancer to local contaminants even where the maps themselves are not published (Feychting and Alhbom 1993). In one study of the possible relation on the effect of power exposure to parents of children with cancer she mapped the homes of young Swedish cancer patients and of a control population of children of a similar age. In topographic maps she then identified local power lines and sources, based on power company data, and measured the distance between the homes of

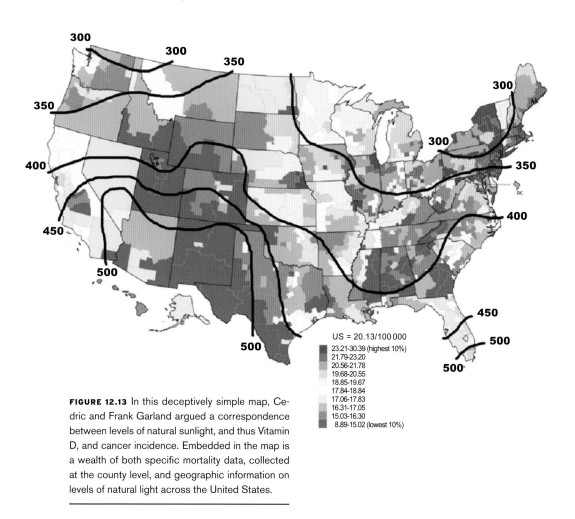

US = 20.13/100 000

23.21-30.39 (highest 10%)
21.79-23.20
20.56-21.78
19.68-20.55
18.85-19.67
17.84-18.84
17.06-17.83
16.31-17.05
15.03-16.30
8.89-15.02 (lowest 10%)

FIGURE 12.13 In this deceptively simple map, Cedric and Frank Garland argued a correspondence between levels of natural sunlight, and thus Vitamin D, and cancer incidence. Embedded in the map is a wealth of both specific mortality data, collected at the county level, and geographic information on levels of natural light across the United States.

cancer patients and their cohorts to local power sources (Feychting et al. 2001). The results were cartographic *sub rosa*, the map workbench the surface on which the statistical analysis was cobbled together, see by the researcher if not the reader of the resulting analysis.

Others similarly struggle to develop other models of natural carcinogenic or anti-carcinonogenic environments. These studies do not employ simple environmental determinism but instead present sophisticated attempts to identify general environmental agents broadly increasing or decreasing the incidence and death rates of particular cancers. While on one scale, cancer etiological research depends on understanding of cellular, genetic, and metabolic processes, on another scale it relies on mapping of the relations between suspected environmental agents and cancer rates.

CANCER AND VITAMIN D

In the early 1980s, for example, Cedric and Frank Garland first argued that there is an inverse relationship between sunlight and colon cancer mortality, theorizing it resulted from the protective effect of higher levels of vitamin D, photosynthesized in the skin from sunlight (Smith 2006, 211. Garland and Garland 1980).[11] A similar

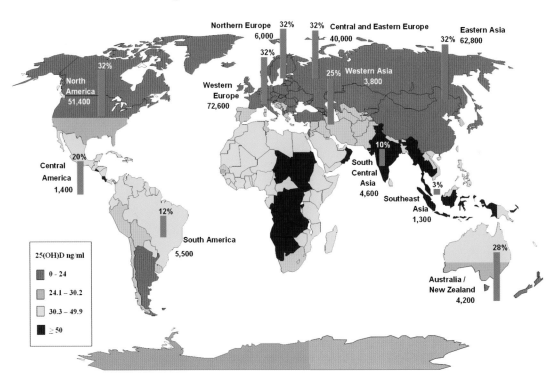

FIGURE 12.14 The end point of the Garlands' work begun in the 1980s on the relation between natural sunlight, and thus Vitamin D, and cancer rates is seen in this map, which predicts the likely decrease in cancer that would result from inclusion of Vitamin D supplements across world populations.

relationship was then seen between sunlight and breast cancer, and later, diabetes. In studying cancer maps, the Garland brothers noticed a strong latitudinal gradient within the United States for colon cancer mortality rates. They observed that mortality rates were much higher in the northeastern and northern parts of the country than in the southern and southwestern parts, where the hours and strength of sunlight are generally greater (Mohr 2009, 80).

The original thesis, since tested "six ways" (Giovannucci 2006), was a basic, prepositional argument: "Given that solar UV-B radiation is the major source of vitamin D for most people," if low levels of vitamin D result in increased cancer then, "one would predict that greater average UV-B radiation in geographical region of residence would correlate with lower risk of colon cancer" (Giovannucci 2006, 21). In the first "hypothesis paper," recently reprinted, the Garlands overlaid gradients of sunlight exposure in the United States on county mortality data for colon cancer. The result was a strong inverse relation between levels of solar intensity and death rates of colon cancer.

Since then researchers have labored to expand the Garland thesis to include other cancers, and diabetes, while developing a biological model to explain their findings. The last thirty years of this research was nicely summarized by Kathleen Egan in an article reviewing this history: "Experimental, clinical, and epidemiological data is pointing to a beneficial role of vitamin D, not only for colorectal malignancy but also potentially for many of the most common and lethal forms of cancer in temperate climates" (Egan 2006, 15). The conclusion of these studies is a model that predicts the potential decrease in cancer rates if vitamin D supplements were used as a preventive (Gorham et al. 2009). The conclusion is compelling. Stated formally it is that "approximately 200,000 cases of breast and 250,000 cases of colorectal cancer annually worldwide would be prevented by raising serum 25(OH)D concentrations to 40–50 ng/mL. This serum 25 (OH)D level is, in general, associated with oral intake of 2000 IU of vitamin D_3 per day and about 10 minutes per day of sunlight around noon, weather allowing" (Gorham et al. 2009). In plain speak: daily usage of vitamin D supplements would over time decrease the world wide cases of breast and colorectal cancer by 450,000 cases a year.

One sees this in a map (fig. 12.14) that argues the effect on colon cancer rates were vitamin D supplements universally used. The map first proposes a world in which the relationship between vitamin D levels and cancer rates—in this case breast cancer—is consistent. If the relationship were distinct for each nation or for various races the world map would not serve. In this world the map aggregates local data into regional and national jurisdictional statements. We see the jurisdictional boundaries of countries but they are insignificant except as data catchments.

The meat of the map is in its coloration aggregating national data to create zones of natural vitamin D, 25(OH)D serum levels. These rates reflect the amount of sunlight available in each jurisdiction with levels being higher in, for example, central Africa being higher than those in Canada and the northern United States. All this projects the argument that decreased cancers would result from the general use of 2,000 international units (IU) of oral vitamin D on a daily basis. *If* the level of vitamin D is inversely related to the incidence of this cancer in the population, *then* use of vitamin D supplements can decrease in a predictable fashion the incidence of this cancer. The effect of vitamin D supplements will differ geographically depending on the amount of natural vitamin received from sunlight and thus the natural 25(OH)D serum levels of persons in different places. Thus the effect of vitamin D supplementation would be greatest in those national regions where average daily sunlight is least.

Over the approximately thirty years of this research Cedric Garland says, in private correspondence, that maps have been indispensable to the research. "Epidemiology *is* maps,"[12] he writes. Maps locate health statistics in a manner that permits their correspondence to potential social and environmental influences. The Garlands' work is an example. It began with the observation of variable cancer rates as a function of latitudinal gradients. Those gradients were used to map hours of sunlight, a natural source of vitamin D. A wealth of work went into the collection of data on serum vitamin D levels and on cancer incidence reported within county resolution health jurisdictions in the United States. The relationship was secured and then data collection was expanded to include a worldview. Finally, a predictive model was refined describing a probably decrease in specific cancer rates were vitamin D supplements to be generally used. Map to map the argument was built, researched, and unfolded.

VERY LOCAL CLUSTERS

More popularly, maps of large-scale, local cancer clusters are springing up on the Internet as local researchers, amateurs in the tradition of Lea of Cincinnati (Fig 7.11), argue the reality of neighborhood cancer concentrations ignored by the official researchers. The work is rarely carried out these days by local physicians, the inheritors of earlier researchers like Arnaudet and Powers, whose epidemiological training favors the more coarse-grained, smaller-scale study. It is the activist resident rather than the clinician or epidemiologist who most typically work first at this scale.

A member of the New Jersey Central Cancer Registries told me in 2007 his organization receives nearly a hundred reports of suspected neighborhood cancer clusters each year.[13] Over the last five years, however, New Jersey cancer registrars have certified at best one such cluster. All the rest have been dismissed as statistical artifacts with no real significance.[14] New Jersey cancer registrars do not dismiss the diagnosis of cancers in these clusters, or their apparent prevalence in the few blocks and streets that seem to be inordinately populated by cancer patients. Whereas historically such "cancer streets" would have been the subject for intense, local medical investigation today they are typically dismissed as anomalies on the basis of statistical tests favoring large populations.

The local cancer maps by citizens who believe their neighborhood "clusters" are not statistical anomalies but real problems caused by local geographies return us to the late-nineteenth- and early-twentieth-century researchers who saw cancer as something to be investigated at the scale of the physician's practice. Those who insist that statistics are unreliable filters incapable of distinguishing a real local outbreak from seemingly random clusters argue the local scale typically ignored by broadly statistical arguments is important.

To say statistical definitions of incidence do not deal well with the local scale of incidence is only to say our knowledge and our methods are incomplete. We are at best in mid-step on the road to a complete understanding of cancers in their genetic, environmental, economic and social complexities. We need the local study as much as the national review. Even our current state of moderate ignorance would not have been possible, however, without the bureaucracies that collect the data of health and disease, map its boundaries and locate it in the collective world of shared experience rather than the abstract world of statistics alone. The local mapping of cancer clusters and the search for environmental determinants in local and regional mapped studies remains a medium of both theory generation and theory testing as well as serving as a corrective to the broadly statistical view of contemporary epidemiologists struggling with disease events that are still imperfectly understood.

"Knowledge is not simply local, it is located. It is both situated and situating. It has place and creates a space" (Turnbull 2000, 19). We see this in the history of nineteenth- and twentieth-century cancer studies, nineteenth-century plague studies, seventeenth-century yellow fever studies, and so on. Each infection, each disease incidence happens to a person in a place. We know this because of the reports of the doctors who treat them and, where necessary, sign the death certificate affirming their demise was a result of this or that disease thing. But those local deaths are located within the communities we build, the governments we encourage and the bureaucracies that result.

There are too many cancer patients, too many multiple sclerosis patients, and too many tuberculosis patients to situate our concern in this town and that neighborhood. As we build our resources across provinces, states, and nations the individual disappears within the mapped statistic of incidence. Even at this level of anonymity the location of the individual case makes a space that is shared by the many who share this or that diagnosis. It is in that space, seen in the map if not in the national registers, that research takes place; in that space that our priorities as a society concerned with disease incidence occurs.

There is no battle between mapped and statistical data. The former is not a mere illustration of the latter. Maps locate numbers that need maps if we are to transform incidence into ideas about causation. Maps presented the statistics of incidence in a manner that encouraged the environmental search for causal elements. Statistics encouraged specific scales of graphic analysis at different points of clinical and political reportage. In the numbers and in the graphics that attempted to make sense of them it was the seeing that was a test of the knowing.

AFTERWORD

We learn as we were taught to learn. We see, we count, we theorize and then as-semble what we think we know to propose a disease in its uniqueness, constructing it as we go. Well water is still tested when diarrheic diseases occur. It is tested when a set of unexplained symptoms cluster in a community (Lagakos, Wessen, and Zelen 1984a, 1984b). We look to sewers generally for contaminated water and industrial plants for contaminated air. Like the Hans Holbein the Younger, we worry that human trade and travel promote the spread of diseases. Like Snow, we worry about the people—in hospital and at home—who do not wash their hands.

And yet knowledge is not a ladder we climb rung by rung, each step once past, never to be trod again. With each new disease we repeat the steps taken before, and perhaps add another one or two if we can. First we see a congress of symp-toms we can describe even if the condition they present is itself unclear. We count the occurrence of these symptoms in a population, and if it is rife we marshal our knowledge resources. Then we seek an explanation in the lexicon of existing theory, assembling our materials to test this or that argument based on past experience. In the testing we find new ways of thinking about disease in populations, and perhaps about ourselves.

Stories are enriched (or impoverished) as our broader knowledge base is changed, transforming in the telling what we thought we knew to be real, what we thought we had seen. In this perpetual reassembling of the known and the newly learned the old is rarely discarded. More commonly it is instead merely . . . reordered. Of *course* we plan each spring and summer for the annual return of influenza in the winter, Thomas Sydenham would say today. His epic study of recurrent disease in the seventeenth century proved that necessary. And of *course* we seek the fouled water when facing a diarrheic outbreak, Snow and Farr would say today (albeit for different reasons): "Didn't we make the reasoning clear?" In the first decade of the twentieth century people *knew* polio was a disease of unsanitary immigrants in unclean tenements (Rogers 1996). After all, other diseases were. Slowly, however, it became something else, a new thing requiring new answers to be sought through new methods of investigation.

At the University of Hawaii's dormitories in 2007 were signs warning residents to wear long sleeves if they went out in the morning and evening to protect themselves from dengue fever and the mosquitoes that are its vector. The caution about dress in the morning and evening hours is one Hillary offered his patients and readers after yellow fever came to Barbados. The rationale is slightly different but the advice the same—one century's knowledge refined by another but both offering a similar prescription.

Identifying the microscopic agent, *seeing* it through the lens, does not eliminate a concern over environmental conditions that might promote its growth or that of its vector, as F. R. Campbell cautioned contemporaries in the 1880s when bacteriology seemed, for a brief period, the answer to everything: "The pathologists who inform us that the microbe of a given disease is shaped like a period, a dish, or a comma, without telling us what meteorological donations are most favorable to its development does only half his duty" (Campbell 1885). Once seen in a matter that brooked no uncertainty the question of the nature of the agent was transposed into another: what geographic, physical, or social conditions—alone or together—promoted a bacterium and its mode of transmission?

It is for this reason that the idea of disease proposed here engages from the start the bureaucratic and social as well as the clinical. The Environmental Protection Agency estimates that at least twenty-three thousand times a year, between three billion and ten billion gallons of raw sewage are discharged into American streams, rivers, and lakes (Wheeler and Smith 2009). Hundreds of government-run sewer system operators have been fined for the resulting discharges, infractions of rules set out to protect water supplies.[1] The sanitary lessons of the nineteenth century are known, and when ignored what results is a burden of disease Victorian researchers like Chadwick and Snow encountered long before we were born.

If more evidence is needed of the importance of this complex conception of disease, one in which history lives into the present, the deaths resulting from tainted water in Walkerton, Ontario, provides a case study our nineteenth-century researchers would have appreciated.[2] The contamination of that community's water started at a pig farm and from there entered the water supply. In a real way it was a nineteenth-century outbreak that twentieth-century researchers did not see (it was not mapped) until widespread illness and some deaths forced officials to investigate a problem that perhaps could have been foreseen (O'Connor 2002, Perkel 2002).

The simple story is a bacterium, a strain of *Escherichia coli*, entered the town's water supply, causing a range of gastrointestinal symptoms. The first defense was a boiled water advisory during an investigation of the source of the infection, eventually found to be farm runoff. The complex story is that people with little training in water purity ran a system they were unable to correctly monitor and that illness and deaths resulted from a situation similar to those mapped, time and again, over the last two hundred years. This was not simply a bacterial but also in origin a bureaucratic disease, one in which safeguards protecting local water supplies were not followed. The clean-up following the deaths was therefore bureaucratic as well as technical.

The greatest failure of the simple history is that it assumes not a clear beginning and a definitive ending. It promises certainty instead of the gray areas of maybe and possibly that are the real domain of disease and its study. With the best of knowledge what typically results is not this or that, but both together. What happened in Walkerton was a bacterial outbreak with bureaucratic and social causes. So too are many of the thousands of outbreaks of gastrointestinal disease like the one in Greater Vancouver in 2002 described in Chapter 2. Time and again, history argues the failure of choice and the complexity that is what we face in disease studies.

Yellow fever did come with the seasons; it did proliferate in the warm moist days in the torrid zone. It was also a creature of the temperate northern city whose piles of odiferous waste served as a fertile center for yellow fever's propagation. And, yes, mosquitoes were especially rife during an outbreak. Cholera was a creature of bad hygiene and of bad urban sanitation; and so are most diarrheic diseases. Hand washing is important, good sewage is important, and safe water—as Snow's miners knew—is a critical component of the disease equation. But hand washing requires money for soap and clean water to wash in. Chadwick understood the relation between health and income. Good sewage disposal requires a communal effort to remake the city, and that demands a bureaucracy up to the task. Good sewage and good water are complex political goals requiring engineering, scientific, and political structures if these systems are to be created.

The real issue is not solely the patient in his or her house. The critical issue, as Alice Hamilton understood, is the means by which the disease spreads from this person to that neighbor, from this neighborhood to another. Disease is rarely simply personal, after all. It is instead almost always a communal experience. Think malaria, that "ragbag of different strains of the parasite(s) and interacting processes" that remains a multifaceted complexity in medicine and science (Turnbull 2000, 162). Yes, its vector is the mosquito that breeds in stagnant water in certain climatic conditions. To control the disease a common approach is to attack the breeding ground of the mosquito, the water in which its larvae hatch. That takes the cooperation of a range of officials and persons in potentially infected malarial areas, too. Wearing long sleeves and staying indoors in the evenings and mornings is rarely sufficient.

The distribution of permethrin-treated bed nets serves not only to protect from malaria the person sleeping at home but contributes as well a protective element for others in his or her immediate vicinity (Hawley et al. 2003). A protective buffer based on the homes thus protected grows with the number of homes whose tenants possess treated bed nets, creating together a broader arena of antimalarial security. When the nets were introduced the first question was whether they worked to protect people against the malarial bite of mosquitoes. The question then became whether denying the mosquito a food source in one hut might put those in other huts at risk. Once the distance between homes in a test village was computed and incidence of malaria ascertained, the result was ideal. There was no added danger to anyone and increased

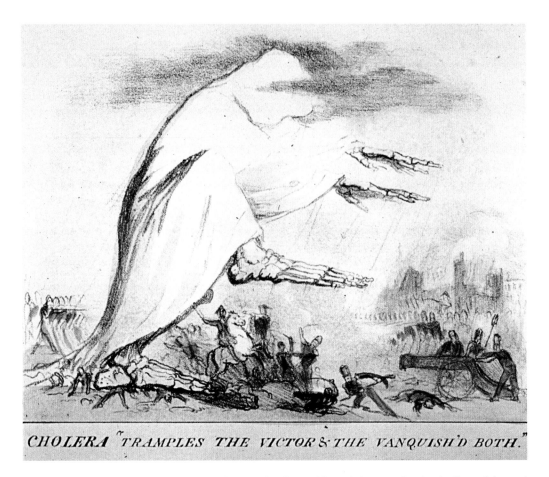

CHOLERA "TRAMPLES THE VICTOR & THE VANQUISH'D BOTH."

FIGURE 13.1 It is not only cholera that tramples the wealthy and the poor, the city dweller and the rural resident. Cholera was simply one of a class of diseases whose signal feature is its epidemic spread.

safety for all in the use of these bed nets (Gimming et al. 2003; Townsend 2005).[3]

Woven into this one small study of malaria in one place were anthropology (of the village), chemistry (the construction of permethrin), ethics ("do no harm," as Hippocrates enjoined), parasitology (a specific type of microscopic agent), public health (the community), geography (location of huts in the village), and statistics (computing risk). The backcloth is the history of disease studies from ancient to modern times, from plague to yellow fever to cholera and typhoid fever. In the *longue dureé*, it builds the matrix of tools (bureaucratic, clinical, social, statistical, visual) that are the way of disease studies today. The lesson is that knowledge, like disease and its treatment, is rarely simple. It is social in its constitution. Another lesson is that the knowledge we make is not permanent but changing, its state at any time a reflection not simply of the past but also of transient social and scientific variables (Turnbull 2000,).

There are four conditions that seem always to promote the development of disease and all four are in full play today, assuring the evolution of our microbial friends will outstrip our own state of development. They are increased trade, rapid urbanization, income inequality (at the scale of cities, nations, and the world), and war. The exploration of these elements together, and the method by which we focus on or ignore them and their effect, is the next reality around which all else will turn. To meet the challenge of the evolving epidemic we need to see at every scale, from that of the electronic microscope to that of the world at large. We will need to see the patient in his or her need and the community in which both the patient and the disease are lodged. In this seeing we will need the mapping that has for centuries been a part of disease studies and the statistics that the map presents, the modeling both engage together.

ACKNOWLEDGMENTS

Books do not appear full-blown in the window of a bookseller or the stacks in a local library. Writers create manuscripts that become a publisher's project, and if all goes well, the result is something both publisher and writer can point to with pride. For the writer the book is the thing but for the publisher it is a piece of the whole, an item in the lists of books in a subject area, that is the result of a publisher's editorial vision.

I have long admired the lists at the University of Chicago Press, especially its offerings in the history of cartography and the sociology of science and technology. It was there, I thought, a project like this, one balanced between both subject areas would find its rightful home. Editorial director Christie Henry championed that hope from the first proposal through the process of manuscript review and into the complexities of production. In the transformation of my manuscript into their project I was privileged to work with editors Abby Collier and Mary Gehl, who guided me through the issues of image permission, textual edits, and production. Without them this book would never have been.

There would have been no manuscript, no work to present without the continuous assistance and collaboration of first, Ken Denike and second, Denis Wood. It was with Ken's guidance that I went from student to teacher to collaborative researcher at the University of British Columbia. For a decade we have labored together to understand aspects of the mapping and statistics that created nineteenth-century medicine and public health. Many of the ideas presented here found their first expression in our joint teaching and work. My friend Denis Wood has been unfailingly generous in the time he gave for critiques of ideas I had half formed, materials I had in draft, half written. More formally, the trace of his writings and thinking about the nature of maps is clearly evident here.

Too few writers or researchers acknowledge the librarians who preserve the works of the past and assure their availability. I have had the assistance of many: Arlene Shaner at the New York Library of Medicine, Crystal Smith at the National Libraries of Medicine, Jim Ackerman and Bob Karrow at the Newberry Library in Chicago. Joe Ditta at the New York Historical Society, Lily Szczygiel at McGill University's Osler Library, and Ralph Stanton at the University of British Columbia are just a few of those whose diligence and knowledge made this book possible. My thanks to them and to the many elsewhere whose assistance is here acknowledged gratefully.

Finally, I am obliged to a number of scholars whose work informed mine and whose correspondence has advanced my understanding. These include, in a very partial list, David Marsh, Dr. Frank Garland, Ralph R. Frerichs, Michael Friendly, Ian Spence, John B. Osbourne, and Peter Vinten-Johansen. I am especially grateful to Professor Peter Haggett and Dame Professor Mildred Blaxter, both now resident in Bristol, United Kingdom, whose seminal work in both medical cartography and statistics, for the former, and medical sociology for the latter, have informed us all. As well I am obliged to Dr. Linda Pickle, whose decades of work on the mapped and statistical studies of cancer in the United States has advanced all our knowledge of both the disease and the methods by which it can be understood.

Books like this are listed by author name but that, like the name on a research paper, is always a fiction. The author draws from so many in building his or her knowledge base; then draws further from others as the ideas form on a page. We build on the past and in the present and hope the result worthy of a future place. For all those who have assisted me in this effort I say, thank you.

NOTES ON THE ILLUSTRATIONS

The charts, diagrams, and maps used in this volume come from a variety of sources, some more obscure than others. While every effort has been made to detail the source of each illustration, its location may be of little use to the reader who seeks to view the images and maps on his or her own, or to use them in his or her research. Where possible, alternate collection sites have been identified using the WorldCat library network (http://worldcat.org) to permit others interested in this work to find the images.

Two libraries with superb collections and extremely complete, online catalogs are the British Library in England and the National Library of Medicine in Bethesda, Maryland. The British Library has placed online its collection of maps of London from the Crace Collection, an inestimable resource (http://www.bl.uk/onlinegallery/index.html). Other libraries where I have been privileged to work, and where research librarians have assisted greatly, include the College of Physicians of Philadelphia, the Newberry Library in Chicago, the New York Academy of Medicine, and my home library at the University of British Columbia.

In an attempt to find the best examples of the work reproduced here I have in some cases ordered two or three different images of a single map from different sources. Here each image is sourced in a manner that attempts to identify an appropriate rights holder. Where possible I have sourced materials to the National Library of Medicine (NLM), which generously does not charge for publication rights, even when in other writings I have listed another source. The U.S. court decision *Bridgeman Art Library, Ltd. v. Corel Corp.* questions the right of libraries to claim ownership of historical images in their collection. That said, where a less expensive and alternative source could not be found—for example, the British Library's Crace Collection—every effort has been made to acknowledge ownership claims and to follow posted library regulations for not only purchase of the images but also their inclusion in this publication.

CHAPTER 1

1.1.-1.3. The images in this chapter are taken from an online exhibition Cholera Online: A Modern Pandemic in Text and Images, produced by the NLM (http://www.nlm.nih.gov/exhibition/cholera/index.html). Illustration 1.1 (image no. A012673 in the online exhibition) was published first in "A History of . . . the . . . Cholera in England and Scotland" in the *Lancet* in 1831–1832. Figures 1.2 (image no. A012121) and 1.3 (A012122) are by Robert Weimar, from Robert Froep's *Symptome der asiatischen cholera*

(1832). English copies can be found at the Wellcome Trust library and Cambridge University. WorldCat also lists a copy in Germany at the Wissenschaftlichen Stadtbibliothek, Mainz.

1.4. This 1832 lithograph by Robert Seaman (image no. A021772 in the online exhibition) fancifully depicts Board of Health physicians searching for the source of cholera. Possible sources included the foul airs of the sewers, the odiferous waste piles in the streets, the discharges of the sick, the close quarters of the poor, and so forth.

CHAPTER 2

2.1. Peter Gould's mapping of AIDS in the United States has been widely reproduced, sometimes with the permission of his estate, often without. It was his goal to use the maps of disease to educate college students about the danger of infection (Gould 1993). The work on diffusion, detailed in technical papers and his book, served to significantly advance understanding of the structure of the virus's diffusion.

2.2. Since the introduction of West Nile Virus in 1999 the Centers for Disease Control and Prevention and other U.S. agencies have continually mapped the presence of the disease at county and state levels. Data on disease incidence in all major species groups affected (humans, mosquitoes, sentinel chickens, veterinary mammals, and wild birds) also is collected and mapped. An index of WNV data maps from 2000-2008 can be found online at http://diseasemaps.usgs.gov/wnv_historical.html.

2.3. The map of a projected influenza epidemic is one of a series modeling the likely effect of a 1918-style event. It is used with the permission of the mapmaker, Eddie Oldfield, and his employer, the New Brunswick Lung Association. The work was done in association with the American Lung Association of Maine.

2.4. The map of air travel from Mexico to world cities was published in the *New England Journal of Medicine*. It was made by Paulo Raposo under the direction of the article's lead author, Dr. Kamran Khan (Khan et al. 2009). Permission for its use was kindly granted by the NEJM.

2.5. These maps of West Nile Virus activity in the United States were built from public databases embedded in maps on a several U.S. government websites. These include the United States Geological Service disease maps Web site (http://diseasemaps.usgs.gov/) and, for the earlier years of the epidemic, http://nationalatlas.gov/natlas/Natlasstart.asp. A perspective on the data was published in Koch and Denike 2007a.

2.6A-D. The maps of the diarrheic outbreak in Greater Vancouver are based on data purchased from the British Columbia Centre for Disease Control in 2001. Earlier versions of these maps were published in Koch 2005 (301–6).

Koch and Denike (2007b) presents a more complex treatment of the data and its mapping.

2.7A-C. These maps of different analytic approaches were developed with Professor Ken Denike as part of a course at the University of British Columbia Department of Geography: Spatial Data Analysis Using GIS. Since their first introduction and publication (Koch 2005), the teaching module has been expanded and modified (Koch and Denike 2007b).

CHAPTER 3

3.1-3.4. Perhaps the best popular volume on Andreas Vesalius is *The Illustrations from the Works of Andreas Vesalius of Brussels* (Saunders and O'Malley 1973). This provides the illustrations and an English-language translation of Vesalius's *De Humani Corporis Fabrica*. I was fortunate to be able to study an early edition of the original text at the Newberry Library in Chicago (http://www.newberry.org). There are various online sites reproducing Vesalius's work. Among th best is a site produced by experts at Northwestern University in which the Latin text is translated (http://vesalius.northwestern.edu/). A surprising number of print copies of Vesalius's anatomy, in any of its many editions, are maintained at a number of major research libraries.

3.5. A wealth of scholarship surrounds the cordiform map, both the methods of its construction and its history. Oronce Finé's famous map is also reproduced on the Web page of the Bibliothèque nationale de France (http://www.bnf.fr/loc/bnf074.jpg) and as an illustration in a recent article by Denis Cosgrove, "Mapping the World" (Ackerman and Karrow 2008, 109). A basic image can be seen on the Wikipedia page for Finé (http://en.wikipedia.org/wiki/Oronce_Finé). The map shown here is a 1560 double-cordiform map by Anton Lafréry and Gerhard Mercator from the Newberry Library's Franco Novacco Map Collection.

3.6A-B. Ortelius's *Theatrum Orbis Terrarum* went through 32 editions after 1570, a testimony to its popularity. This figure was originally plate 2.4 in this work. A surprising number of the earliest editions are in libraries. The U.S. Library of Congress has a copy and a good online description of it (http://hdl.loc.gov/loc.gmd/g3200m

.gct00003). Another copy exists at Chicago's Newberry Library that provides as well a limited set of print-slides online (http://www.newberry .org/smith/slidesets/ss02.html). Both the British Library (http://catalogue.bl.uk/) and the Folger Library in Great Britain have copies (http://shake speare.folger.edu). There are, in addition, fac-similes published by Finns Books (http://www .finns-books.com/contdocs.htm) and another by a Dutch firm, *Theatrum Orbis Terrarum* Ltd.

3.7A-B. Braun and Hogenberg's *Civitates Orbis Terrarum* exists in its many editions at a startling number of research libraries. I was fortunate to be able to peruse several, includ-ing an original Latin and a later English version, at the Newberry Library in Chicago. My favorite was a 1601 version whose London had been lightly crayoned by a child, swaths of light color ripping across the surface. It is only partially and inexpertly done, however. Presumably a parent stopped the artist from his or her rendering. A complete and detailed digital version of *Civitates* can be found at the Historic Cities Web site created by the Hebrew University Department of Geography in conjunction with the Jewish National and University Library (http://historic cities.huji.ac.il/mapmakers/braun_hogenberg .html).

CHAPTER 4

4.1-4.2. The forty-one woodblocks from Holbein's *Dance of Death* have been made avail-able online, including a downloadable version from Project Guttenberg (http://www.gutenberg .org/etext/21790). A good site with bilingual text (French and English) is http://www.lamortdansl art.com/danse/Manuscrit/Holbein/dd_holbein .htm. A low-cost facsimile of a copy owned by the Library of Congress, with the addition of an English translation, is available from Dover Publi-cations (http://store.doverpublications.com/ 0486228045.html). Eleven editions of the *Dance of Death* were published between 1538 and 1562, and as many as a hundred unauthor-ized editions and imitations may have been is-sued by the end of the century (Gundersheimer 1971, x). Copies of the original, or later editions and copies, are lodged in various art and re-search libraries, including an early edition I was

privileged to study at the Newberry Library. The images from Holbein included here are reprinted with permission from the Newberry Library.

4.4-4.5B. The only version of Arrieta's maps of which I am aware (figs. 2.7, 2.9) aside from the one at the New York Academy of Medicine Library (NYAM) is a reported copy in the Brit-ish Library (shelfmark: 663.f.3). Undoubtedly copies reside in Italian libraries as well. Jarcho (1970) used the NYAM edition while researching his 1970 article on the beginnings of medical cartography. It was this copy I used as well, first in *Cartographies of Disease* and again in this volume.

4.7. The portrait of Thomas Sydenham is widely reproduced online (e.g., http:// en.wikipedia.org/wiki/Thomas_Sydenham).

4.8. This version of the famous seventeenth-century Bill of Mortality is based on a version by Edward R. Tufte (2006), who granted permis-sion for its use here. Tufte straightened, repaired, and reproduced the oversize original so it could be seen on a modern-sized page.

4.9A-C. It appears that only two copies of the Faithorne-Newcomb map have survived One is at the Bibliothèque nationale and the other, "discovered" in 1854, is in the British Library's Crace Collection (shelfmark: Maps. Crace 1, 35). The British Library catalog also lists an 1857 tracing of the original.

CHAPTER 5

5.1, 5.2, 5.4, 5.5. Both Mathew Carey's and William Hillary's yellow fever studies are part of the online digital library at the College of Physi-cians of Philadelphia. This service provides a partial, but not necessarily the complete histori-cal document. Individual pages can be copied from the online resource. The index for the online service is at http://www.collphyphil.org/HMDL Subweb/copyright.htm. A bibliographic search shows numerous editions in over fifty libraries of a 1970 reissue of Carey's study by Arno Press. Hillary's work is available in some libraries in later editions published between the 1860s and 1880s (http://www.worldcat.org).

5.3. The illustration of Playfair's chart is taken from a recent reissue of his *Commercial and Political Atlas and Statistical Breviary*, with a forward by Howard Wainer and Ian Spence (2005). The whole is a trove of material for historians of science, medicine, and technology. The image used here was obtained from the editors, whose facsimile edition was based on a copy of the original owned by the Annenberg Rare Book and Manuscript Library of the University of Pennsylvania. The image is also online at Michael Friendly's Gallery of Data Visualization (http://www.math.yorku.ca/SCS/Gallery/).

5.6–5.7. Copies of the premier issues of the *Medical Repository*, in which Seaman's maps of yellow fever were printed, are lodged both at McGill University's Osler Library (Montreal, Canada) and the NYAM. The maps are reproduced in a number of books and articles, including Koch 2005 and Shannon 1981. For the classical discussion of Seaman's mapping, see Stevenson 1965.

5.8. Pascalis's map of the 1819 yellow fever outbreak was found at the New York Library of Medicine. The NLM also has this work in its collection, albeit as microfilm. This copy was reproduced with permission of the NLM.

CHAPTER 6

6.1 Dr. Jameson's 1820 *Report on the epidemick cholera morbus* resides today in the British Library collection (shelfmark: I.S.be.249). Another print copy is reported at the University of Utrecht, the Netherlands. A number of North American research libraries similarly report it in their collections, typically among the rare books of their medical libraries. A sample of these libraries includes those at McGill University in Montreal, Harvard in Boston, and Rush and the University of Chicago Library in Chicago. In some of these holdings the book is available in microfilm only and in others it is found in print.

6.2A-B. Brierre de Boismont's *Relation historique et médicale du choléra-morbus de Pologne* can be found both in the NLM (ID: 4720230R) and in the British Library (shelfmark: 1168.g.23). Other copies are lodged at the Bibliotheek Van de Universitat Van Amsterdam and the Universiteit Leiden, as well as a number of U.S. research libraries.

6.3. The tables of cholera cases reported to

Hamett by Dantzick hospital officials are in his 1832 report, as is his map. Hamett's 189-page publication is available both through the British Library (shelfmark: RB.23.a.2460) and the NLM (ID: 34720760R). Copies are available at various other research libraries, including McGill University and in a surprising number of U.S. medical libraries. Recently, it was added to the collection at Google Books (http://books.google .com).

6.4A-B. This map of the "progress of cholera in Asia, Europe, Africa, etc." was published in the November 19, 1831, issue of the *Lancet*. The 245 × 210-millimeter map is cataloged separately at the British Library (shelfmark: Maps. 46840,21). A number of research libraries have a collection of the *Lancet* from its inception to the present. Unfortunately, the 1831 volume is reported as missing from the NLM. Other research libraries have it, however, including the NYAM. Impressively, it is also available in the online collection of *Lancet* articles beginning in 1823 at Elselvier's ScienceDirect Web site (http://www.ScienceDirect.com).

6.5 The copy of Friedrich Schnurrer's *Die Cholera Morbus* I reviewed is in the collection of the College of Physicians of Philadelphia, the library created by Benjamin Rush. Other copies can be found in various European libraries while in the United States it is cataloged at the NLM (ID: 64750290R), as well as a handful of medical libraries. Schnurrer's *Chronic der Suchen* is also available at the NLM.

6.6A-B. Alexander Christie's "Map of the Countries Visited by Epidemic Cholera" is found in his *A Treatise on Epidemic Cholera*. The copy I first saw is in the British Library (shelfmark: T1463.[13].), but later I used a microfilm copy from the NLM catalog (ID: 34631290R). Others copies exist at Harvard's Conway Library and at the University of Chicago.

6.7A-B. Fredrick Corbyn's map is located in his 348-page *Treatise on Epidemic Cholera* and resides at both the British Library (shelfmark: Humanities. 1168.g.19) and the NLM (ID: 34631370R). At least nine other U.S. university libraries (but no Canadian ones) are said to contain copies of this book.

6.8A-B. Amariah Brigham's map of cholera

was folded into his 368-page treatise, which is in collections at the NYAM, whose copy I first used, and the NLM (ID: 34630680R). Other copies exist at a number of research libraries. The map is also discussed in Koch 2005 (62–68).

CHAPTER 7

7.1. Lithographer James Basire's 686 × 838-millimeter postal reform map (fig 7.1), commissioned in 1838 by the House of Commons for use in its debates over a national system of postal reform. The map shows interconnected routes of mail transmission within different cost areas that increase with distance from its center (shelfmark: Maps. Crace XIX, 48). It can be seen online at: http://www.bl.uk/onlinegallery/onlineex/crace/s/007000000000019u00048000.html.

7.2. This 1838 map by James Wyld shows the delivery range for two-penny stamps across Metropolitan London (shelfmark: Maps. Crace XIX, 49). Its use of bold lines of transportation and a circular imagery gives the sense of the city as an organic entity seen through a microscope. I know of no other source for this map.

7.3. This 584 × 635-millimeter map of London's metropolitan boroughs is one of two made in 1832 and has been preserved in the British Library's Crace Collection (shelfmark: Maps. Crace XIX, 53). This map, by cartographer Robert K. Dawson is easier to read and to reproduce than the larger, black-and-white map by an anonymous cartographer and engraver. "The Metropolitan Boroughs as Defined by the Reform Bill" (shelfmark: Maps. Crace XIX, 52) is also available in the Crace collection's online catalog (http://www.bl.uk/onlinegallery/onlineex/crace/t/largeimage88397.html).

7.4. This map of London registration districts in 1850 was kindly provided by the Arts and Humanities Data Service History archive, University of Essex, which has recreated boundaries in a digital gazetteer (Burton, Westwood, and Carter 2004). The project's Web site is located at http://ahds.ac.uk/history/creating/index.htm.

7.5. The map of cholera in New York in 1832 is from a book by David Meredith Reese (1833). New-York Historical Society reference librarian

Joseph Ditta kindly brought this map to my attention. It can also be found in the NLM collection (ID: 64740930R) and at several other North American libraries. At the time of this writing, antiquarian booksellers were advertising multiple copies for sale.

7.7A-B. Professor Emmanuel Elliot of the University of Le Havre provided me with a digital copy of the C. E. Helis map of cholera in Rouen in 1832. Elliott and his colleagues used the map to recreate the path of cholera over time in that outbreak (Elliott, Daude, and Bonnet 2008). Professor Elliot informs me the original, reproduced in Helis's *Souvenirs du choléra en 1832 à Rouen* (1833), is archived in both the municipal archives of the town of Rouen (the source of this copy), and separately at the Bibliothèque nationale de France in Paris.

7.8. Rothenburg's lithograph of cholera in Hamburg was published in his 1836 *Uber die Cholera-Epidemie des Jahre 1832 in Hamburg* and is currently available at the British Library (shelfmark: 7561.bbb.49.[3.]) and on microfilm at the NLM (ID: 100910779). Grainger's version of the map was included in his 1850 report to Parliament. It was reproduced in A. H. Robinson's widely available *Early Thematic Mapping in the History of Cartography* (1982, 170–71) and my *Cartographies of Disease* (2005).

7.9A-B. Chadwick's sanitary map of Leeds, embedded in his1842 report, is widely available on both sides of the Atlantic at a range of libraries, most importantly in the United States at the NLM (ID: 3140740R) and in the UK at the British Library (shelfmark: G13877–80). There the report is also collected in the *House of Lords Session Papers*, 1842, vols. 26–28; and session 1843, vol. 32. In the more than 160 years since its original submission it has been republished many times by varying writers, including Robinson (1982, 186–87) and Koch (2005, 56–59).

7.10. Grainger's map of cholera in London was included in his *Report of the General Board of Health on the epidemic cholera of 1848 & 1849* (appendix B). A good reproduction is included in my *Cartographies of Disease* where I mistakenly discounted Grainger's allegiance to a broadly miasmatic theory of cholera (Koch 2005,

870–89).

7.11A-C. Petermann's 1848 cholera map is preserved, with his notes, at the British Library (shelfmark: 1101. [01.]). A detail of the map in black and white, used to show its shading technique, is included in Robinson (1982, 181–82). Gilbert included a version of the Petermann map in a 1958 article on early British maps of health and disease. There the map is simplified, presumably to make journal production easier, but not discussed. Robinson details these changes in his work (1982, 180–81nn331, 235).

7.12. John Lea's monograph is cataloged at the NLM (ID: 101194450) and on microfilm at the University of British Columbia (Call no.: AW5. S45). The *Western Lancet* for these years is at the NLM (ID: 0031305) as well as the NYAM.

CHAPTER 8

8.1. The Crace Collection contains a number of maps commissioned by the Metropolitan Commission of Sewers as part of its program to upgrade the metropolitan system. This map shows the grid overlaid on the city to create the sewer commission survey. The dark area in the center is the area of work. The work can be found online at http://www.bl.uk/onlinegallery/onlineex/crace/i/007000000000007u00264000.html.

8.2A-B. In this map for an 1849 survey, engineers used observation posts to set up a precise triangulation system to survey location and relative altitude in the planning of drainage systems. In the detail, one can see in red altitude markings used by the commission in its design of the new system. The map was published by James Wyld in 1850 at a scale of 1:17600 (3 5/8": 1 statute mile). The map can be see online at http://www.bl.uk/onlinegallery/onlineex/crace/1 /007000000000007u00261000.html.

8.3 The portrait of a thirty-four-year-old Snow was painted in 1847 by Thomas J. Barker, whose family was cared for by Snow. The image is generally available on Wikipedia (http://en.wikipedia.org/wiki/John_Snow); a good copy of the painting—and the story of production—is available on the UCLA School of Epidemiology Web site (http://www.ph.ucla.edu/epi/snow/barker_thomas.html).

8.4A-B. Vinten-Johansen and his coauthors have kindly made publicly available online all the illustrations published in their *Cholera, Chloroform, and the Science of Medicine: A Biography of John Snow* (2003). These images, including the Albion Terrace schematic, are now lodged at The John Snow Archive and Research site (http://www.matrix.msu.edu/~johnsnow/book_images8.php). Permission for their use here was given by Professor Vinten-Johansen.

8.5, 8.6A-B. Thomas Shapter's *The History of Cholera in 1832 in Exeter* is widely available at North American and British research libraries, including the British Library (shelfmark: 7560.d.59.), the NYAM, and the NLM (ID: 64750630R). A reproduction of the map is also found in Koch 2005 (49–52).

8.7. The portrait of William Farr is available on Wikipedia (http://en.wikipedia.org/wiki/William_Farr). An equally forbidding portrait is in the online Portraits of Statisticians gallery, maintained by the University of York in Heslington, UK (http://www.york.ac.uk/depts/maths/histstat/people/farr.gif).

CHAPTER 9

9.1. The portrait of Dr. John Sutherland was taken from his obituary in the *Illustrated London News* (August 1, 1891) and was more recently used in Smith 2002.

9.2-9.8. Farr's 1852 report, with its maps, charts, and tables, is available at the NLM (ID: 101305438), NYAM (New Stacks S.72), and of course, the British Library (shelfmark: B.S.34/12.). Other copies are available at the Wellcome Trust in London and in the United States at Chicago, Harvard, and Minnesota university libraries, among others.

9.9A-C. Snow's South London registration map is in the second edition of *On the Mode of Transmission of Cholera* but not in the 1836 reprint edited by W. H. Frost. *On the Mode of Transmission of Cholera* is collected in the British Library (shelfmark: 7560.e.67), The Johns Hopkins University Institute of Medicine Rare Books Room (control no.: 15650776), the NYAM, and other libraries. A reproduction of the South London map can be found in Robinson 1982 (179) and Koch 2005 (93).

9.11. This portrait of John Simon is a copy of a photograph reprinted in Frazer 1950.

9.12. Simon's *Report on the Last Two Cholera-Epidemics of London as Affected by the Consumption of Impure Water* is located in the NLM (call number: WCB S594r 1856), the Wellcome Trust Library (record number: b10479557), and at least five U.S. university libraries. I first read it at the British Library in London.

CHAPTER 10

10.1. The map of St. James parish in 1720 by engraver Richard Blome was based on a 1683 map of the area. It was then updated for the 1750 imaging of the area. These maps are available from the Crace Collection. The 1720 map can be seen online at http://www.bl.uk/onlinegallery/onlineex/crace/t/007000000000015u000370a0.html.

10.2. Whitehead's map of St. Luke parish is in his 1854 pamphlet *The Cholera in Berwick Street*. Copies exist in the British Library (shelfmark: 7560.b.76) and the NLM (ID: 101169980), as well as a few other libraries.

10.3. was constructed using ArcGIS 9.1 software (ESRI), and later modified using Arc-GIS 9.2. First, St. Luke Parish boundaries were transferred from Whitehead's parish map to a digital version of Snow's 1854 map from *On the Mode of Transmission of Cholera*. Then Whitehead's area of intense cholera activity, described by street in his report, was used to create the area of intense cholera activity.

10.4. Snow's iconic map of the Broad Street outbreak in *On the Mode of Transmission of Cholera* is available at the British Library, the NLM (ID: 101094474), and other research libraries, including the library of the College of Physicians of Philadelphia, the NYAM, and so forth. One of the few reproductions of the map in its entirety is in Robinson 1982 (177). It has been redrawn by a number of authors, whose alterations have been dramatic (Koch 2005, 96).

10.5. Snow's map for the parish inquiry report is in the collections of a number of research libraries and has been reproduced in a number of publications, including Koch 2005. Its irregular polygon, based on walking distance,

is not, as others have claimed, either a Thiessen polygon or a forerunner of them (McLeod 2000). It was, however, an immensely innovative attempt to create a nearest neighbor-style district within which incidence could be calculated.

10.6. Cooper's cholera map can be found in a number of locations. I first encountered it at the library of the Philadelphia College of Physicians, which kindly provided me with a copy. I later reviewed a better preserved copy at the British Museum. It has been variously published, in part or whole, in Koch 2005 and other texts.

10.7A-B. Reverend Henry Whitehead's version of the Board of Inquiry map is located in the British Library (shelfmark: 7560.c.58). Details of the map have been reproduced previously in Koch 2005 (111–13).

10.8. For information on T. Heber Jackson's study of cholera in the 1850s I am indebted to historian John B. Osborne, who shared both references and his own published works in this area.

10.9. Acland's study of cholera in Oxford is relatively famous. It is described in a number of studies of cholera in this era and has been elsewhere reproduced in, for example, Koch 2005. As a cartographic argument relating altitude to disease incidence using contour lines, it is a marvelous example of the medical cartography of the period.

10.10. This portrait of Whitehead is copied from a late-nineteenth-century book on his life and contributions (Rawnsley 1898). There appear to be no portraits of him in his early years or with John Snow.

CHAPTER 11

1.1. Sedgwick recreated a map of the Broad Street outbreak, after the original map by Snow, for the 1911 edition of his *Principles of Sanitary Science and the Public Health*. This map appears to be consistent across all editions of his textbook, published between 1901 and into the 1920s. The copy I used was obtained from the Woodward Library at the University of British Columbia. Sedgwick's text is generally available in a range of medical and health-related libraries in England and in North America. For a discussion of the map and its use by others see Koch

2004.

11.2, 11.3, 11.6-11.8. Frost's maps, graphs, and some of his papers are collected in Maxcy's 1941 *Papers of Wade Hampton Frost, M.D.* Maxcy's book is widely available at numerous research libraries, including, of course, the NLM (ID: 29420550R). Most of Frost's work cited here was first published in the *Bulletin of the Hygienic Laboratory*, whose 154 issues ran from 1900 to 1930 and are archived in the NLM (ID: 21420100R).

11.4, 11.5. Alice Hamilton's typhoid fever research, published in 1903 in the *Journal of the American Medical Association*, was done during her tenure at Chicago's Hull House. Rina Schultz (2007) recently gathered, with a good introduction, a collection of Hull House maps and papers that serve as an introduction to the sociology the Hull House initiative championed in its intensive study of Chicago's Nineteenth Ward. Hamilton's paper was an enormous triumph, combining bacteriology, etymology, and public health methodologies in a single argument.

11.9-11.11B. These maps were all made using ArcGIS software (ESRI), version 9.2. Waldo Tobler was among the first to develop a popularly available GIS version of Snow's streets, pumps, and deaths data sets (http://www.ncgia.ucsb.edu/pubs/snow/snow.html). My work began in the 1990s with a data set provided by ESRI, Inc., which appears identical and is likely derivative of Tobler's published work. In addition, the Epi Info 2000 data set from the U.S. Centers for Disease Control and Prevention has a partial version of Snow's data set, albeit with an incomplete record of deaths, pumps, and streets. In preparing the work for this chapter, I have been able to compare carefully the respective data sets of Cooper, Whitehead, and Snow for the first time. For a general statement on this mapping, done by hand, see Koch and Denike 2009.

CHAPTER 12

12.1, 12.2. Haviland's *The Geographical Distribution of Disease in Great Britain*, the source of the maps of the Cumberland and Westmorland districts, is available at the NLM (ID: 64011350R). In Great Britain it is reported to be at libraries at Cambridge, Oxford, and London universities, the British Library, and others. The copy from which these illustrations were taken is at the University of British Columbia's Woodward Library.

12.3A-B. D'Arcy Power's articles, and maps were found in issues of *The Practitioner* at the NLM (ID: 0404245). While some libraries have microfilm of the medical journal, which was published in London, no comprehensive list of other locations has been compiled.

12.4. Arnaudet's Plan de Cormeilles is one of several maps in his study of cancer in Normandy from *La Normandie médicale*. It is collected at the NLM (ID: 19210310R). In Rouen, and in some libraries, it appears to be listed under the title *Societe de medicine de Rouen*.

12.5-12.8. Four editions of *Charles E. Green's Cancer: A Statistical Study* were published between 1911 and 1917. Figures 10.5–10.7 were taken from the 1917 edition at the NLM (ID: 12911480R). The early editions were all between 89 and 94 pages, while this edition was 140 pages. The 1911 edition seems to be the only one at the British Library (shelfmark: 7422.3.2). No serious attempt has been made to find other copies. Permission to reproduce these maps was graciously provided by the publisher, Green and Son.

12.9, 12.10A-B. Percy Stock's 1928 report to the International Conference on Cancer is collected at the British Library (shelfmark: W38-8225-1938.DSC). I obtained the 1936 report from which figures 10.9 and 10.10 were taken through interlibrary loan at the University of British Columbia from the NLM (where the annual reports published between 1928 and 1942 are collected under the ID 101251480). Attempts to identify the current rights holder for Stock's work have been unsuccessful.

12.11. The early National Cancer Institute atlases are widely available in most academic and some public libraries. Dr. Linda Pickle, a senior statistician at the NCI for thirty years who worked on atlases from the late 1970s to 2010, provided the maps included here. In producing this series, she transformed the 1970s atlas page, which used a different coloration scheme, to assure a visual consistency across the map set. The maps were based on NCI data and produced with an ESRI mapping program.

AFTERWORD

13.1. This image from an 1832 issue of *McLean's* magazine can be found on the NLM's Web site on the history of cholera (http://www.nlm.nih .gov/exhibition/cholera/images.html). With other images collected from that site and used in this book (for example, those in chapter 1), they are provided for researchers as part of the NLM mandate. I am obliged, however, to the NLM's Crystal Smith for her help in tracking down and obtaining images at a publishable resolution.

ILLUSTRATION CREDITS

3.5, 3.7*A*, 3.7*B*: Photo courtesy of the Edward E. Ayer Collection, The Newberry Library, Chicago; 3.6*A*, 3.6*B*, 4.3: Photo courtesy of the Franco Novacco Map Collection, The Newberry Library, Chicago; 4.1, 4.2: Photo courtesy of the John M. Wing Foundation, The Newberry Library, Chicago; 4.4, 4.5*A*, 4.5*B*, 5.8, 11.1: Courtesy of the New York Academy of Medicine Library; 4.8: Courtesy of Edward Tufte ; 4.9*A*, 4.9*B*, 4.9*C*, 7.1, 7.2, 7.3, 7.11*A*, 7.11*B*, 7.11*C*, 8.1, 8.2*A*, 8.2*B*, 10.1*A*, 10.1*B*: © The British Library; 5.6, 5.7, 11.4, 11.5: Osler Library of the History of Medicine, McGill University, Montréal; 6.1: Courtesy of the Boston Medical Library, Francis A. Countway Library of Medicine; 6.2*A*, 6.2*B*, 6.3*A*, 6.3*B*, 6.4*A*, 6.4*B*, 6.6*A*, 6.6*B*, 6.7*A*, 6.7*B*, 6.8*A*, 6.8*B*, 7.5, 7.6, 7.8, 7.9*A*, 7.9*B*, 7.12, 8.5, 8.6*A*, 8.6*B*, 9.2, 9.3, 9.4, 9.5, 9.6*A*, 9.6*B*, 9.7*A*, 9.7*B*, 9.12, 10.2, 12.3*A*, 12.3*B*, 12.4, 13.1: Courtesy of the National Library of Medicine; 6.5: Courtesy of the College of Physicians of Philadelphia; 8.4*A*, 8.4*B*: Courtesy of Peter Vinten-Johansen; 10.10, 12.1, 12.2: University of British Columbia Library, Rare Books and Special Collections; 12.11: From *National Atlas of Disease Mortality in the United Kingdom* (after G. M. Howe 1963); 12.12: Courtesy of Linda W. Pickle.

NOTES

CHAPTER 1

1. Nor is the history of visualization as a means of disease definition limited to the historical. For a history of the importance of seeing and the progressively complex technologies that enabled disease states to be visualized in the twentieth century, see Kevles 1997.

2. Gunnar Olsson's 2007 *Abysmal: Critique of Cartographic Reason* makes a similar argument across a vast spectrum of social histories. It attempts to merge the histories of art and history with that of cartography, with not simply the boundaries of the landforms of the world but the content that exists within those boundaries. In reading Olsson I have been helped by John Pickles's 2007 review of Olsson's work.

CHAPTER 2

1. For example, consider the mapping—and the map-intensive news stories—that results in neighborhoods like Sayreville, New Jersey, reporting cancer clusters local officials deny exist (Sloan 2007). Specific questions related to cancer mapping and issues of clusters are discussed here in chapter 12.

2. "Tracking the trail of BSE fears," *Vancouver Sun*, May 24, 2003.

3. The World Health Organization offers a series of publication on the varying forms of tuberculosis. All are collected on their Web site at http://www.who.int/tb/challenges/en/. The CDC reports on tuberculosis outbreaks as well in its *Morbidity and Mortality Weekly Report* (http://www.cdc.gov/mmwr/).

4. While asserted here as obvious, the idea of mapping as a propositional argument is relatively new. In 2007 I was fortunate to have the opportunity to read proofs of Wood and Fels's *The Nature of Maps* (2008). Their analysis struck me as immediately and directly applicable to the subject of disease studies I was then struggling with. I am therefore deeply obliged to them for that opportunity.

5. An earlier description of this outbreak was included in Koch 2005, 301–6. The Salmonella database is based on records at the British Columbia Centre for Disease Control. The data set, in which cases were randomized to within two blocks, was purchased from an employee at the BCCDC. For a discussion of its utility in teaching and as an analytic example, see Koch and Denike 2007a.

6. Theodore Smith, one of America's first research scientists, isolated the salmonella bacillus in pigs. Mistakenly believing it was the cause of a porcine cholera, he named it *Salmonella choleraesuis*. In 1900, the entire family to which this bacillus belongs was named Salmonella in honor of David Elmer Salmon, a veterinarian with whom Smith worked. For a brief review of the bacterium and the history of its discovery, Wikipedia is a useful lay source (http://en.wikipedia.org/wiki/Salmonella).

7. "Richmond bakery probed after 47

cases of food poisoning," *Vancou-ver Sun*, September, 12, 2000.

8. Know in French as *le grippe*, the type A influenza was actually two outbreaks. The first originated in U.S. training camps and a second, even more virulent strain manifested simultaneously in 1918 in Brest, France, Freetown, Sierra Leon, and Boston, Massachusetts. It was called "Spanish" because Spain did not partici-pate in World War I and did not impose press censorship during the period of the pandemic. A very complete popular review of the epidemic can be found on Wikipedia entry (http://en.wikipedia.org/wiki/1918_flu_pandemic).

CHAPTER 3

1. Finé's *Recens et integra orbis de-scriptio cordiform* world map is beauti-fully reproduced in Akerman and Karrow 2007. I am obliged to Ruth Watson for her discussion of the relationship between the cordiform world map and the history of sixteenth-century anatomy. That personal discussion went well beyond her recent paper on the cordiform map (Watson 2008) and the earlier work of Mangani (1998).

2. The 1606 English-language version of Ortelius's *Theatrum* was printed by John Norton of London. It is one of several copies of the *Theatrum* in the Newberry Library's extraordinary historical car-tography collection. I am indebted to the Newberry librarians and staff for making it and several other editions in various languages available to me.

3. There is a huge volume of scholarly writ-ings that have grown around this and other early atlases and the printing tech-nology that permitted their production. Varying editions of *Civitates* are preserved at a range of different research libraries. I was fortunate to view several, including a 1600 English-language edition, at the Newberry Library in Chicago (http://www.newberry.org/smith/slidesets/ss04.html).

4. Two histories focus on the imaging of human anatomy, and anatomy text-books, over time. Rifkin and Ackerman (2006) present a near-encyclopedic history of anatomy and anatomical texts from the Renaissance into the current digital age. Mayor (1984), an art his-torian, considers instead the classical tradition through the 1600s with only a modest nod at more recent work. Both are, in their own ways, useful.

5. This is a point nicely made in essays by unnamed authors that accompany an online version of *de Fabrica* produced by Northwestern University. See http://vesalius.northwestern.edu/flash.html.

6. The Library of Congress lists in its ex-traordinary map and rare book collection a number of examples of Mercator's work, including his 1595 atlas. An inexpensive CD-ROM of Mercator's atlas, with excel-lent commentary, was recently issued (Mercator, Jackson, and Karrow 2000).

CHAPTER 4

1. The biblical quote accompanying Hol-bein's woodblock print of Death and the merchant warns against, as a 1971 trans-lation puts it, "The getting of treasures by a lying tongue" (Proverbs 21:6). The text accompanying the image warns, "The Merchant's wealth's a worthless thing, of others, won by lies, the spoils; But Death will sure repentance bring, Snaring the snarer in his toils" (Holbein 1971, 134). The text therefore is a caution against the lies that were the salesman's, and a caution against the mercantile search for profit over the search for virtue. That said, the idea that death was in the merchant's bale, brought by ship, also illustrated the belief in plague as a portable good that traveled in and with the ships of the day.

2. Saul Jarcho was a medical practitioner and historian. The author of over five hun-dred scholarly articles and a specialist in tropical medicine, he also edited the *Bul-letin of the New York Academy of Medi-*

cine for over a decade. A bibliography of his work is available online at http://www.whonamedit.com/doctor.cfm/3151.html.

3. I am obliged to research librarian Arlene Shaner of the New York Academy of Medicine's library, who helped me translate the maps' legends and review salient parts of Arrieta's text. It was at the NYAM that I have twice had the opportunity to both study Arrieta's maps and ponder his text.

4. The National Library of Medicine lists 117 items under "Sydenham, Thomas, author." While some are later editions of his work, or commentaries on it by others, an impressive number of titles are original to Sydenham. These writings appear to be in English, German, and of course, Latin. Thus some are simply the translation of his work into another language, although there is an astonishing range to the works he produced. Of equal monument is the degree to which by the period of his greatest productivity, the mid-seventeenth century, printed materials were already items of general knowledge translated and produced internationally for broad consumption.

5. William Hillary's text is one of a number of eighteenth- and nineteenth-century publications now available, in part or in whole, through online digital libraries. The preface to Hillary's *Observations on the changes of Air* . . . was found in the online offerings of the College of Physicians of Philadelphia, the oldest medical library in the United States (http://www.collphyphil.org/HMDLSubweb/catalogue.htm#Hillary%20W).

6. Slack is at some pains to both describe the various sources he used in reconstructing Norwich mortality figures and the parish populations they are based on. He is confident in the general correlation between mortality from plague and general socioeconomic status, parish to parish, if not in the accuracy of the specific numbers reported for individual parishes (Slack 1985, 138n68, 372nn662–71).

7. It would not be until the first years of the twentieth century that these three epistemic things, the different plagues, were joined into a single disease. In the last great pandemic of the 1890s the response in many countries was to spread lye as a cleansing agent in the homes of those affected or to burn the houses entirely. For a review of the history of plague, see Marriott 2003.

8. This point is made by J. A. I. Champion in an article reproduced online as "Epidemics and the Built Environment in 1665." The entire edited collection, *Epidemic Disease in London* (Champion 1993) was published by the Centre for Metropolitan History (http://www.history.ac.uk/cmh/epipre.html).

9. Porter's interest resides in the plague experience, and especially its occurrence in London. In service of that focus his knowledge of plague as an international pandemic that periodically visited England, where it was endemic, is vast. Hacking's interest is principally in the history of mathematics and the idea of probability that arose from the work of Graunt and others.

10. There is a very good basic discussion of Graunt's work, and of this period, in Ian Hacking's discussion of Petty's political arithmetic and the perspective it spawned (see Hacking 2006, 105–10).

11. The general history of the growth from simple counts to population ratios and denominators and from there to the world of more complex statistics is beautifully told in Hacking's books on the history of chance and probability (Hacking 1990, 2006). Here the broad history he traces is summarized in its application to disease studies.

12. For a discussion of Foucault's conception of "bio-power" and the importance of the generation of populations see Shelly Tremain's brief discussion and excellent references (Tremain 2005, 4–5).

13. I am obliged to historian David Marsh for a copy of his 2007 University of London

lecture on Faithorne and Newcourt's map of London. Presented in the Maps and Society lecture series directed by Catherine Delano-Smith, it was through Marsh that I first learned of the use of this map and its survey by early demographers. At the time of this writing Marsh's lecture is scheduled as a future publication in the journal *Imago Mundi*. Delano-Smith's 2005 paper on the use of stamps in Renaissance maps provided another layer of understanding of the authoritative look of maps like Arrieta's and the complex production materials assembled to assure the publication of maps in this era.

14. White and Hardy (1970) give the source of both Huygens's graph and the letter accompanying it as Huygens, *C. Ouvres Complètes*, 19 vols. La Haye, 1888–1937. The graph has been variously reproduced by a range of authors, including Spense, sometimes in what appears to be a copy of the original and other times as a contemporary, modernly constructed graph.

CHAPTER 5

1. "Black water fever" is, in fact, a complication of malaria in which the blood breaks down and kidney failure occurs. Symptoms include chills, high fever, jaundice, vomiting, passage of black urine and stool . . . and death. Its symptoms are similar, but it is distinct from yellow fever. The differences were not necessarily obvious, however, to the eighteenth-century physician.

2. Ian Spence at the University of Toronto's Department of Psychology kindly made available to me both Playfair's graph of trade between Barbados and England and notes on the psychology and utility of early graphs in the eighteenth century.

3. The National Library of Medicine lists over thirty citations for works produced by Mathew Carey as author, printer, or bookseller. Like Noah Webster and others of this era, the conjunction of printing and publishing, on the one hand, and medicine as well as other sci-

ences, on the other, seemed natural. It would only be later, in the nineteenth century, that the idea of a publisher writing on science, or a medical investigator who was also a book printer, would come to be seen as strange.

CHAPTER 6

1. I am very obliged to John B. Osborne for drawing my attention to Tanner's writing and to his maps. An urbanist, Osborne has written carefully on the cholera experience in U.S. cities in 1832, and it was in those publications I found a reference to Tanner's maps. Unfortunately, I have to date only read Tanner's work in microfilm copies in which the map is absent, and thus, here I rely on his own description of its coloration as a method of argument.

CHAPTER 7

1. I am obliged to Professor Michael Friendly of York University in Toronto for this point: that medicine developed in tandem with the bureaucracy and the sense of general political concern, not as a science divorced from government or the citizenry.

2. The British Library's Crace Collection holds a trove of nineteenth-century postal maps that serve both to document the stages of the system's development and the importance of the mapped medium in the process. Other maps similar serve to explore the expansion of the London sewer system in the 1840s and 1850s (http://www.collectbritain.co.uk/collections/crace/). For a review of the collector's maps of London see Whitfield 2007.

3. In the United States it would not be until 1906 that large-scale, permanent geographic areas were first proposed as catchments for health and socioeconomic data. These would replace the sometimes eccentrically gerrymandered wards whose boundaries changed, election to local election, and whose differences in area and population made

them difficult to analyze (Krieger 2006).

4. The official citations for these acts are An Act for Marriages in England. 6&7 Wm. IV. C. 85 (1836) and An Act for Registering Births, Deaths, and Marriages in England. 6&7 Wm. IV. C. 86 (1836): 6 & 7 Will. 4 C A P. LXXXVI.

5. John B. Osborne's research on the reaction of United States and Canadian cities to cholera in the early 1830s fills in what was a significant gap in research on nineteenth-century responses to cholera. Uniquely, he emphasizes both the specific and the general bureaucratic responses of urban boards of health and the method by which they sought to first gain data on cholera and then apply that data to their own locales (Osborne 2008).

6. I am very much obliged to Professor Emmanuel Elliott of the Université du Havre, France, for both this copy of the Hellis map of cholera in Rouen and for his translation of sections of the text. Elliott's reading of the map and the text that accompanied it was presented at the 2007 meeting of the American Association of Geographers and again in a more detailed discussion at the 2008 meeting (Elliott, Daude, and Bonnet 2008).

7. As Dale Porter notes, the meaning of sewage changed from the 1830s through the 1850s. Before the 1830s it was "any surface runoff involving liquids" and assumed to be a likely source of miasmatic disease. In the 1840s and 1850s it was defined as generally but considered by most to be more a nuisance than something dangerous (Porter 1998, 54). By the 1850s an average of 260 tons of sewage passed daily through 71 main sewer outlets in London, all leading to the Thames.

8. Petermann's 1848 maps of cholera are typically discussed as if only the national map was produced. This may be because Jusatz published a facsimile of the national map alone in an article in his journal *Petermann Geographische Mitteilunger*. It was this map, not the original, that Gilbert referenced in his distorted version of Petermann's map in a well-known article

(E. W. Gilbert 1958). Robinson (1982) draws attention to these distortions in his writing, citing a copy of the original, which is archived at the British Library.

9. The interlibrary loan system that located for me a copy of Lea's article listed the year of publication as unclear. The dates given in the bibliography are the most likely based on the writing of others, including a relatively recent 2002 article by G. Davey Smith in the *International Journal of Epidemiology*.

10. The Salford study arguing the relation cited by Lea was included in Dr. John Sutherland's report to the General Board of Health on the 1849–1849 British cholera epidemic (Smith 2002). It was likely from this report that the New York edition of the *London Quarterly Review* extracted its material.

CHAPTER 8

1. Morris (1976) gives a number of sources both for Lord Morpeth's address to the House of Commons and for a general review of the sense of both Parliament and the public in 1848. These include *Hansard's Parliamentary Debates*, 3rd series, vols.96–101 (1848); *People's Journal*, vol. 6 (1848), 213; *Tait's Edinburgh Magazine*, vol. 16 (1849), 1, 21–22. For scholarly completeness and a clear style of address few books have yet to match Morris in his consideration of cholera in its first pandemic appearance. The relevant section of Morris's epilogue, including Morpeth's address, has been posted on the resource pages developed by UCLA's Professor Ralph R. Frerichs (http://www.ph.ucla.edu/epi/snow/morris/morris_publichealthacts_a.html).

2. The acts of 1847 and 1848 were joined and then cited as:11 & 12 Vict. Cap. 123 ; 12 (1849).

3. Why an invisible, waterborne agent would be any less dismal than one that was airborne is unclear. Certainly Snow was not opposed to the idea of microscopic, disease-causing agents. And

as an anesthesiologist he was certainly comfortable and family with gaseous activity. Still, the rhetoric allowed him to forcefully distance himself from airborne theories of communicable disease.

4. The use of E.A. Parke's 1855 critique of Snow's 1849 monograph rather than an 1849 review serves to make a critical point. Snow's general approach and thesis and much of his data were carried from the 1849 monograph directly into the expanded second edition published in 1855. In that transposition Snow did not materially alter its presentation in light of comments and criticisms argued after the first publication. The length and detail of the 1855 critique exceeds that of Park's 1849 review of Snow's short monograph and thus provides a better understanding of contemporary readings of Snow's work from 1849 through 1855.

5. In table 8.1, Snow's total population and total mortality figures result from summing the columns of his data for all rows. Snow calculated total "mortality per 1,000 persons" by dividing the total deaths from cholera by the total population and then multiplying by 1,000 persons. This is certainly fair enough. But this approach loses the specificity gained by the regional divisions. Another and better approach would have been to take the mean of all the mortality ratios in his table. Snow handled regional data in a similar fashion in his 1856 study of cholera South London's registration districts (Koch and Denike 2006).

6. I am deeply obliged to Professor Peter Vinten-Johansen for bringing these papers, and their use of maps, to my attention.

7. Snow never married and did not leave family members with tales of his early years and his medical training. Nor was he a diarist; he kept only a casebook that recorded his procedures. Most of our knowledge of Snow comes from the reminiscences written after his death by friends like Richardson (1858) and Whitehead (1866). These materials, and what

might be gleaned from elsewhere, were used for a recent biography of Snow (Vinten-Johansen et. al 2003). Since then, other writers (Koch 2005, Johnson 2006, Hempel 2006) have used that biography as a resource in their writings on Snow.

8. In their biography of Snow, Vinten-Johansen et al. (2003) provide a complete listing of Snow's papers and, in the book's first chapter, the best summary to date of Snow's early years.

9. Ian Hacking provides a very useful and readable primer of the history of this transformation, its meaning for the nature of acceptable evidence, and the manner in which statistical applications changed the thinking about realities, in his *The Taming of Chance* (1990).

10. I am indebted to Professor Ralph Freich for his online reproduction of this section of the registrar general's *First Annual Report* (1838, xiv).

CHAPTER 9

1. John Sutherland was appointed an inspector for the then newly constituted Board of Health in 1848, and in this capacity contributed to the *Report of the General Board of Health on the Epidemic Cholera of 1848–1849* (Smith 2002, 924; Snow 2002). With others of his era he was willing to admit that water might be a predisposing condition for the introduction of cholera but not the principal medium of its generation and diffusion.

2. An appendix to the second edition of Snow's *On the Mode of Communication of Cholera* included a table of cases collected in South London during the 1854 outbreak. For these cases he included the definition given on the death certificate by a local physician. Over fourteen different diagnoses were reported for choleric and dysenteric disorders.

3. A very good, very simple description of the mathematics of coxcomb graph design is posted online at http://understandinguncertainty.org/node/214.

4. I am obliged to Professor Michael Friendly of York University, Toronto, for information on Guerry's early use of the polar projection in the presentation of meteorological data in *Annales d'Hygiène Publique et de Mèdecine Lègal*.

5. The "rice-water discharges" were a symptom distinguishing Asiatic cholera from English cholera caused by food poisoning. The rice-water appearance, occurring in violent and sustained diarrhea, results from the discharge of intestinal epithelial cells that to the naked eye appeared to resemble rice grains.

6. Tithe commissioners visited parishes where monetary rents had not been agreed to and attempts to adjudicate disputes had been unsuccessful. As part of the review process parishes were mapped, often with altitude data, in reports filed with the government.

7. In a recent reworking of Farr's data using logistic regression, British epidemiologists have shown that elevation above high water, water supplier, and income each had independent and statistically significant effects on registration district cholera mortality rates (Bingham, Verlander, and Cheal 2004). The result, which uses modern statistical techniques unavailable to Snow or Farr, improves contemporary studies of this historical data but does not bear upon Farr's assertion, based on mid-nineteenth-century statistical analytics, that the only strong correlation in his data was between altitude and mortality.

8. I owe this analysis, and the use of the scatter plots, to my friend and frequent coauthor Professor Kenneth Denike of the University of British Columbia. He and I together analyzed Simon's report and his data, but it was Ken Denike who saw the problem of resolution and scale. The scatter plot is based on his use of Simon's data. So too is the suggestion that only a modern Bayesian analytic using a quadratic regression analysis would be sufficient to solve the problem Simon saw. Work on that solution is currently underway.

CHAPTER 10

1. The number of pumps varies across the nineteenth century maps of the Broad Street investigations from eleven to fifteen. Snow mapped thirteen and fourteen public pumps, respectively, in his two maps of the study area. Whitehead mapped fifteen public pumps, while Cooper mapped only eleven. The matter is further confused by later appropriations of the map by contemporary authors who typically include only eleven pumps in their iterations of what they call John Snow's map (Koch 2004).

2. There are a number of versions of Snow's Broad Street map available online. Many—indeed, most—are variations on Snow's map drawn by twentieth-century mapmakers that change the symbology of the original map and the data in it. Perhaps the best digital presentation of this map can be found on Ralph Frerich's John Snow Web site at the University of California, Los Angeles: http://www.ph.ucla.edu/epi/snow/snowmap1_1854_lge.htm.

3. Most historians have assumed the number of deaths was the same in Snow's two maps of Broad Street. Overlaying the two in a GIS and comparing them uncovers a previously unreported difference in Snow's two mapped data sets.

4. It was a forerunner of what today is called a Thiessen polygon (Bailey and Gatrell 1996, 156). They are built from a database in which the nearest member of one set, in this case pumps, is measured from the target set, in this case the cholera deaths.

5. "Letter—The Cholera in St. James Parish," *Bell's New Weekly Messenger*, September 7, 1854.

6. "Decline of the Cholera and the Infamous Gully-Holes in the Street—Life Destroyed by Exhalations from the Sewers," *Bell's New Weekly Mes-*

senger, September 24, 1854.

7. See also Johnson 2006, 52. I am obliged to Steve Johnson for copies of newspaper stories from the period of the Broad Street inquiry.

8. The cholera—this day. *Bell's Life in London*, September 7, 1854, 3.

9. As Peter Vinten-Johansen has pointed out to me privately, what I sometimes have called "Whitehead's map" is technically a collaboration between both the parish Cholera Inquiry Committee and the Board of Health. In the parish report a radius of cholera activity. The map in the Board of Health report was identical except for the exclusion of the radius of cholera activity centered on the Broad Street pump. I am vastly obliged to Professor Vinten-Johansen for his clarification and his history of this map.

10. On both Snow and Whitehead there are both Whitehead's writings from the 1860s and the work of Whitehead's biographer who, writing two years after the priest's death, described a strong relationship between the doctor and the priest (Johnson 2006, 181; Rawnsley 1898).

11. John Simon's quote on mapping is included as well in P. N. Gilbert 2004, albeit with a different interpretation.

12. I am grateful to Pennsylvania historian John B. Osborne for a photocopy of the Jackson report. Osborne also kindly sent me the draft of an article he has been preparing on the Columbia, Pennsylvania, outbreak and its request to the College of Physicians of Philadelphia for medical aid. As a result, fellows of the college traveled to Columbia to assess the situation and attempt to discover its origin, and thus its relation to different theories of the disease. Jackson's study stemmed from the request to the college and its attempt to provide assistance (Osborne 2009).

CHAPTER 11

1. What Snow wrote was: "but the attacks had so far diminished before the use of the water was stopped, that it is impossible to decide whether the well still contained the cholera poison in an active state, or whether, from some cause, the water had become free from it" (1855a, 51). Whitehead, in the 1860s, thought it possible that the removal of the pump handle stopped a second wave of infection from occurring.

2. For a detailed discussion of the manner in which Sedgwick transformed studies of typhoid fever into instructional cases that served to emphasize a single methodology, see Koch 2005, chap. 8.

3. Following Sedgwick, Frost's promotion of heroic Snow was very much an American fiction. British and European medical and public health communities were slower to adopt this story of Snow as a foundation myth of their health sciences. See, for example, Vandenbroucke, Eelkman, and Beukers 1991 for a continental perspective. By the end of the twentieth century, however, the Snow story had gained global prominence.

4. There are obvious parallels with the famous *E. coli* outbreak in Walkerton, Ontario, in 2000. There the contamination was traced to a single well contaminated by a local pig farm, but only a month after the outbreak began. In this recent Ontario outbreak seven died and hundreds became ill, a reminder that because the epidemiology is simple and well known does not mean it isn't still useful.

5. Many physicians have spoken fondly of the inspiration of de Kruif's *The Microbe Hunters*. Abraham Verghese, on the other hand, credits writers like A. J. Cronin's *The Citadel*, and especially Somerset Maugham's *On Human Bondage* for his interest in medicine (Verghese 2005).

6. Huge literatures have evolved that ask about the nature of a host of health states. For example, is obesity a matter of simple willpower (Just say No!), the result of social conditions, or perhaps, a matter of genetically informed patterns of metabolism? Is it the same thing, defined by body mass, in

Polynesians as it is in smaller-boned persons of Japanese ancestry?

7. The question and its answer were both suggested by participants in two lectures I presented based on earlier drafts of this material. The first was at Texas A&M University where I was the Hayes lecturer in medical geography in November 2007. Audience feedback from that presentation permitted a tighter focus of this argument in a 2008 lecture for the seventeenth annual Maps and Society series hosted by the Wartburg Institute University of London. I am obliged to the audiences of both communities for their questions and suggestions.

8. There is no way to determine if Snow knew about Thiessen polygons, although it is clear they were first argued as an analytic during his working lifetime. Some have suggested, incorrectly, that the walking area Snow distinguished around the Broad Street pump *was* a Thiessen polygon (McLeod 2000). Polygons are regularly sided and based on a simple algorithm; while Snow's walking area was eccentric and irregular. Another way of generating the pump catchments, available to us but not then available to Snow, employs a nearest-neighbor analysis (Bailey and Gatrell 1996). By first associating each death with a pump nearest to it and then calculating the distance between each case and its nearest pump, a more precise set of associations is rendered. Using nearest-neighbor analysis the number of deaths assigned to the Broad Street is the same as those identified in Snow's irregular walking area.

9. Contemporary statisticians would require other tests in considering the data Snow argued and I present. They would also suggest different calculations, including risk ratios. Because these techniques were not in wide use in Snow's day they are not used here, with the exception of risk ratios, whose utility is so great it seemed wise to present them without dwelling on their significance.

10. In his 1852 study Farr included a meteorological table that includes wind speed and direction for London generally in 1849. That data was collected as a matter of course by technicians at the Royal Greenwich Observatory.

CHAPTER 12

1. Materials in this chapter expand upon a plenary lecture I gave in 2007 at the annual meeting of the North American Association of Central Cancer Registries (NAACCR). In Canada and the United States all provinces and states have local registrars whose responsibility is the collection and distribution of data on cancer incidence in their health jurisdictions. My knowledge of cancer's transformation into a public health disease has been deepened through discussions with members of NAACCR and other researchers in the field. My acquaintance with Dr. Linda Pickle, for more than thirty years a researcher and statistician at the U.S. National Cancer Institute, has been especially helpful and informative.

2. For a discussion of Malgaigne's map of hernias among military recruits in France see Koch 2005, 45–48.

3. I am obliged to Charlotte Sparham of Vancouver for her assistance in translating the Bakker's Dutch journal paper into English.

4. Founded in 1923 "to attack and defend the disease of cancer in all its forms," the British Empire Cancer Campaign was in almost constant rivalry with the then well-established Imperial Cancer Research Council, whose members were concerned the new agency would jeopardize its funding and its standing. Despite this, the Campaign became a hugely successful organization and a primary granting agency for researchers. In 1963 "Research" was added to its title, and in 1970 the title "Cancer Research Campaign" was adopted. Details of this history are included on the organization's Web site (http://www.crc.org.uk) and on a Wellcome Library Web page (http://

www.aim25.ac.uk/cats/20/4605.htm).

5. Howe's 1963 atlas is available at a range of research libraries including the National Library of Medicine (NLM unique ID: 28911340R). For a more general discussion of Howe's atlas and the evolution of computerized medical cartography see Koch 2005 (227–37).

6. Nixon's declaration of the "war on cancer" in his 1971 State of the Union address began with the request for a modest $100 million in funding. Over the years that figure has grown as the National Cancer Institute (NCI) grew into a critical research institution. Currently all states have cancer registries whose members report county-level incidence in annual reports that are collected nationally and used by academic and official researchers. The most popular venue for the data they collect has been the national cancer atlases published by NCI.

7. Dr. Linda Pickle, a consummate authority on both the history of cancer atlases and the statistics they present, has my gratitude for the help she provided in procuring maps for this section and, more generally, for her continuing friendship. Her most recent paper on the history of the cancer atlas is a critical study on the history of spatial mapping and statistics (Pickle 2009).

8. Materials on the "Navajo neuropathy" and its histories were drawn from the work by L.A. Times reporter Judy Pasternak's series, "Blighted Homeland" (http://www.latimes.com/news/nationworld/nation/la-na-navaj019nov19,0,1645689.story). EPA maps of the areas are available at the EPA Web site (http://yosemite.epa.gov/r9/sfund/mappicsx.nsff). The online maps did not have adequate resolution for reproduction here and interested readers are urged to find them online.

9. Openshaw's use of a Monte Carlo simulation to test relevance of data in the mapped data, and to draw the conclusions upon the map surface, was in its day an extraordinarily innovative and complex piece of computer programming. In recent years the basic technique has been improved and made generally available in SatScan software developed by Martin Kuldorff and made available for free at http://www.satscan.org/. The map has been widely reproduced and can be found in, among other sources, Koch 2005.

10. I am obliged to Dr. Feychting, who was kind enough to explain the importance of mapping to me in a series of 2007 emails. The scores, perhaps hundreds, of maps used included both topographic maps and maps from local power companies. Our discussions were in the context of the 2007 lecture I gave to the NAACCR.

11. I am deeply obliged to Frank Garland both for permission to use maps from his published work and for his careful discussion of the relationship between the mapping and the research he, his brother, and others have carried out. He was kind enough to review my basic description of the almost thirty years of research into the relationship between specific cancers and vitamin D level.

12. Cedric Garland, personal communication.

13. When I was invited to speak to the annual meeting of the NAACCR I was told the history began after World War II. And yet, in *Cartographies of Disease* I had a cancer map from Britain in the 1890s. That said, I was as surprised as anyone when I discovered, through preparing my lecture, a history of cancer mapping beginning at the latest in the mid-nineteenth century. As often happens, the opportunity to focus a general lecture for a specific audience was the impetus for a whole new line of investigation.

14. For an example of the history of one such cluster dismissed by statisticians during the period when the subject was a matter of intense public debate, see Koch 1991 (150–60). There are, of course, many such stories in the public literature. In the professional literature there are problem-

atic clusters that appear consistent, and perhaps significant, but for which there is little clear explanation. The cancer clusters in Woburn, Massachusetts, are one such well-studied example (Lagakos, Wessen, and Zelen 1984a, 1984b).

AFTERWORD

1. The Gannett News Service investigation of Environmental Protection Agency (EPA) reports of violations like this provides a useful example of the problems of "simple" diseases with complex bureaucratic histories. Their data was drawn principally from the EPA Enforcement & Compliance History Online database (http://www.epa-echo.gov/echo/). Translating these incident reports into a solid social narrative was the work of Wheeler and Smith (2008, 2009) and their editors.

2. I am obliged to the unnamed reviewer of this manuscript who insisted it was necessary to at least mention the Walkerton case in the context of this argument. As he or she noted, the failure was not simply bureaucratic but visual: the failure to map and thus see the location of the wells that combined to make the town's water supply, the failure to map the increasing incidence of gastrointestinal disease until a full-blown and deadly outbreak resulted.

3. The work done on this project was principally statistical, using SAS software to compute distances and then to calculate the effect of the treated nets. The goal of the study was not to see if the treated nets would diminish malarial incidence but rather whether that incidence came at the cost of greater risk to others in the community. I am grateful to one of the researchers, Dr. Allen Hightower of the CDC, for information on this work and its methodologies.

WORKS CITED AND CONSULTED

Ackerknecht, E. H. 1948a. Hygiene in France: 1815–1848. *Bulletin of the History of Medicine* 22, no. 2:117–55.

———. 1948b. Anticontagionism between 1821 and 1867. *Bulletin of the History of Medicine* 22 (Sept.–Oct. 19): 562–93. Repr. in *International Journal of Epidemiology* 2009, no. 38: 7–21.

Acland, H. W. 1856. *Memoir of the cholera at Oxford, in the year 1854, with considerations suggested by the epidemic.* London: J. Churchill.

Akerman, J. R., and R. D. Karrow, eds. 2007. *Maps: Finding our place in the world.* Chicago: University of Chicago Press.

Anderson, L. 1987. *Research contributions made possible by the NCI cancer atlases published in the 1970s.* Bethesda, MD: National Cancer Institute Office of Cancer Communications.

Anonymous [E. A. Parkes]. 1849. *London Medical Gazette* 9:466.

Anonymous. 1855. Bibliographical record. *British Foreign Medical Chirurgical Review* 16:148.

Arikha, N. 2007. *Passions and tempers: A history of the humours.* New York: HarperCollins.

Arnaudet, A. 1890. Nouvelle contribution a l'étude du Cancer en Normandie. *La Normandie Médicale* 4, no. 7:105–11.

Arrieta, F. 1694. *Raggualio historico del contagio occurso della provincia de Bari negli anni 1890, 1891, e 1892.* Naples: Dom. Ant. Aparrino e Michele Luigi Mutii.

Ashton, J. 1974. *The epidemiological imagination.* Philadelphia: Open University Press, 1974.

Athanasios, D. A., G. C. Pavlos, and K. I. Theodoros. 2007. Early evidence-based medicine: Clues on statistical analysis in medicine from Galen's writings. *American Statistician* 61, no. 2:154–58.

Austin, J. L. 1975. *How to do things with words*, ed. J. O. Urmson and M. Sbisà. 2nd ed. Cambridge, MA: Harvard University Press.

Bailey, T. C., and A. C. Gatrell. 1995. *Interactive spatial data analysis.* New York: John Wiley & Sons.

Bakker, C., J. Van Dam, and C. Bonne. 1926. Waar Woonden onze Kankerpatiën? *Nederlandsch Tijdschrift vor Geenskunde* 70, no. 17:1698–1702.

Bernstein, E. 2002. John Bellers, champion of the poor and advocate of a League of Nations. In *Cromwell and Communism*, chapter 17. http://www.marx.org/reference/archive/bernstein/works/1895/cromwell/17-bellers.htm#top.

Binding, P. 2003. *Imagined corners: Exploring the world's first atlas.* London: Hodder Headline Book Publishing.

Bingham, P. N. Q. Verlander, and M. J. Cheal. 2004. John Snow, William Farr and the 1849 outbreak of cholera that affected London: a reworking of the data highlights the importance of the water supply. *Public Health* 118:387–94.

Blot, W. J., J. M. Harrington, A. Toledo, R. H. Hoover, and J. F. Fraumeni. 1978. Lung cancer after employment in shipyards during World War II. *New England Journal of Medicine* 299, no. 12:620–24.

Booth, C. 1902. *Life and labour of the people.* 17 vols. London: Macmillan.

Braudel, F. 1980. History and the social sciences: The longue durée. In *On history.* Trans. Sarah Mathews. Chicago: University of Chicago Press.

Braun, G., and F. Hogenberg. 1966. *Civitates Orbis Terrarum.* Intro. by R. A. Skelton. Cleveland. World Publishing Co.

Brierre de Boismont, A. 1832. *Relation historique et medicale du cholera-morbus de Pologne, comprenant l'apparition de la maladie, sa marche, ses progres, ses symptomes, son mode de traitement et les moyens preservatifs.* Paris: Germer-Bailliere.

Brigham, A. 1832. *A treatise on epidemic cholera: Including an historical account of its origin and progress, to the present period.* Hartford, CT: H. and F. J. Huntington.

Brodeur, P. 1989. *Currents of death: Power lines, computer terminals, and the attempt to cover up their threat to your health.* New York: Simon and Schuster.

Brody, H., M. R. Rip, P. Vinten Johnsen, et al. 2000. Map-making and myth-making in Broad Street: The London cholera epidemic, 1854. *Lancet* 356:54–58.

Brömer R. 2000. The first global map of the distribution of human diseases: Friedrich Schnurrer's "Charte über die geographische Ausbreitun der Krankheiten" (1827). *Medical History Supplement* 20:176–85. http://www.ncbi.nlm.nih.gov/pmc/articles/ PMC2530995/.

Brown, L. A. 1956. The longitude. In *The world of mathematics*, vol. 2, ed. J.R. Newman, 780–821. New York: H. Wolff Books.

Brown, P. E. 1961. John Snow—The autumn loiterer. *Bulletin of the History of Medicine* 35:519–28.

Budd, W. 1849. *Malignant cholera: Its mode of propagation and its prevention.* London: John Churchill.

———. 1873. *Typhoid fever.* London: Longmans, Green, and Co.

Burbank, F. 1971. *Patterns in cancer mortality in the United States, 1950–1967.* National Cancer Institute Monograph No. 33. Bethesda, MD: National Cancer Institute.

Burton, N., J. Westwood, and P. Carter. 2004. *GIS of the ancient parishes of England and Wales, 1500–1850.* Colchester, UK: UK Data Archive.

Callahan, G. N. 2006. *Infection: The uninvited universe.* New York: St. Martin's Press.

Campbell, E. M. T. 1949. An English philosophico-chorographical chart. *Imago Mundi* 6:79–84.

Campbell, F. R. 1885.The relation of meteorology to disease. *Buffalo Medical and Surgical Journal* 26:193–214.

Carey, M. 1793. *A short account of the malignant fever lately prevalent in Philadelphia with a statement of the proceedings that took place on the subject in different parts of the United States.* Philadelphia. http://www.collphyphil.org/HMDLSubweb/Pages/C/CareyM/ malfevPgAccess.htm.

Carr, D., J. F. Wallin, and A. Carr. 2000. Two new templates for epidemiology applications: Linked micromap plots and conditioned choropleth maps. *Statistics in Medicine* 19:2521–38.

Carr, D., D. White, and A. MacEachren. 2005. Conditioned choropleth maps and hypothesis generation. *Annals of the American Association of American Geographers* 95:32–53.

Carrell, J. L. 2003. *The speckled monster: A historical tale of battling smallpox.* New York: Dutton.

Carvalho, F. M., F. Lima, and D. Kreibel. 2004. Re: on John Snow's unquestioned long division. *American Journal of Epidemiology* 159:422–23.

CDC [US Centers for Disease Control and Prevention]. 1999. Cholera. *Weekly epidemiological record* 31:249–256. Geneva: World Health Organization.

———. 2000. *Epi info manual*. Atlanta: CDC.

———. 2001. Special issue, *Emerging Infectious Diseases* 7, no. 4. http://www.cdc.gov/ncidod/EID/v017n04/contents.htm.

———. 2002. Cholera epidemic associated with raw vegetables—Lusaka, Zambia, 2003–2004. (CDC Editorial Note). *Journal of the American Medical Association* 17:2077–78.

———. 2003. West Nile virus statistics, surveillance, and control: Maps and data. http://www.cdc.gov/ncidod/dvbid/westnile/Mapsactivity/surv&contr0103Maps.htm.

———. 2004. Bovine Spongiform Encephalopathy in a Dairy Cow—Washington State, 2003. *Morbidity and Mortality Weekly Report* 52, no. 53: 1280–85. http://www.cdc.gov/mmwr/preview/mmwrhtml/mm5253a2.htm.

Central Board of Health (England). 1831a. *Papers relative to the disease called* Cholera spasmodica *in India now prevailing in the North of Europe*. London: Winchester and Varnham.

———. 1831b. *Rules and regulations proposed by the Board of Health for the preventing the introduction and spreading of* Cholera morbus. London: Burgess and Hill, 1831.

———. 1831c. *Rules and regulations proposed by the Board of Health for the preventing the introduction and spreading of . . .* Cholera morbus, *by a committee of the Lords of his Majesty's Privy Council, to be added, A warning to the British public*. London: Burgess and Hill.

Chadwick, E. 1842. *Report to Her Majesty's principal Secretary of State for the Home Department, from the Poor Law Commissioners, on an inquiry into the sanitary condition of the labouring population of Great Britain; with appendices*. London: Clowes.

Champion, J. A. I. 1993. Epidemics and the built environment in 1665. In *Epidemic disease in London*, a collection of working papers given at the symposium "Epidemic disease in London: From the Black Death to cholera.", London: Centre for Metropolitan History, 1999. http://www.history.ac.uk/cmh/epipre.html.

Chave, S. P. W. 1958. Henry Whitehead and cholera in Broad Street. *Medical History* 2: 92–109.

Choi, T. Y. 2003. Narrating the unexceptional: The art of medical inquiry in Victorian England and the present. *Literature and Medicine* 22, no. 1:66–83.

Chrisman, N. 1997. *Exploring geographic information systems*. New York: John Wiley & Sons.

Christie, A. T. 1833. *A treatise on epidemic cholera; containing its histories, symptoms, autopsy, etiology, causes and treatment*. London: J & C Adlard.

Colten, C. E. 2006. *An unnatural metropolis: Wrestling New Orleans from nature*. Baton Rouge: Louisiana State University Press.

Cooper, E. 1854. *Report on an enquiry and examination into the state of the drainage of the homes situate in that part of the Parish of St. James, Westminster*. Metropolitan Commission of Sewers 478/21. London Metropolitan Archives.

Cooter, R. 2006. A liking for Snow. *Lancet* 367, no. 9523:1647–48.

Corbyn, F. 1832. *Treatise on the epidemic cholera, as it has prevailed in India; together with the reports of the medical officers, made to the medical boards of the presidencies of Bengal, Madras, and Bombay*. Philadelphia: Carey, Lea, and Carey.

Cosgrove, D. 2007. Mapping the world. In *Maps: Finding our place in the world*, ed. J. R. Akerman and R. D. Karrow, 57–114. Chicago: University of Chicago Press.

Crosby, A. W. 1972. *The Columbian exchange: Biological and cultural consequences of 1492*. Westport, CT: Greenwood Press.

Davenhill, W. 2005. Access to better information for public health. *ArcUser*, April–June. http://www.esri.com/news/arcuser/0405/umbrella_30.html.

Davis, D. 2007. *The secret history of the war on cancer.* New York: Basic Books.

Defoe, D. 1969. *A journal of the plague year, being observations or memorials of the most remarkable occurrences, as well as publick as private, which happened in London during the last Great Visitation in 1665,* ed. Louis Landa. London: Oxford University Press.

de Jong, R. J. 1926. Modern Opvattingen over het kankerproblem. *Nederlandsch Tijdschrift voor Genneskunde* 70, no. 2: 1964–83.

De Knecht-van Eekelen, A. 2000. The debate about acclimatization in the Dutch East Indies (1840–1840). *Med Hist Supplement* 20:70–85.

de Kruif, P. H. 1939. *The microbe hunters.* New York: Harcourt.

Delano-Smith, C. 2005. Stamped signs on manuscript maps in the Renaissance. *Imago Mundi* 57, no. 1:59–62.

Denike, K., and T. Koch. 2007. *The statistics in the map: Rewriting John Snow's South London study.* Seminar in biostatistics, University of British Columbia.

Desowitz, R. S. 1997. *Who gave Pinta to the Santa Maria? Tracking the devastating spread of lethal tropical diseases into America.* New York: Harcourt.

Doen, E. H., and E. Elliot. 2007. Circulation and diffusion of the 1831–32 epidemic of cholera in Normandy (France). Paper presented at the annual meeting of the American Association of Geographers, Boston.

Dorling, D. 1999. Book review: Visual explanations: images and quantities, evidence and narrative. *Progress in Human Geography* 23, no. 1:127–31.

Drew, J. F. 1970. *A history of bubonic plague in the British Isles,* Cambridge: Cambridge University Press.

Dubuc, E. 1832. *Rapport adressé à l'intendance sanitaire de Rouen, sur le choléra-morbus, observé à Sunderland, Newcastle et les environs.* Rouen: D. Brière

East London Water Company. 1866. *Medical Times and Gazette,* September 8.

Edney, M. H. 2005. The origins and development of J. B. Harley's cartographic theories. *Cartographica,* monograph 54.

Egan, K. M. 2006. Sunlight, vitamin D, and the cancer connection revisited. *International Journal of Epidemiology* 35, no. 2:227–30.

Elliot, E., E. Daude, and E. Bonnet. 2008. Spatial dynamics and place effects of the epidemics of cholera in Normandy (France) (1832–1892). Paper presented at the annual meeting of the American Association of Geographers, Boston.

Elliott, P., and D. Wartenberg. 2004. Spatial epidemiology: current approaches and future challenges. *Environment Health Perspectives* 112, no. 9:998–1006.

Evans, R. J. 1992. Epidemics and revolutions: Cholera in nineteenth-century Europe. In *Epidemics and ideas: Essays on the historical perception of pestilence,* ed. T. Ranger and P. Stack, 149–74. Cambridge: University of Cambridge Press.

Eyler, J. M. 1973. William Farr on the cholera. *Journal of the History of Medicine* 28, no. 2:79–100.

———. 1979. *Victorian social medicine: The ideas and methods of William Farr.* Baltimore, MD: Johns Hopkins University Press.

———. 1997. Review of John Snow, anaesthetist to a queen and epidemiologist to a nation: a biography. *Bulletin of the History of Medicine* 71, no. 4:716–17.

———. 2001. The changing assessments of John Snow's and William Farr's cholera studies. *Präventivmed* 46:225–32.

Farr, W. 1852a. *Report on the mortality from cholera in England, 1848–1849.* London: Her Majesty's Stationery Office. (Also published as Farr, W. 1852. *Registrar General's Report on Cholera in England 1849–1850.* London: W. Clowes & Son).

———. 1852b. Influences of elevation on the fatality of cholera. *Journal of the Statistical Society of London* 15:153–83.

———. 1853. Cholera and the London water supply. *Registrar General's weekly return of births and deaths in London*, November 19, 401–6.

———. 1855. *Letter of the president of the General Board of Health to the right honorable Viscount Palmerston accompanying a report from Dr. Sutherland on epidemic cholera in the metropolis in 1854.* London: George E. Eyre and William Spottiswoode.

Feychting, M., and M. Alhbom. 1993. Magnetic fields and cancer in children residing near Swedish high-voltage power lines. *American Journal of Epidemiology* 138, no. 7:467–81. http://aje.oxfordjournals.org/cgi/content/abstract/138/7/467.

Feychting M., N. Plato, G. Nise, and A. Ahlbom. 2001. Paternal occupational exposures and childhood cancer. *Environmental Health Perspectives*, February. http://ehp.niehs.nih.gov/members/2001/109p193-196feychting/feychting.pdf.

Forry, S. 1842. *The climate of the United States and its endemic influences.* New York: J. & H. G. Langley.

Fotheringham, A. S., and F. B. Zhan. 1996. A comparison of three exploratory methods for cluster collection in spatial point patterns. *Geographical Analysis* 28, no. 3:200–18.

Foucault, M. 1972. *The archeology of knowledge; and, the discourse on language.* Trans. A. M. Sheridan Smith. New York: Pantheon.

———. 1973. *The birth of the clinic: An archaeology of medical perception.* New York: Pantheon.

———. 1978. An Introduction. Vol. 1. of *The history of sexuality*. Trans. Robert Hurlye. New York: Random House.

Foucault, M. 1980. Truth and power. In *Power/knowledge: Selected interviews and other writings 1972–1977*, ed. Colin Gordon, trans. by Gordon et al. New York: Pantheon.

Frazer, W. M. 1950. *A history of English public health: 1834–1939.* London: Ballière, Tindall, and Cox.

Friendly, M., and D. Denis. 2005. The early origins and development of the scatterplot. *Journal of the History of the Behavioral Sciences*, 41, no. 2: 103–30.

Friendly, M., and G. Palsky. 2007. Visualizing nature and society. In *Maps: Finding our place in the world*, ed. J. R. Akerman and R. W. Karrow Jr. Chicago: University of Chicago Press.

Froriep, R. 1832. *Symptome der asiatischen cholera im November und December 1831 zu Berlin.* Weimar: Landes-Industrie-Comptoir.

Frost, W. H. 1910. The water supply of Williamson, West Virginia, and its relation to an epidemic of typhoid fever. *Hygienic Laboratory Bulletin* 72:55–90.

———. 1912. Epidemiologic studies of acute anterior poliomyelitis. *Hygienic Laboratory Bulletin* 90:9–105, 234–52.

———. 1927. *Epidemiology: Public health prevention medicine* Vol. 2. New York: Thomas Nelson & Sons.

———. 1936. Introduction to *Snow on Cholera: Being a reprint of two papers by John Snow, M.D., together with a biographical memoir by R. B. W. Richardson, M.D.* New York: Commonwealth Fund.

———. 1941. *Papers of Wade Hampton Frost, M.D. A contribution to epidemiological method*, ed. Kenneth F. Maxcy. New York: Commonwealth Fund, 1941.

Garland C. F., and F. C. Garland. 1980. Do sunlight and vitamin D reduce the likelihood of colon cancer? *International Journal of Epidemiology* 9:227–31. Repr. in *International Journal of Epidemiology* 35, no. 2 (2006): 217–20

General Board of Health. 1855. *Report of the committee for scientific inquiries in relation to the cholera-epidemic of 1854.* London: Her Majesty's Stationary Office.

Gilbert, E. W. 1958. Pioneer maps of health and disease in England. *Geographical Journal* 124:172–83.

Gilbert, P. N. 2004. *Mapping the Victorian social body.* Albany, New York: SUNY Press.

Gilbert, P. N. 2008. *Cholera and nation: Doctoring the social body in Victorian England.* Albany, New York: SUNY Press.

Gimming, J. E., M. S. Kolczak, A. W. Hightower, et al. 2003. Effect of permethrin-treated bed nets on the spatial distribution of malaria vectors in Western Kenya. *American Journal of Tropical Medicine and Hygiene* 68 (Supplement 4): 115–20.

Giovannucci, E. 2006. Vitamin D and colorectal cancer—twenty-five years later. *International Journal of Epidemiology* 35, no. 2:222–24.

Gorham, E. D., S. G. Mohr, F. C. Garland, and C.F. Garland. 2009. Vitamin D for cancer prevention and survival. *Clinical Reviews in Bone and Mineral Metabolism* 7, no. 2:159–75.

Gould, P. 1993. *The slow plague: A geography of AIDS.* Oxford: Blackwell.

Grainger, R. 1850. Appendix B: Reports from commissioners. In *Report of the General Board of Health on the epidemic cholera of 1848 & 1849.* Vol. 21: 199, Parliamentary Session January 31–August 15.

Graunt, J. 1662. *Natural and political observations mentioned in a following index, and made upon the bills of mortality . . . With reference to the government, religion, trade, growth, ayre, and diseases of the said city.* London.

Green, C. E. 1917. *The cancer problem: A statistical study.* Edinburgh: W. Green & Son.

Guerry, A.-M. 1829. Mèmoire sur les variations mèterologiques comparèes aux phènomènes physiologiques. *Annales d'Hygiène Publique et de Mèdecine Lègal* I.

Gundersheimer, W. L. 1971. Introduction to *The Dance of Death by Hans Holbein the Younger.* New York: Dover publications.

Hacking, I. 1990. *The taming of chance.* New York: Cambridge University Press.

———. 1999. *The social construction of what?* Cambridge, MA: Harvard University Press.

———. 2006. *The emergence of probability: A philosophical study of early ideas about probability, induction, and statistical inference.* 2nd ed. New York: Cambridge University Press.

Halliday, S. 2002. Commentary: Dr. John Sutherland, *Vibrio cholera* and "predisposing causes." *International Journal of Epidemiology* 31:912–14.

Hamilton, A. 1903. The fly as a carrier of typhoid: An inquiry into the part played by the common housefly in the recent epidemic of typhoid fever in Chicago. *Journal of the American Medical Association* XL, no. 9:576–83.

Hamett, J. 1832. *The substance of the official medical reports upon the epidemic cholera. Which prevailed among the poor at Dantzick between the first of July and the first part of September, 1831.* London: S. Highley, Fleet Street.

Hamlin, C. 2007. Two tales of a city. *American Scientist.* September 2, 2008. http://www .americanscientist.org/bookshelf/pub/2007/3/two-tales-of-a-city.

Hargrove, T. 2007. Fatal outbreak: CDC reports on food-borne illnesses provide treasure trove of state data. *IRE Journal* 30, no. 1: 30.

Harley, B. 1988. Maps, knowledge, and power. *Cartographica* 24:59–68.

Harris J., and S. Holm. 1993. If only AIDS were different. *Hastings Centre Report* 23, no. 6:6–13.

Haviland, A. 1869. *Abstracts of two papers: I. On the geographical distribution of heart disease and dropsy in England and Wales. II. On the geographical distribution of cancer in England and Wales.* London.

———, A. 1875. *The geographical distribution of health disease and dropsy, cancer, in females and phthisis in females, in England and Wales.* London: Smith, Elder.

312

———. A. 1892. *The geographical distribution of disease in Great Britain and Wales.* London: Swan Sonneschein and Co.

Hawkins, F. B. 1831. *History of the epidemic spasmodic cholera of Russia; including an account of the disease which has prevailed in India and which has traveled . . . from Asia into Europe, etc.* London.

Hawley, W. A., P. A. Phillips-Howard, F. O. Ter Kuile, et al. 2003. Community-wide effects of permethrin-treated bed nets on child mortality and malaria morbidity in western Kenya. *American Journal of Tropical Medicine and Hygiene* 68 (suppl.4): 121–27.

Hayes, E. B., N. Komar, R. S. Nasci, et al. 2005. Epidemiology and transmission dynamics of West Nile virus disease. *Emerging Infectious Diseases* 11, no. 8:1167–73.

Healy, M. 2003. Defoe's journal and the English plague writing tradition. *Literature and Medicine* 22, no. 1:25–44.

Henderson, D. A. 1999. Eradication: Lessons from the past. *Mortality and Morbidity Weekly Report* 48(SU01), 16–22. http://www.cdc.gov/mmwr/preview/mmwrhtml/su48a6.htm

Hellis, E. C. 1832. *Réflexions sur le choléra-morbus observé par M. E. Dubuc en Angleterre.* Rouen: Nicétas Periaux.

———. 1833. *Souvenirs du choléra à Rouen et dans le Département de la Seine-Inférieure en 1832; ornés d'un plan de rouen, Indiquant la marche de l'Epidémie dans la ville.* Paris: Ballière: Delaunay.

Hempel, S. 2006. *The medical detective: John Snow and the mystery of cholera.* London: Granta Publications.

Hightower, A. 2005. Spatial analysis optimizes malaria prevention measures. *ArcUser Online* (April–June). http://www.esri.com/news/arcuser/0405/malaria1of2.html.

Hillary, W. 1759. *Observations on the changes of the air and the concomitant epidemical diseases in the island of Barbados. To which is added a treatise on the putrid bilious fever, commonly called the yellow fever, and such other diseases as are indigenous or endemical in the West India island, or in the torrid zone.* London: C. Hitch and L. Hawes.

Hoffman, F. L. 1915.*The mortality from cancer throughout the world.* Newark, NJ: Prudential Press.

Holbein, H. 1971. *The dance of death: A complete facsimile of the original 1538 edition of Les simulachres & historiees faces de la mort.* New York: Dover Publications.

Holt, H. M. E. 1984. Assistant commissioners and local agents: their role in tithe commutation, 1836–1854. *Agricultural History Review* 32:189–200.

Howe, G. M. 1963. *National atlas of disease mortality in the United Kingdom.* 2nd ed. London: Thomas Nelson and Sons, Ltd.

Hull House. 1895. *Hull-House maps and papers. A presentation of nationalities and wages in a congested district of Chicago, together with comments and essays on problems growing out of the social conditions.* By residents of Hull-House, a social settlement at 335 South Halsted Street, Chicago, Ill. New York: Crowell.

Humphreys, N. 1885. Biographical sketch. In *Vital statistics: A memorial volume of selections from the reports and writings of William Farr*, ed. N. Humphreys. London: Office of the Sanitary Institute.

Hunter, D. 1978. *Papermaking: The history and technique of an ancient craft.* New York: Dover Publications.

Hunter, J. M., and J. Young. 1971. Diffusion of influenza in England and Wales. *Annals of the Association of American Geographers* 61:637–53.

Ingram, D. 1755. *An historical account of the several plagues that have appeared in the world since the year 1346. With an enquiry into the present prevailing opinion that the plague is a contagious distemper, capable of being transported in merchandise from one country to another. . .* London: R. Baldwin and J. Clark.

Jackson, T. H. [1855] 1958. Report of T. Heber Jackson, MD: Cholera in Lancaster and Columbia in 1854. *Lancaster County Historical Journal*. April, 123–31.

Jameson, J. 1819. *Report on the epidemick cholera morbus, as it visited the territories subject to the presidency of Bengal, in the years 1817, 1818 and 1819. Drawn up by order of the government, under the superintendence of the medical board.* E. Balfour: Government Gazette Press.

James, R. 1764. *A dissertation on fevers and inflammatory distempers. Wherein an expeditious method is proposed of curing those dangerous disorders. To which is added, an account of the success with which the fever powder has been given in the small-pox, yellow fever, slow fever, and rheumatism.* London.

Jarcho, S. 1970. Yellow fever, cholera, and the beginnings of medical cartography. *Journal of the History of Medicine and Allied Science* 25:131–42.

———. 1978. Christopher Packe (1868–1949): Physician-cartographer of Kent. *Journal of the History of Medicine and Allied Science* 33, no. 1:44–52.

Jekel, J. F., J. G. Elmore, and D. L. Katz. 1996. *Epidemiology, biostatistics, and preventive medicine.* Philadelphia: W. B. Saunders Co.

Johnson, G. 1855. On Epidemic diarrhoea and cholera, their pathology and treatment; with a record of cases. London.

Johnson, S. 2006. *The ghost map: The story of London's most terrifying epidemic—and how it changed science, cities, and the modern world.* New York: Riverhead Books.

Jusatz, H. 1940. Die geographisch-medizinische Erforschung von Epidemien. *Petermann Mitteilungen* 86, 201–4.

Kain, R. J. P., and Richard, O. 2001. *Historic parishes of England and Wales: An electronic map of boundaries before 1850 with a gazetteer and metadata.* Colchester: UK Data Archive.

Kearns, G. 1985. Urban epidemics and historical geography: Cholera in London, 1848–9. *Historical Geography Research Series* 15:1–48.

Kevles, B. H. 1997. Naked to the bone: Medical imaging in the twentieth century. New Brunswick, NJ: Rutgers University Press.

Khan, K., J. Arino, H. Wei et al. 2009. Spread of a novel influenza A (H1N1) virus via global airline transportation. *New England Journal of Medicine* 361:212–14.

Kish, G. 1965. The cosmographic heart: Cordiform maps of the 16th century. *Imago Mundi* 19:13–21.

Koch, T. 1991. *Journalism for the 21st century: Online information, electronic databases, and the news.* Westport, CT: Greenwood Press.

———. 2004. The map as intent: Variations on the theme of John Snow. *Cartographica* 39, no. 4:1–13.

———. 2005. *Cartographies of disease: Maps, mapping, and medicine.* Redlands, CA: ESRI Press.

———. 2006. "False truths": Ethics and mapping as a profession. *Cartographic Perspectives* 54:4–15.

Koch, T., and K. Denike. 2004. Medical mapping: The revolution in teaching—and using—maps for the analysis of medical issues. *Journal of Geography* 103, no. 2:76–85.

———. 2006. John Snow's South London study: A Bayesian evaluation and recalculation. *Social Science & Medicine* 63, no. 1:271–83.

———. 2007a. Certainty, uncertainty, and the spatiality of disease: A West Nile virus example. *Stochastic Environmental Resource Risk Assessment* 21:523–31.

———. 2007b. Aaron's solution, instructor's problem: Teaching surface analysis using GIS. *Journal of Geography* 106, no. 2:69–78.

———. 2009. Crediting his critic's concerns: Remaking John Snow's map of Broad Street cholera, 1854. *Social Science & Medicine* 69:1246–51.

Krieger, N. 2000.Epidemiology and social sciences: Toward a critical reengagement in the 21st century. *Epidemiologic Review* 22, no. 1:155–63.

———. 2006. A century of census tracts: Health & the body politic (1906–2006). *Journal of Urban Health: Bulletin of the New York Academy of Science* 83, no. 3:355–61.

Krieger, N., and A. Birn. 1998. A vision of social justice as the foundation of public health: Commemorating 150 years of the spirit of 1848. *American Journal of Public Health* 88, no. 11:1603–6.

Krieger, N., with the Coordinating Committee of Spirit of 1848. 1997. Spirit of 1848: A network linking politics, passion, and public health. *Radical Statistics* 66:22–32.

Kuhn, 1962. *The structure of scientific revolutions.* Chicago: University of Chicago Press.

Lagakos, S. W., B. J. Wessen, and M. Zelen. 1984a. An analysis of contaminated well water and health effects in Woburn, Massachusetts, SIMS technical report 3. *Journal of the American Statistical Association* 81 (395): 583–96. http://www.hopkintonschools.org/hhs/departments/math/statisticswebsite/woburnstudy.pdf.

———. 1984b. The Woburn health study: An analysis of reproductive and childhood disorders and their relation to environmental contamination. Technical report, Harvard School of Public Health, Dept. of Biostatistics, Boston, MA.

Lancet. 1831. History of the rise, progress, ravages, etc. of the blue cholera of India. *Lancet* 17, no. 429:241–84.

———. 1832. Narrative of the rise, progress, and ravages of the malignant cholera from its origin in 1817 in India to its appearance in London, in 1832.

———. 1848. Medical news. *Lancet* 52, no. 1302: 195–96.

———. 1853. Report on cholera. *Lancet* 62, no. 1573: 393–94.

Langmuir, A. D. 1961. Epidemiology of airborne infection. *Bacteriological Reviews* 25:173–181.

Latour, B. 1988. *The pasteurization of France.* Trans. Alan Sheridan. Cambridge, MA: Harvard University Press.

———. 1993. *We have never been modern.* Trans. Catherine Porter. Cambridge, MA: Harvard University Press.

Law, J. 2004. *After method: Mess in social science research.* New York: Routledge.

Lea, J. 1850. *Cholera, with reference to the geological theory: A proximate cause—a law by which it is governed—a prophylactic.* Cincinnati: Wright, Ferris, & Co. Gazette Office.

———. 1851. Cholera—the geological theory. *Western Lancet*: 89–97.

Lee-Feldstein A. 1983. Arsenic and respiratory cancer in humans: follow-up of copper smelter employees in Montana. *Journal of the National Cancer Institute* 70, no. 4:601–9.

Lilienfeld, D. E. 2000. John Snow; the first hired gun? *American Journal of Epidemiology* 152, no. 1:4–9.

Lining, John. 1799. *A description of the American yellow fever, which prevailed at Charleston, in South Carolina, in the year 1748. By Doctor John Lining, physician . . .* Philadelphia: Thomas Dobson.

MacDonald, G. 1890. Cancer statistics in New Zealand. *New Zealand Medical Journal* 3:252.

Mack, M. P. 1962. *Jeremy Bentham: An odyssey of ideas 1748–1792.* London: Heinemann.

MacDonald, G. 1890. Cancer statistics in New Zealand. *New Zealand Medical Journal* 3:252.

Malgaigne, J. F. 1840. Recherches sur le fréquence des hernies selon les sexes, les ages, er relativement à la population. *Annales d'Hygiène et de Médicine Légale* July: 1–50.

Mangani, G. 1998. Abraham Ortelius and the hermetic meaning of the cordiform projection. *Iamgo Mundi* 50:59–83.

Markel, H. 2004. *When germs travel.* New York: Pantheon.

Marks, G., and W. K. Beatty. 1976. *Epidemics.* New York: Charles Scribner's Sons.

Marriott, E. 2003. *The plague race: A tale of fear, science, and heroism.* London: Picador.

Marsh, D. 2007. Maps, myths, and gardens: Faithorne and Newcourt's map of London. Lecture, Maps and Society Lecture Series, Warburg Institute, School of Advanced Study, University of London.

Mason, T. J., and F. W. McKay. 1974. *U.S. cancer mortality by county, 1950–1969.* Washington DC: US Government Printing Office.

Mason, T. J., F. W. McKay, R. N. Hoover, W. J. Blot, and J. F. Fraumeni Jr. 1975. *Atlas of cancer mortality for U.S. counties, 1950–1969.* Bethesda, MD: U.S. Department of Health, Education, and Welfare.

———. 1976. *Atlas of cancer mortality among U.S. nonwhites, 1950–1969.* Bethesda, MD: U.S. Department of Health, Education, and Welfare.

Maxcy, K. F. 1941. Introduction to *Papers of Wade Hampton Frost, M.D. A contribution to epidemiological method.* New York: Commonwealth Fund, 1941.

Mayor, A. H. 1984. *Artists & anatomists.* New York: Metropolitan Museum of Art.

McLeod, K. 2000. Our sense of Snow: The myth of John Snow in medical geography. *Social Science & Medicine* 50, no. 7–8:923–36.

Melosi, M. 2000. *The sanitary city: Urban infrastructure in America from colonial times to the present.* Baltimore, MD: Johns Hopkins University Press.

Mercator, G., I. Jackson, and B. Karrow. 2000. *Atlas sive Cosmographicae Meditationes De Fabrica Mundi Et Fabricati Figura.* Oakland, CA: Octavio Press.

Miller, H. J., and E. A. Wentz. 2003. Representation and spatial analysis in geographic information systems. *Annals of the Association of American Geographers* 93, no. 3:574–94.

Mitman, G., and R. L. Numbers. 2003. From miasma to asthma: The changing fortunes of medical geography in America. *History and Philosophy of the Life Sciences* 25, no. 3: 391–412.

Mohr, J. C. 2005. *Plague and fire: Battling black death and the 1900 burning of Honolulu's Chinatown.* New York: Oxford University Press.

Mohr, S. B. 2009. A brief history of vitamin D and cancer prevention. *Annals of Epidemiology* 19:79–83.

Mohr, S. B., C. F. Garland, E. D. Gorham, and F. C. Garland. 2008. The association between ultraviolet B irradiance, vitamin D status and incidence rates of type 1 diabetes in 51 regions worldwide. *Diabetologia* 51, no. 8:1391–98.

Monmonier, M. 2002. *Spying with maps.* Chicago: University of Chicago Press.

Morabia, A. 2006. Review: S. Hempel, the medical detective: John Snow and the mystery of cholera (Granta Books, London). *British Medical Journal* 332:1220.

Moretti, F. 1998. *An Atlas of the European novel: 1800–1900.* New York: Verso.

———. 2005. *Graphs, maps, trees: Abstract models for a literary history.* New York: Verso.

Morris, R. J. 1976. *Cholera in 1832: The social response to an epidemic.* London: Croom Helm.

Nasci R. S., D. J. White, H. Stirling, et al. 2001.West Nile virus isolates from mosquitoes in New York and New Jersey, 1999. *Emerging Infectious Diseases* 7:626–29.

Neta, R. 2008. What evidence do you have? *British Journal of the Philosophy of Science* 58:89–119.

Newsholme, A. 1927. *Evolution of Preventive Medicine.* Baltimore, MD: Williams & Wilkins Co.

Nuland, S. 1988. *Doctors: The biography of medicine.* New York: Vintage Books.

O'Connor, D. R. 2002. *Report of the Walkerton Inquiry, Parts 1 and 2*. Government of Ontario Commission of Inquiry. Toronto: Publications Ontario. http://www.attorneygeneral.jus
.gov.on.ca/english/about/pubs/walkerton/.

Olsson, G. 2007. *Abysmal: A critique of cartographic reason*. Chicago: University of Chicago Press.

Openshaw, S., M. Charlton, C. Wymer, and A. Craft. 1987. A mark 1 geographical analysis
machine for the automated analysis of point data sets. *International Journal of Geographical Information System* 1, no. 4:335–58.

Openshaw, S., M. Charlton, and A. Craft. 1988. Searching for leukemia clusters using a geographical analysis engine. *Papers of the Regional Science Association* 64:95–106.

Orent, W. 2004. *Plague: The mysterious past and terrifying future of the world's most
dangerous disease*. New York: Free Press.

Ortelius, A. 1606. *de Theatrum terrarum*. English language ed. London: John Norton.

Osborne, J. B. 2008. Preparing for the pandemic: City boards of health and the arrival of
cholera in Montreal, New York, and Philadelphia in 1832. *Urban History Review* 36, no.
2:29–42.

Osborne, J. B. 2009. The Lancaster County cholera epidemic of 1854 and the challenge to
the miasma theory of disease. *Pennsylvania Magazine of History and Biography* 133.1:
5–28.

Palmer, R. 1993. In bad odor: Smell and its significance in medicine from antiquity to the
17th century. In *Medicine and the five senses*, ed. W. F. Bynum and R. Porter, 61–68.
New York: Cambridge University Press.

Paneth, N. 2004. Assessing the contributions of John Snow to epidemiology 150 years after
removal of the Broad Street pump handle. *Epidemiology* 15, no. 5:514–16.

Paneth, N., P. Vinten-Johansen, and H. Brody. 1998. A rivalry of foulness: Official and unof-
ficial investigations of the London cholera epidemic of 1854. *American Journal of Public
Health* 88, no. 10:1545–53.

Parkes, E. A. 1855a. Review: Mode of communication of cholera by John Snow. *British and
Foreign Medico-Chiurgical Review* 15:449–56.

———. 1855b. The public health and nuisances removal bill: Dr. Snow's evidence. *Lancet*
XV: 634–35.

Pascalis, F. 1796. *Medico-chymical dissertations on the cause of the epidemic called yellow
fever, and in the preparation of the best antinomial preparations for the use of the medicine*. Philadelphia: Snowden and M'Corkle.

———. 1798. *An account of the contagious epidemic yellow fever, which prevailed in Philadelphia in the summer and autumn of 1797: Comprising the questions of its causes and
domestic origin, characters, medical treatment, and preventives*. Philadelphia: Snowden
& M'Corkle.

———. 1819. *A statement of the occurrence of a malignant yellow fever, in the city of
New-York, in the summer and autumnal months of 1819; and of the check given to its
progress . . . foreign ports*. New York: W. A. Mercein.

———. 1820. A statement of the occurrences during a malignant yellow fever in the city of
New-York. *Medical Repository*: 229–256.

Pasternak, J. 2007. Official indifference: Uranium mines reveal environmental crisis for Navajos. *IRE Journal* 30, no. 2:30–33.

Pattison, John. 1998. The emergence of bovine spongiform encephalopathy and related
diseases. Special issue, *Emerging Infectious Diseases* 4, no. 3:390–94.

Perkel, C. 2002. *Well of lies: The Walkerton water tragedy*. Toronto: McClelland & Stewart.

Petermann, A. 1848. *Statistical notes to the cholera map of the British Isles showing the
districts attacked in 1831, 1832, and 1833*. London: John Setts.

Pickle, J. 2007. Radical thought-in-action: Gunnar Olsson's "critique of cartographic reason." *Geographisker Anneler* 89, no. 4:394–97.

Pickle L. W. 2009. A history and critique of U.S. mortality atlases. *Spatial and Spatio-temporal Epidemiology* 1, No. 1, 3-17.

Pickstone, J. V. 2000. *Ways of knowing: A new history of science, technology, and medicine.* Manchester: Manchester University Press.

Picon, A. 2003.Nineteenth-century urban cartography and the scientific ideal: The case of Paris. *Osiris* 18:135–49.

Playfair, W. 2005. *William Playfair's Commercial and Political Atlas and Statistical Breviary*, ed. H. Wainer and I. Spence. New York: Cambridge University Press.

Porter, D. H. 1998. *The Thames embankment: environment, technology, and society in Victorian London.* Akron, OH: University of Akron Press.

Porter, R. 1988. *The greatest benefit to mankind: A medical history of humanity.* New York: W. W. Norton.

Porter, S. 1999. *The great plague.* London: Sutton Pub.

———. 2005. *Lord have mercy upon us: London's plague years.* Stroud, Gloucestershire: Tempus Publishing.

Power, D'Arcy. 1899. The locational distribution of cancer and cancer houses. *Practitioner* LXII: 415–29.

———. 1903. A further contribution to the distribution of cancer. *Practitioner* LXX: 1–20.

Public Health Agency of Canada. 2000. An outbreak of *Salmonella enteritidis* linked to baked goods from a local bakery in lower mainland British Columbia: Preliminary Report, October. *Canada Communicable Disease Report* 26–20:175.

Pullman, B. 1992. Plague and perceptions of the poor in early modern Italy. In *Epidemics and ideas: Essays on the historical perception of pestilence*, ed. T. Ranger and P. Slack. Cambridge: University of Cambridge Press.

Pyle, G. F. 1986. *The diffusion of influenza: Patterns and paradigms.* Lanham, MD: Rowman and Littlefield.

Rawnsley, H. D. 1898. *Henry Whitehead: 1825–1896: A memorial sketch.* Glasgow: James MacLehose and Sons.

Reese, D. M. 1833. *A plain and practical treatise on the epidemic cholera, as it prevailed in the city of New York, in the summer of 1832 . . .* New York: Conner & Cooke.

Rheinberger, H. 1997. *Toward a history of epistemic things: Synthesizing proteins in the test tube.* Palo Alto, CA: Stanford University Press.

Rhodes, R. 1997. *Deadly feasts: Tracing the secrets of a terrifying new plague.* New York: Simon and Schuster.

Richardson, B. W. 1858. *On chloroform and other anaesthetics: Their action and administration—edited, with a memoir of the author by Benjamin Ward Richardson..*London.

———. 1936. Introduction to *Snow on cholera: Being a reprint of two papers by John Snow, M. D.*, ed. W. H. Frost. New York: Commonwealth Fund.

Rifkin, B. J., and M. J. Ackerman. 2006. *Human anatomy (from the Renaissance to the digital age).* New York: Abrams.

Robinson, A. H. 1982. *Early thematic mapping in the history of cartography.* Chicago: University of Chicago Press.

Rogers, N. 1996. *Dirt and disease: Polio before FDR.* New Brunswick, NH: Rutgers University Press.

Rosen, G. 1993. *A history of public health.* Expanded ed. Baltimore, MD: Johns Hopkins University Press.

Rothenburg, J. N. 1836. *Die Cholera-Epidemie des Jahre 1832 in Hamburg.* Perthes and Besser.

Roueche, B. 1991. Introduction to *The microbe hunter*. Rev. ed. New: Penguin Plume Books.

Rush, B. 1793. *An account of the bilious remitting yellow fever, as it appeared in the city of Philadelphia, in the year 1793*. Philadelphia: Thomas Dobson.

Rush, B. 1809. *Medical inquiries and observations*, vol. 1. 3rd ed. Philadelphia: Thomas and William Bradford.

Sadegh-Zadeh, K. 2008. The prototype resemblance theory of disease. *Journal of Medicine and Philosophy* 33:196–239.

Salim, A., L. Ruiting, P. R. Reeves et al. 2005. *Vibrio cholerae* pathogenic clones. *Emerging Infectious Diseases* 11, no. 11:1758–60.

Saunders K. B., and C. D. O'Malley. 1973. *The illustrations from the works of Andreas Vesalius of Brussels*. New York: Dover Publications.

Schatz, G. S. 2005. *Introductory note to World Health Organization: Revision of the international health regulations* (44 ILM 1011). Geneva: World Health Organization.

Schnurrer, F. 1825. *Chronik der Seuchen, In Verbindung mit den gleichzeitigen Vorgängen in der physischen Welt und in physischen Welt und in der Geschichte der Menschen*. 2 vols. Tubingen: Osiander.

———. 1831. *Die Cholera morbus, ihre Verbreitung ihre Zufälle, die versuchten Heilmethoden, ihre Eigenthümlichkeiten und die im Grossen dagegen anzuwendenden Mittel*. n.p.

Schultz, R. L., ed. 2007. *Hull-House maps and papers: A presentation of nationalities and wages in a congested district of Chicago*. Urbana: University of Illinois Press.

Seaman, V. 1793. *A dissertation on the mineral waters of Saratoga: Containing, a topographical description of the country, and the situation of the several springs ; an analysis of the waters, as made upon the spot, together with remarks on their use in medicine, and a conjecture respecting their natural mode of formation: also, a method of making an artificial mineral water, resembling that of Saratoga, both in sensible qualities and in medicinal virtue*. New York: Samuel Campbell.

———. 1796. *An account of the epidemic yellow fever, as it appeared in the city of New-York in the year 1795: containing, besides its history, &c., the most probable means of preventing its return, and of avoiding it, in case it should again become epidemic*. New York: Hopkins, Webb & Co.

———. 1798. Inquiry into the cause of the prevalence of yellow fever in New York. *Medical Repository* 1, no. 3:303–23.

Sedgwick, W. T. 1911. *Principles of sanitary science and the public health with special reference to the causation and prevention of infectious disease*. New York: Macmillan.

Shannon, G. W. 1981. Disease mapping and early theories of yellow fever. *Professional Geographer* 33, no. 21:221–27.

Shapin, S. 1994. *A social history of truth: Civility and science in seventeenth-century England*. Chicago: University of Chicago Press.

———. 1996. *The scientific revolution*. Chicago: University of Chicago Press.

———. 2006. Sick City. *New Yorker*. November 6. http://www.newyorker.com/archive/2006/11/06/061106crbo_books.

Shapin, S., and S. Schaffer. 1985. *Leviathan and the air-pump: Hobbes, Boyle and the experimental life*. Princeton, NJ: Princeton University Press.

Shapter, Thomas. 1849. *The history of cholera in 1832 in Exeter*. London: John Churchill.

Shephard, D. A. E. 1995. *John Snow: Anaesthetist to a queen and epidemiologist to a nation—a biography*. Chapel Hill, NC: Professional Press.

Simon, J. 1856. *Report on the last two cholera-epidemics of London as affected by the consumption of impure water*. London: George E. Eyre and William Spottiswoode printers.

———. 1890. *English sanitary institutions, reviewed in their course of development, and in some of their political and social. relations.*

Skelton, R. 1958. *Explorer's maps: Chapters in the cartographical record of geographical discovery.* London: Routledge & Kegan Paul.

———. 1964. *Theatrum Orbis Terrarum.* Amsterdam: N. Israel.

———. 1965. *Civitates Orbis Terrarum. Amsterdam:* N. Israel.

Slack, P. 1985. *The impact of plague in Tudor and Stuart England.* London: Routledge & Kegan Paul.

———. 1992. Introduction to *Epidemics and ideas: Essays on the historical perception of pestilence,* ed. T. Ranger and P. Slack. Cambridge: University of Cambridge Press.

Sloan, C. 2007. CBS2 exclusive leads to cancer investigation in N.J. Town. June 5. CBS 2 News (New York). http://www.topix.com/city/sayreville-nj/2007/05/cbs-2-exclusive-sayre ville-cancer-mystery.

Smith, D G. 2002. Commentary: Behind the Broad Street pump: aetiology, epidemiology, and prevention of cholera in mid-19th century Britain. *International Journal of Epidemiology* 31:92–93.

———. 2006. Cultural Climate, physical climate, life, and death. International Journal of Epidemiology 35:2, 1–2.

Snow, J. 1847. *On the inhalation of the vapour of ether in surgical operations: Containing a description of the various stages of etherization, and a statement of the result of nearly eighty operations in which ether has been employed.* London: Churchill.

Snow, J. 1849a. *On the mode of communication of cholera.* London: Churchill.

———. 1849b. On the pathology and mode of transmission of cholera [part 1] *Medical Times and Gazette* November 2, 745–52.

———. 1849c. On the pathology and mode of transmission of cholera [part 2]. *Medical Times and Gazette* November 30, 923–20.

———. 1851. On the mode of propagation of cholera. *Medical Times and Gazette* 24:559–662, 610–12.

———. 1853. On the prevention of cholera. *Medical Times and Gazette,* 367–69.

———. 1854. The cholera near Golden-square and at Deptford. *Medical Times and Gazette* 9:321–22.

———. 1855a. On the mode of transmission of cholera, second edition. Repr. in *Snow on Cholera: A reprint of two papers by John Snow, M.D.,* ed. W. H. Frost. New York: Commonwealth Fund, 1936.

———. 1855b. Dr. Snow's Report. In *Report of the cholera outbreak in the parish of St. James, Westminster, during the autumn of 1854,* 97–120. London, J. Churchill.

———. 1856. Cholera and the water supply of the south districts of London in 1854. *Journal of Public Health* 2:239–247.

Snow, S. J. 2002. Commentary: Sutherland, snow and water: The transmission of cholera in the nineteenth century. *International Journal of Epidemiology* 31:908–11.

Sontag, S. 1977. *Illness as metaphor.* New York: Vintage Books.

Spence, I. 2006. *William Playfair and the psychology of graphs.* Paper presented at Joint Statistical Meetings, Seattle, Washington. http://www.psych.utoronto.ca/users/spence/ Spence_JSM_2006.pdf.

Spielman, A, and M. D'Antonio. 2001. *Mosquito: A natural history of our most persistent and deadly foe.* New York: Hyperion.

Stern, A. M., and Markel, H. 2004. International efforts to control infectious diseases, 1851 to the present. *Journal of the American Medical Association* 292, no. 12:1474–79.

Steuart, R. and B. Phillips. 1819. *Reports on the epidemic cholera which has raged throughout the Peninsula of India since August 1817.* Bombay: Jesus.

Stevenson, L. G. 1965. Putting disease on the map: The early use of spot maps in the study of yellow fever. *Journal of the History of Medicine and Allied Sciences* 20:226–62.

Stocks, P. 1924.Cancer and Goitre. *Biometrika* XVIV: 364–401.

———. 1928. "On the evidence for a regional distribution of cancer prevalence in England and Wales." *Report of the International conference on Cancer, London, 17th–20th, July 1928. British Empire Cancer Campaign.* Bristol: J. Wright.

———. 1936. Distribution in England and Wales of cancer in various organs. In *British Empire Cancer Campaign Annual Report*, 239–80. n.p.

Strauss, B. M. Fyfe, K. Higo et al. 2005. Salmonella enteritidis outbreak linked to a local bakery, British Columbia, Canada. *Canadian Communicable Disease Report* 31, no. 7. http://www.phac-aspc.gc.ca/publicat/ccdr-rmtc/05v0131/dr3107ea.html.

Sutherland, J. 1850. *Appendix (A) to the report of the general board of health on the epidemic cholera of 1848 & 1849.* London: Her Majesty's Stationary Office.

Sydenham, T. 1676. *Observationes medicae circa morborum acutorum historiam et curationem.* London: Typis A. C. impensis Gualteri Kettilby.

———. 1680. *Epistolae responsoriae duae . . . Prima de morbis epidemicis ab anno 1675.* London: Typis A. C. impensis Gualteri Kettilby.

Tanner, H. S. 1832. *A geographical and statistical account of the epidemic cholera, from its commencement in India to its entrance into the United States . . . Compiled from a great variety of printed and manuscript documents.* Philadelphia.

Thacker, S. B., and D. F. Stroup. 1998. *Deciphering global epidemics: Analytical approaches of the disease records of world cities, 1888–1912.* Foreword by A. Cliff, P. Haggett, and M. Smallman-Raynor. Cambridge: Cambridge University Press.

Tomalin, C. 2002. *Samuel Pepys: The unequalled self.* New York: Penguin Books.

Townsend, A. 2005. Spatial analysis optimizes malaria prevention measures. *ArcUser*, June. http://www.esri.com/news/arcuser/0405/malaria1of2.html.

Tremain, S. 2005. Foucault, governmentality, and critical disability theory. In *Foucault and the government of disability*, ed. S. Tremain, 1–26. Ann Arbor: University of Michigan Press.

Tufte, E. R. 1983. *The visual display of quantitative information.* Cheshire, CT: Graphics Press.

———. 1997. *Visual explanations.* Cheshire, CT: Graphics Press.

———. 2006. *Beautiful evidence.* Cheshire, CT: Graphics Press.

Turnbull, D. 2000. *Masons, tricksters, and cartographers.* Abington, UK: OPA Routledge.

Twyman, M. 1998. *The British library guide to printing: History and techniques.* London: The British Library.

US Bureau of the Census. *Statistical Abstract of the United States, 1946*, ed. M. H. Hansen. Washington DC: US Government Printing Office.

USGS [U.S. Geological Survey]. 2006. West Nile virus Maps: Historical. http://diseasemaps .usgs.gov/wnv_historical.html.

US Geological Survey 2007. Disease maps 2007. http://diseasemaps.usgs.gov/index .html/.

Valenčius, C. B. 2002 *The health of the country: How American settlers understood themselves and their land.* New York: Basic Books.

Vandenbroucke, J. P. 2001. Changing images of John Snow in the history of epidemiology. *Präventivmed* 46:288–93.

Vandenbroucke, J. P., H. M. Eelkman, and H. Beukers. 1991. Who made John Snow a hero? *American Journal of Epidemiology* 1233:967–73.

Verghese, A. 2005. The calling. *New England Journal of Medicine* 352, no. 18:1844–47.

Verghese, A., S. L. Berk, and F. Sarubbi. 1989. Urbs in rure: Human immunodeficiency virus infections in rural Tennessee. *Journal of Infectious Diseases* 160, no. 6:1051–55.

Vinten-Johansen, P., H. Brody, N. Paneth, S. Rchman, and M. Rip M. 2003. *Cholera, chloroform, and the science of medicine: A life of John Snow.* New York: University of Oxford Press.

Wainer, H., and I. Spence. 2006. Introduction to *William Playfair's commercial and political atlas and Statistical Breviary*, ed. H. Wainer and I Spence. New York: Cambridge University Press.

Watson, R. 2008. Cordiform maps since the sixteenth century: The legacy of nineteenth-century classificatory systems. *Imago Mundi* 60, no. 2:182–94.

Webster, N., ed. 1796. *A collection of papers on the subject of bilious fevers, prevalent in the United States for a few years past.* New York: Hopkins, Webb & Co.

Wells, S. 1888. The Morton lecture on cancer and cancerous diseases. *British Medical Journal* 2, no. 2:1201–5.

Wertheimer, N. et al. 1979. Electrical wiring configurations and childhood cancer. *American Journal of Epidemiology* 109:273–84.

Wheeler, L. and G. Smith. 2008. Aging systems releasing sewage into river, streams. *USA Today*, August 12. http://www.usatoday.com/news/nation/2008-05-07-sewers-main_N.htm.

———. 2009. Cleaning EPA's dirty sewer data. *IRE Journal*, Nov.–Dec. 31.

White, C., and R. J. Hardy. 1970. Hygens' graph of Graunt's data. *Isis* 61:107–8.

Whitehead, H. 1854. "The cholera in Berwick Street." 2nd ed. London: Hope & Co.

———. 1855. Mr. Whitehead's report. In *Report of the cholera outbreak in the parish of St. James, Westminster, during the autumn of 1854*, 120–67. London: J. Churchill.

———. 1865. The Broad Street pump: An episode in the cholera epidemic of 1854. *MacMillan's Magazine*, December, 113–22.

Whitfield, P. 2007. London: A Life in Maps. Chicago: University of Chicago Press.

Winn D. M., W. J. Blot, C. M. Shy, L. W. Pickle, and J. F. Fraumeni Jr. 1981. Snuff dipping and oral cancer among women in the southern United States. *New England Journal of Medicine* 304:745–49.

Williamson, T. 2000. *Knowledge and its limits.* Oxford: Oxford University Press.

Wood, D. 1992. *The Power of maps.* New York: Guilford Press.

———. 2004. Notes for a history of the term "atlas." Paper presented at the Unsettling Archives Symposium, Whitechapel Gallery, London.

———. 2006. Map art. *Cartographic Perspectives* 53:5–14

Wood, D., and J. Fels. 1986. Designs on signs: Myth and meaning in maps. *Cartographica* 23, no. 3:54–103.

———. 2008. *The natures of maps: Cartographic constructions of the natural world.* Chicago: University of Chicago Press.

Wood, D., and J. Krygier. 2009. Maps. In *International Encyclopedia of Human Geography*, ed. R. Kitchen and N. Thrift. Oxford: Elselvier.

World Health Organization. 2003. Cholera. *Weekly Epidemiological Record* 78:269–76. http://www.who.int/wer/2003/en/wer7831.pdf.

Woodworth, J. M. 1875. The cholera epidemic of 1873 in the United States. House of Representatives. 43rd Cong., 2nd sess. House Ex. Doc. No. 95. Washington DC: US Government Printing Office.

York, J. 1855. Mr. York's Report. In Report of the cholera outbreak in the parish of St. James, Westminster, during the autumn of 1854, 168–72. London: J. Churchill.

INDEX

Account of the Malignant Fever Lately Prevalent in Philadelphia (Carey), 77

Acland, H. W., *Memoir on the Cholera at Oxford*, 213–14

Africa, sunlight and cancer rates in, 271–72

AIDS. *See* HIV/AIDS, mapping of, 284n2.1

airborne diseases, 60–61, 231; cancer theories, 254–57; cholera theories, 5, 106, 126, 131, 134–35, 136, 144, 145, 147, 151, 152, 156, 163, 176, 178, 187, 197–98, 211; plague theories, 49, 56–57, 57, 61–63; yellow fever theories, 80, 83–84, 200

air pollutants. *See* airborne diseases

air quality, epidemiological model of, 240

air travel, and H1N1 influenza, 14–16

altitude: and cancer incidence, 48–49, 254; and cholera incidence, 135, 138, 175, 176, 178, 179–80, 182, 213–14, 301n7; Halley's graph of, 69

anatomical studies and teaching texts, 2, 31–36, 72, 296n4; map frame for, 36–37; technologies of, 45–47

animalcule, as disease agents, 2, 3

Ankographia (Packe), 73

Annals d'Hygiène Publique et Medicine Legal, 119

Arnaudet, A., 251

Arrieta, Filippo, 51–54, 55, 56, 63, 71, 81, 146, 283n4.4–4.5B

Atlas of Cancer Mortality for the United States (National Cancer Institute), 264–65

Atlas sive Cosmographicae Meditationes de Fabrica Mundi et Fabricati Figura (Mercator), 47

bacteria, 2, 248, 276–77. *See also names of specific diseases*

Barbados, yellow fever in, 73, 82

Bari, Italy: *cordon sanitaire*, 53, 56; mapping of, 51–54; plague in, 51–58; quarantine of plague victims, 51, 54–55, 57

Barker, Thomas J., 288n8.3

Basire, James, 120, 287n7.1

Beijerinck, Martinus, 248

Bellers, John, 95

Bell's Life in London, cholera reports in, 203

Bell's New Weekly Messenger, cholera reports in, 203

Bengal, cholera outbreaks in, 96–98

Bennett, Park, 74

Bentham, Jeremy, 161–62

Berengaria da Carpi, Jacopo, 31, 34

Berghaus, Henrich, *Physikalischer Atlas*, 135

Bernoulli, Daniel, 65–66, 67

Bill of Mortality, 285n4.8

Binding, Paul, 37

birth records, 72

Black Death. *See* plague

black water fever, 74, 298n1 (ch. 5)

Booth, Charles, *Life and Labour of the People*, 125

Boston, yellow fever outbreaks in, 79, 82

bovine spongiform encephalopathy, mapping of, 8

Boyle, Robert, 26, 69–70, 72, 73, 120

Braudel, Ferdinand, 45

Braun, Georg, 39, 42–44, 47, 51, 232, 285n3.7A–B; *Civitates Orbis Terrarum*, 39, 42–44, 45, 46, 51

Brierre de Boismont, Alexandre: *la march du cholera-morbus dans L'Indie et Dans l'Asie Central*, 98, 99

Brigham, Amariah: *Treatise on Epidemic Cholera*, 112, 114

Bristol, England, typhoid fever in, 154–55

British Columbia Centre for Disease Control, 284n2.6A–D, 295n5; diarrhea outbreak report, 25

British Empire Cancer Campaign, 259, 260, 303n4

Broad Street, cholera studies of: and airborne disease theory, 5; Cooper's study, 204–5, 207; critique of Snow's study, 202–4, 207–8; quantifying map studies, 233–42; Snow's investigation, 193, 198–203, 209–11; and waterborne disease theory, 4–5; Whitehead's investigation, 194–98, 205–9

Brodeur, Paul, 269

Brown, P. E., 154

BSE. *See* bovine spongiform encephalopathy, mapping of

Budd, William, 153, 154, 155

Buffalo, New York, polio outbreaks in, 224

Calcutta, cholera outbreaks in, 91

California, West Nile virus in, 18
Campbell, F. R., 276
Canada: BSE in, 8; disease resulting from contaminated water supply, 276–77; sunlight and cancer rates in, 271–72; West Nile virus in, 19
Canada Communicable Disease Report, 25
cancer: as airborne disease, 254–57; altitude and incidence rates, 253, 254; breast cancer, 270–71 colorectal cancer, 270–72; as contagious disease, 254; diet's role in, 252; economic costs, 258; in England, 248–51, 254, 257, 261, 263; environmental cancers, 268–72, 272–73; environmental causes posited, 251, 259, 263, 265; esophageal, 266; ethnicity of patients, 259; in France, 251–52, 255; Green's study, 254–57; local and regional studies, 249, 252, 258, 259, 263, 267, 272–73, 295n1 (ch. 2); lung cancer, 261, 263, 267; mapping cases, 249–57, 260, 261, 270–72, 304n13; in Massachusetts, 304n14; microscopy investigating, 248; mortality rates, 247, 249, 255, 258, 260, 266; national studies, 259, 263, 268–69; Navajo neuropathy, 268, 304n8; occupational causes of, 265, 268; oral cancers, 266–67; and power lines and transformers, 269–70; and public health, 246–47; racial differences in incidence, 264–65; and radiation exposure, 268–69; social and occupational differences in incidence, 259, 261; statistical studies, 258–59, 263, 269; in Sweden, 269–70; tobacco usage, 265, 266; in United States, 262–63; and vitamin D, 270–72; as waterborne disease, 252
Canterbury, England, mapping of, 73
Carey, Matthew, 76, 77; A Short Account of the Malignant Fever, Lately Prevalent in

Philadelphia, 80, 81
Centers for Disease Control and Prevention, 16, 19, 284n2.2; food-related diarrhea outbreak report, 25
Chadwick, Edwin, 144, 232, 276, 277; Report on the Sanitary Condition of the Labouring Population of Great Britain to the Poor Law Commissioners, 132–35
Charleston, South Carolina, yellow fever outbreaks in, 79
Cheffins, C. F., 200
Chi, theory of, 2
Chicago, typhoid fever in, 11.5, 221–23, 290n11.4
China, theory of Chi, 2
cholera, 277; Asiatic cholera, 27, 98–100, 167; bacillus identified, 22; British cholera, 98; cholera morbus, 98, 99–100, 101; El Tor serotype, 10; English cholera, 27, 155, 167; institutionalized victims, 143, 199; microscopy in investigation of, 209, 214, 218; mutation of, 6; plague compared to, 172; and poverty, 107, 116, 182, 210; quarantine of victims, 106–7; and social unrest, 142, 143
The Cholera in Berwick Street (Whitehead), 195, 197
cholera (sites): altitude and incidence, 135, 138, 175, 176, 178, 179–80, 182, 213–14, 301n7; Calcutta, 91; Cincinnati, 139–41; Columbia, Pennsylvania, 211–13; Dantzick, 100–4; Exeter, 156–57; Glasgow, 149; Great Britain, 108, 135–39; Hamburg, 130–31; India, 91, 96–98, 110, 111, 115; international map of cholera, 109–10, 114; local studies of, 4–5; London, 4–5, 138–39, 149–54, 158, 164, 169; Macao, 110; Malacca, 110; mapping of, 10, 104–7; Montreal, 126; Newcastle, 159–60; New York City, 115, 125–27; Paris, 119; Philadelphia, 115; Poland, 98–100; Rouen, 128–30;

Russia, 98; Salford, 141, 164; Scotland, 149; Siam, 110; St. Louis, 140; trade routes, 101, 106, 110, 114, 129, 136; urban areas, 108, 110, 135–39
cholera (theories), 3; as airborne disease, 5, 106, 126, 131, 134–35, 136, 144, 145, 147, 151, 152, 156, 163, 176, 178, 187, 197–98, 211; climactic conditions, 106, 112, 116, 131, 168, 204; as fever, 147; miasmatic disease theory, 116, 136, 145, 151, 197–98, 210; mobile nature of disease, 106; natural law theory, 164–65, 178–80; portable nature of, 146; and poverty, 103–4, 130, 174; sanitary conditions, 103–4, 107, 116, 130, 145, 147–48, 147–51, 153–54, 160–61, 205, 207–9, 211, 237, 239–40; as waterborne disease, 4, 139–41, 147–54, 163, 174, 180, 185–86, 187, 191, 198–203, 210, 241, 301n7
Christie, Alexander Turnbull, A Treatise on Epidemic Cholera, 110
Chronik der Seuchen (Schnurrer), 109–10
Cincinnati, Ohio, cholera outbreaks in, 139–41
Civitates Orbis Terrarum (Braun and Hogenberg), 39, 42–44, 45, 46, 51, 232, 285n3.7A–B
climactic conditions promoting disease, 59–60, 63, 69, 74–77, 80–81, 98, 116; cancer, 246–47; cholera, 106, 112, 116, 131, 168, 204; mortality tables demonstrating, 169, 171–72; plague, 63; yellow fever, 74–77, 80, 82, 90
College of Physicians of Philadelphia, 90, 285n5.1–5.5
Colorado, cancer incidence in, 2669
Columbia, Pennsylvania, cholera outbreaks in, 211–13
Commercial and Political Atlas and Statistical Breviary (Playfair), 77–78, 77–79
Constantinople, plague of Justinian in, 48

contagium vivum fluidum, 248

Cooper, Edmund, 204–5, 207, 234; Broad Street mapping by, 5

Corbyn, Frederick, *The Epidemic of Cholera as it prevailed in India*, 111, 112

cordon sanitaire, 146; in Bari, 53, 56; mapping, 3

Creutzfeldt-Jakob disease, 8

Crosby, Alfred, 72

Daily News (London), cholera reports in, 204

Dance of Death (Holbein the Younger), 49–50, 285n4.1–4.2, 296n1

Dantzick, 133; cholera outbreaks in, 100–4

da Vinci, Leonardo, 31, 46

Dawson, Major, 174

Defoe, Daniel, 62

De Humani Corporis Fabrica (Vesalius), 31, 34, 35, 37, 41, 45

de Ketham, Johannes, *Faciulo de medicina*, 46

de Kruif, Paul, *The Microbe Hunters*, 230

dengue fever, 275–76

Denike, Ken, 24

Desowitz, Robert, 73

De usu partium (Galen), 34

diabetes, and vitamin D, 270

diarrhea outbreak, 301n5 (ch. 9); cholera (*see* cholera); food poisoning causing, 22–25; as naturally occurring event, 168; and sewer lines, 22; summer diarrhea, 59; in Vancouver, 19–25, 26, 29, 84, 277

Die Cholera morbus (Schnurrer), 109

diet, and cancer, 252

Dirichlet, J. P. Lejeune, 234

disease theories, 2, 3, 26–29, 57–61 (*see also specific diseases*); contagious and anticontagonist, 82; and globalization of mercantilism, 72–73; and mapping, 5–6, 73; maps published as evidence testing, 84, 87; population-based analysis, 65, 67

Dubuc, M. E., 128

dysentery: English dysentery, 58; food poisoning and, 96; summer dysentery, 58, 59, 96

eastern equine encephalitis, 9

economic effects: of cancer, 258; of yellow fever epidemics, 77–78, 77–79

elevation. *See* altitude

Elizabeth I, 47

Elliot, Emmanuel, 287n7.7*A–B*, 299n6

empirical method, 69–70

England. *See* Great Britain

environmental causes, of cancer, 251, 263, 265, 268–73

Environmental Protection Agency, 268, 276, 305n1

The Epidemic of Cholera as it prevailed in India (Corbyn), 111, 112

epidemic profile, 64; and accurate reporting of cases, 260

epidemics: defined, 1; economic effects of, 77–78; human environments promoting, 109–10; as spatial phenomena, 2

Epidemiological Society of London, 154

Epidemiology: Public Health Prevention Medicine (Frost), 228–29

Epistola responsoria (Sydenham), 59

ethnicity, of cancer patients, 259

Eustachi, Bartolemo, *Tabulae anatomicae*, 36–37

Exeter, England, cholera outbreaks in, 156–57

Eyler, John M., 185

Faciulo de medicina (de Ketham), 46

Faithorne and Newcourt's *Exact Delineation of London*, 67–68, 71

Farr, William, 5, 125, 154, 161–63, 164–65, 186, 214, 240, 247, 253, 274; *A Report on the Morality of Cholera in England, 1848-1849*, 166–80; on Snow's cholera theories, 172

febrile diseases, theories of, 59–61, 74, 119, 147; cholera, 147; miasmatic theory, 82

fetid air causing disease. *See* airborne diseases

Feychting, Maria, 269–70, 304n10

Finé, Oronce: *Recens et integra orbis descripto*, 37

Florida, West Nile virus in, 18

food poisoning: diarrhea outbreak caused by, 22–25, 29; and dysentery, 58, 96

Forry, S., 141

Foucault, Michel, 34, 45

foul odors. *See* airborne diseases; miasmatic disease theory

France: cancer rates in, 251–52, 255; cholera outbreaks in, 128–30; national collection of public health data, 123; Royal Academy of Medicine, 119

Friendly, Michael, 298n1 (ch. 7)

Frost, W. H., 218–21, 245, 302n3; *Epidemiology : Public Health Prevention Medicine*, 228–29; influenza pandemic of 1918 studied by, 228; polio investigated by, 224–29; Snow studied by, 228–29, 230

Galen of Pergamon, 31, 33, 34, 36, 46; *De usu partium*, 34

Garland, C. F., 270, 272

Garland, F. C., 270, 304n11

Genoa, plague in, 55

A geographical and statistical account of the epidemic cholera (Tanner), 114

Geographica (Ptolemy), 38

geographic proximity of cases, and causality, 24, 25

Gilbert, E. W., 138

Gilbert, Pamela, 163

Glasgow: cholera outbreaks in, 149; influenza outbreak in, 131–32

Gould, Peter, 284n2.1; AIDS mapping by, 9

Grainger, Richard, 130, 131, 134–35, 158, 232

Grainger, William, 144

Graunt, John, 64–65, 67, 68, 71, 72, 269; *Table of Casualties*, 64–65, 246

Great Britain: Board of Health, 130, 134; cancer mortality rates in, 259–61; cancer rates

Great Britain (continued)
in, 257, 263; Central Board
of Health, 107, 108; cholera
in, 27; cholera outbreaks
in, 108, 135–39; General
Board of Health, 143, 166;
General Register Office (see
registration districts); influenza
outbreaks in, 143; Nuisances
Removal and Contagious
Disease Prevention Act,
143–44; plague outbreaks
in, 48–49; postal service in,
120–22; Public Health Act,
143; smallpox pandemic in, 58
Green, Charles Edward,
251, 253, 254–57
Grey, George, 166
Guadeloupe, yellow fever in, 73
Guerry, André-Michel, 169

Halifax, Nova Scotia, yellow
fever outbreaks in, 79
Halley, Edmund, 69
Hamburg, Germany, cholera
outbreaks in, 130–31
Hamett, John, 100–4,
108, 127, 133
Hamilton, Alice, 221–22, 277
Hamlin, Christopher, 230
Hassall, Arthur Hill, 209
Havana, yellow fever in, 73
Haviland, Alfred, 246–48, 263
Hawaii: quarantine of
plague victims in, 49
Heine, Jakob, 223
Hellis, Eugène-Clément,
Souvenirs du Choléra,
128–30, 131
hemogastric pestilence, yel-
low fever termed, 74
Hempel, Sandra, 233; The
Medical Detective, 230
Hightower, Allen, 305n3
Hillary, William, 61, 74–76, 276
Hippocrates, 31, 58, 59, 61, 76
HIV/AIDS, 3, 246; mapping
of, 8–9; retroviruses, 9
H1N1 influenza, 3; and air
travel, 14–16; pandemic
of 2009, 14–16
Hobbes, Thomas, 26, 70, 72, 73
Hoffman, F. L., 261
Hogenberg, Franz, 43, 44, 47,
51, 232, 285n3.7A–B
Holbein the Young, Hans, 49–50,

285n4.1–4.2, 296n1 (ch. 4)
Holland: cancer inci-
dence studies, 258
hospital admission tables,
used in mapping, 103
Howe, G. M., 263
humours, 60; symptoms
grouped by, 2
Hurdwar: cholera outbreaks in, 91
Huygens, Christian, 65, 69

India: cholera in, 27, 96–98
inductive thinking, and disease
theories, 147, 217–18, 231
influenza, 6, 58; in Glasgow,
131–32; in Great Britain, 143;
H1N1 (see H1N1 influenza);
mapping of, 10, 284n2.3;
1918 pandemic, 10, 228,
296n8; Spanish flu, 27
Ingram, Dale, 50, 57, 74
International Air Transport
Association, 14, 15
International Sanitary Conference:
of 1851, 164; of 1874, 214
Iowa, polio outbreaks
in, 224, 225–28
Italy, plague outbreaks in, 51–58

Jackson, T. Heber, 211–13
Jamaica, yellow fever in, 77
James, Robert, 77
Jameson, James, 96, 168; Report
on the Epidemic Cholera
Morbus, 96–98, 286n6.1
Japan, theory of Chi, 2
Jarcho, Saul, 51, 296–97n2
Journal of Public Health, 191
Journal of the Statistical
Society of London, 161
Jusatz, H., 136

Kansas, influenza in, 27
Kellwaye, Simon, 61
Koch, Robert, 214; cholera
bacillus identified by, 4, 22

La Crosse encephalitis, 9
La march du cholera-morbus dans
L'Indie et Dans l'Asie Central
(Brierre de Boismont), 98, 99
The Lancet, 154, 160, 180,
231, 286n6.4A–B; mapping
of cholera in, 104–8, 114
Latour, Bruno, 70
Lea, John, 139–41, 157, 249

Leeds: income and disease mortality mapping in, 132–35
Leprosy, 48
leukemia, in Denver, 269
Leviathan and the Air-Pump
(Schapin and Schaeffer), 70
Life and Labour of the
People (Booth), 125
life expectancy tables,
64–65, 67, 69
local disease studies, 3–4, 84,
118, 241, 268; of cancer incidence, 249, 258, 259, 263,
272–73, 295n1 (ch. 2); of cholera incidence, 4–5, 126, 127,
139–41, 210–11; disease in
place, 58–61; of HIV/AIDS, 9
London: Bills of Mortality, 64,
71, 268; Board of Health,
210; Braun's map of, 43, 44;
cancer incidence in, 253, 254,
261; cholera outbreaks in,
138–39, 149–54, 158, 164,
169; mapping of, 67–68, 70;
Metropolitan Commission of
Sewers, 174, 194, 204–5,
288n8.1; pest-houses and
pest-fields, 192–93, 207,
237, 240; plague in, 51, 64;
population explosion in, 47;
registration districts in, 123,
125; Tithe Commission,
174. See also Broad Street,
cholera studies of; South
London, cholera studies of
London Medical Gazette, 160
London Quarterly Review, 141
London Sewer Commission,
mapping of cholera deaths, 5
lymphoma, in Denver, 269

MacMillan's, cholera re-
port in, 207, 214
Maine, predictive map-
ping of influenza in, 10;
West Nile virus in, 18
malaria, 58, 277–78
mapping, 2–4, 8; in anatomical
teaching texts, 36–37; atlases, 37–41, 39–40, 42–44,
46–47; cancer cases, 249–57,
260, 261, 270–72, 304n13;
computers in aid of, 6–7,
263–65; cordiform map, 37,
39, 40, 284n3.5, 296n1 (ch.
3); coxcomb projection, 169,

172; and data catchments, 118, 135; evidentiary nature of, 5, 19; exercise of power, 54; as experimental system, 13–14; geographical matrix in, 14; GIS versions, 289n10.3, 290n11.9–11.11B, 301n3; of hospital records, 103; as method of assemblage, 13; Monte Carlo simulation, 304n9; morbidity and mortality tables, 13, 232–33, 260–67; numerical information, 5, 13, 37, 119, 253; ordinance survey maps, 120, 174, 248–49; and postal services, 120; and predictive models, 9, 10; projections, 37, 39, 46–47; quantifying maps, 233–42; risk ratios in, 240; scatter plots, 188; and seeing, 3, 26, 69; spatial dimensions, 13, 55, 201–2; statistical information of, 5, 6–7, 18, 25, 141, 167, 169, 190, 252, 253, 273; technologies of, 30, 31, 39–40, 45–47, 263; temporal dimensions, 5, 16, 55, 99, 110, 115, 168, 179, 219, 224, 226, 254; testing a propositional disease theory, 84, 87; and theories of origin and diffusion, 5–6, 73; Thiessen polygons, 234, 289n10.5, 301n4 (ch. 10), 303n8; as visual statement of data, 12, 13

Margarita Philosophica (Reisch), 34
Marsh, David, 297n13
McDonald, Gordon, 247
measles, 72
The Medical Detective (Hempel), 230
Medical Gazette, 145
Medical Inquiries and Observations (Rush), 80
medical journals, maps published in, 84
Medical Repository (magazine), 84, 87, 88, 286n5.6–5.7
Medical Society of London, 245
Medical Times and Gazette, 214
Medico-Chymical Dissertation (Pascalis), 87–90
Memoir on the Cholera at Oxford (Acland), 213–14
mercantilism. See trade routes, and disease transmission
Mercator: atlas, 294n6; Atlas sive Cosmographicae Meditationes de Fabrica Mundi et Fabricati Figura, 47
Mexico: H1N1 influenza in, 14–16; West Nile virus in, 19; yellow fever in, 73
miasmatic disease theory, 5, 49, 60–63, 77, 85–87, 103–4, 116, 119, 126, 135, 143, 145, 156, 176, 231; cancer theory, 247; cholera, 106, 116, 136, 145, 151, 197–98, 210; febrile diseases, 82; yellow fever, 85–87, 90, 200
The Microbe Hunters (de Kruif), 230
microscopy, 276; in cancer investigations, 248; in cholera investigations, 209, 214, 218
mobility of population, 261–63. See also trade routes, and disease transmission
Mondino, dissection manual of, 31
Monmonier, Mark, 201
Montreal: cholera outbreaks in, 126
morbidity rates, 72, 118; mapping, 5, 13, 232–33, 260–67
Moretti, Franco, 12
Morgagni, Giovanni, 2, 36, 72
Morpeth, Lord, 143
Morris, R. J., 299n1
The Mortality from Cancer through the World (Prudential Insurance Company), 258
mortality rates, 64, 72, 118; cancer mortality, 247, 249, 255, 258, 260, 266; and climactic conditions, 169, 171–72; Farr's cholera study, 165–80; ligne de vie, 69; in London, 64, 71; mapping, 5, 13, 232–33, 260–67; national collection of, 123; and population analytics, 153; spatial descriptions of, 156; yellow fever victims, 80, 87
mosquitoes: and spread of yellow fever, 73–74, 85–86
multifactoral disease causes, 5, 24; yellow fever, 83

naming, 26–29; and seeing, 2
National Cancer Institute, 265, 268, 304n6; Atlas of Cancer Mortality for the United States, 263–64
national disease studies, 118, 241; of cancer incidence, 259, 263, 268–69; of cholera incidence, 126; of encephalitis incidence, 9; and mobility of population, 261–63
natural law theory: of cholera outbreaks, 164–65, 167, 178–80; of diarrhea outbreaks, 168
Navajo neuropathy, 268, 304n8
neighborhood disease studies. See local disease studies
New Brunswick, predictive mapping of influenza in, 10
Newcastle, England, cholera outbreaks in, 159–60
Newcourt, Richard, 67–68, 71
New England Journal of Medicine, 14, 15
New Jersey: cancer incidence in, 295n1 (ch. 2); Central Cancer Registries, 272
New Orleans, yellow fever outbreaks in, 79
New Slip, New York, yellow fever mapping in, 84–87
New York Academy of Medicine, 90, 285n4.4–4.5B
New York City: Board/Department of Health, 83, 89, 95; cholera outbreaks in, 115, 125–27; West Nile virus in, 18; yellow fever outbreaks in, 79, 80, 82, 83–90, 119, 200
New Yorker magazine article regarding power lines and cancer incidence, 269
New York Library of Medicine, 286n5.8
New York State, West Nile virus in, 9, 10
New Zealand, cancer mortality in, 247–48
Nightingale, Florence, 169
Nixon, Richard, 304n6; "war on cancer," 264
North American Association of Central Cancer Registries, 303n1
Norwich, England, plague outbreaks in, 62–63
nuclear power sites. See radiation exposure

obesity, studies of, 302–3n6

Observationes medicae (Sydenham), 59

occupational differences in disease incidence, 118, 277; of cancer, 259, 261, 265, 268

odors causing disease. *See* airborne diseases

Ohio, polio outbreaks in, 224

Old Slip, New York: cholera outbreak in, 127; yellow fever mapping in, 87, 89

Olsson, Gunnar, 2, 295n2 (ch. 1)

On the Mode of Communication of Cholera (Snow), 144–54, 187, 192, 200, 209, 211, 229, 230–31, 233, 300n2

Openshaw, S., 268, 269, 304n9

Ortelius, Abraham, 37, 38, 39, 40–42, 43, 44, 47, 52, 284n3.6A–B; *Theatrum Orbis Terrarum*, 37–38, 39, 40–42, 43, 44, 46, 52

Osborne, John B., 126, 298n1 (ch. 6), 299n5, 302n12

outbreaks, defined, 1

Packe, Christopher, 73, 120; *Ankographia*, 73; Philisophico-Chorographical Chart of East Kent, 73

Pan-American Health Organization, mapping by, 8

pandemics: defined, 1; as spatial phenomena, 2

parasitic infection, cancer as, 248, 251

Paré, Ambroise, 61–62

Paris, cholera outbreaks in, 119

Parkes, Edmund Alexander, 149, 151, 152, 164, 182, 186, 231, 300n4

Pascalis, Felix, 91; *Medico-Chymical Dissertation*, 87–90

Pecquereau (mapmaker), 128

Pennsylvania, cholera outbreaks in, 211–13

Pepys, Samuel, 64

Periaux, A., 128

Perry, Robert, 131–32

Petermann, Augustus, 135–39, 158, 169, 299n8

Petty, William, 64–65, 72

Philadelphia: cholera outbreaks in, 115; yellow fever outbreaks in, 77, 80–81, 82

Philisophico-Chorographical Chart of East Kent (Packe), 73

phthisis, death from, 250

physician surveys, 96–97

Physikalischer Atlas (Berghaus), 135

Pickle, Linda, 303n1, 304n7

Pickstone, Andrew, 31

Picon, Antoine, 67

plague, 48–50, 58; Bills of Mortality, 268; and cholera outbreaks, 172, 203, 205, 207–9, 237; pest-houses and pest-fields, 95, 193; quarantine of victims, 49, 50–51, 63; yellow fever compared to, 74

plague (sites): in Bari, 51–58; Constantinople, 48; in England, 48–49; in London, 51, 64; trade routes, 50–51; as urban disease, 61–63; in Venice, 49, 51

plague (theories), 3; as airborne disease, 49, 56–57, 57, 61–63; climatological conditions, 63; as divine punishment, 49, 56–57; and mapping, 3; as portable disease, 49–50, 56, 60; poverty, 49, 62–63; sanitary conditions and, 62–63

Plain and Practical Treatise on Cholera (Reese), 126–27

Playfair, William, *Commercial and Political Atlas and Statistical Breviary*, 77–78, 77–79

Poland, cholera outbreaks in, 98–100

poliomyelitis, 273; in Buffalo, 224; Frost's investigation of, 224–29; in Iowa, 224, 225–28; in Ohio, 224; portability of, 227–28; and poverty, 223; and sanitary conditions, 223

political arithmetic, 64

population data, graph of, 69, 153, 233–34

Porter, Dale, 299n7

postal reform map, 287n7.1, 287n7.2

postal services, and dissemination of information, 120–22

poverty and social class, 61, 277, 279; of cancer victims, 258; and cholera outbreaks, 116, 130, 174, 182; in cholera outbreaks, 103–4, 107,

210; and disease outbreaks, 132–35, 160; and plague outbreaks, 49, 62–63; and poliomyelitis outbreaks, 223

Powassan virus, 9

Power, D'Arcy, 248–51

power lines and transformers, and cancer incidence, 269–70

predictive models: and mapping, 9, 10; of spread of AIDS, 9

prion, 2, 8

proteinaceous infectious particle. *See* prion

Prudential Insurance Company, *The Mortality from Cancer through the World*, 258

Ptolemy, 39; *Geographica*, 38

public health, and cancer, 246–47

Public Health and Marine Hospital Service, 218

public health data, and mobility of population, 261–63

public health services, 95, 114, 117, 118–19; origins of, 95–96; physician surveys, 96–97

quarantine: of Bari plague victims, 51, 54–55, 57; of cholera victims, 106–7; of plague victims, 49, 50–51, 63

Quarterly Review, 108

Quetelet, Adolphe, 135, 142

racial differences in disease incidence: of cancer, 264–65; cancer studies, 263

radiation exposure: and cancer, 268–69

Rawnsley, H. D., 215

Recens et integra orbis descripto (Finé), 37

Reese, David Meredith, 131, 287n7.5; *Plain and Practical Treatise on Cholera*, 126–27

regional disease studies, 118, 241; of cancer incidence, 249, 258, 259, 264–65, 267; of cholera incidence, 126; of HIV/AIDS, 9; of West Nile virus, 16–19

registration districts, 123, 125, 134, 152, 162, 165–66, 174, 246, 253, 258, 260, 261

Reisch, Gregor, 34; *Margarita Philosophica*, 34

Report on the Epidemic

Cholera Morbus (Jameson), 96–98, 286n6.1

Report on the Morality of Cholera in England, 1848–1849 (Farr), 166–80

Report on the Sanitary Condition of the Labouring Population of Great Britain to the Poor Law Commissioners (Chadwick), 132–35

retroviruses, HIV as, 9

Richardson, Benjamin, 246

Richardson, Benjamin Ward, 216

risk ratios, in mapping, 240

Robinson, A. H., 136

Rouen, France, cholera outbreaks in, 128–30

rurality and urbanity: in cancer incidence studies, 259; cholera outbreaks, 108, 110, 135–39; in disease transmission, 277–78; plague theories, 61–63

Rush, Benjamin, 77, 84; *Medical Inquiries and Observations*, 80

Russia, cholera outbreaks in, 98

Salford, England, cholera outbreaks in, 141, 164

Salmon, David Elmer, 295n6

salmonella: bacillus identified, 22

Salmonella choleraesuis, 22, 295n6

Salmonella enteritidis, 25

sanitary conditions, 275, 299n7; and cholera outbreaks, 103–4, 107, 116, 127, 130, 145, 147–51, 153–54, 160–61, 205, 207–9, 211, 237, 239–40; and diarrhea outbreaks, 22; and disease outbreaks, 133–35, 276; Metropolitan Commission of Sewers, 288n8.1; Nuisances Removal and Contagious Disease Prevention Act, 143–44; and plague outbreaks, 62–63; and polio, 223, 275; public sanitation as a clinical prophylaxis, 119; and typhoid fever, 221–23; and yellow fever, 77, 80–81, 82, 85–87, 89

SARS. *See* severe acute respiratory syndrome (SARS)

Schaeffer, Simon: *Leviathan and the Air-Pump*, 70

Schnurrer, Friedrich: *Chronik der Seuchen*, 109–10; *Die*

Cholera morbus, 109

Scotland, cholera outbreaks in, 149

Seaman, Valentine, 83, 84, 85, 86, 87, 91, 200, 249

Sedgwick, William Thompson, 216–18, 219, 229

seeing, 295n1; and anatomical studies, 31, 33, 34, 37; and mapping, 3, 26, 69; and naming, 2; and trade growth charts, 78

Senefelder, Alois, 96

severe acute respiratory syndrome (SARS), 3

sewer systems. *See* sanitary conditions

Shapin, Steven, 45, 60, 232; *Leviathan and the Air-Pump*, 70

Shapter, Thomas, 156–57

A Short Account of the Malignant Fever, Lately Prevalent in Philadelphia (Carey), 80, 81

Simon, John, 210, 211, 214, 253, 268; Snow's rebuttal of, 190–91; South London cholera studies, 186–90

Skelton, Raleigh A., 30

Slack, Paul, 62

slavery, and yellow fever outbreaks, 73–74

smallpox, 58, 72, 143; population-based analysis, 65, 67

Smith, Theobold, salmonella bacillus identified by, 22

Smith, Theodore, 295n6

Snow, John, 4, 139, 164, 214, 253, 275; Broad Street cholera outbreak investigation, 193, 198–203, 209–11; critique of Broad Street study, 202–4, 207–8, 232–33; Farr's comments on theories, 172; Frost's study of, 228–29, 230; *On the Mode of Communication of Cholera*, 144–54, 187, 192, 200, 209, 211, 229, 230–31, 233, 300n2; Parke's critique, 300n4; rebuttal to Simon's study, 190–91; Sedgwick's study of, 216–18; South London cholera studies, 180–86; writings on, 229–32

social differences in disease incidence. *See* poverty and social class

Social Medicine (journal), 142

South London, cholera studies of, 158; and airborne disease theory, 5; by John Simon, 186–90; by John Snow, 180–86

Souvenirs du Choléra (Hellis), 128–30, 131

St. Christopher: yellow fever in, 73

St. Louis, Missouri, cholera outbreaks in, 140

St. Louis encephalitis, 9

Statistical Abstracts of the United States, 261–62

statistical disease studies, 161–63, 260–61; cancer incidence studies, 258–59, 263, 269; and mapping, 5, 6–7, 18, 25, 141, 167, 169, 190; and national health data, 258, 259; national health statistics, 268–69

Stocks, Percy, 258, 259–61, 263

Sunderland, England, 108

Sutherland, John, 164, 300n1

Sweden, cancer incidence in, 269–70

Sydenham, Thomas, 58, 59, 60–61, 64–65, 69, 71, 72, 74, 275; *Epistola responsoria*, 59; *Observationes medicae*, 59

syphilis, 6, 72

Tabulae anatomicae (Eustachi), 36–37

Tanner, Henry Schenck, *A geographical and statistical account of the epidemic cholera*, 114

Theatrum Orbis Terrarum (Ortelius), 37–38, 39, 40–42, 43, 44, 46, 52, 284n3.6A–B

Thomas Morton Lecture, Wells's lecture on cancer, 247

Thomson, R. D., 154

tobacco usage: and cancer, 263, 265, 266; and lung cancer, 261

topographic survey techniques, 121–22

trade routes: and disease transmission, 39, 72, 77–78, 80, 81, 82, 90, 106, 116, 278–80; cholera, 99–100, 101, 106–7, 110, 112, 114, 116, 129, 136; plague, 50–51, 53, 54, 56, 57, 63; yellow fever,

trade routes (continued)
78–79, 81, 82–83, 88, 90
*Transactions of the
Epidemiological Society
of London*, 214
Treatise on Epidemic Cholera
(Brigham), 112, 114
*A Treatise on Epidemic
Cholera* (Christie), 110
tuberculosis, 6; mapping of, 10, 12
Tufte, Edward R., 285n4.8
Turnbull, David, 104
typhoid fever, 58, 72; in Chicago, 11.5, 221–23, 290nn11.4; and sanitary conditions, 221–23; as waterborne disease, 154–63, 219; in West Virginia, 219–21
typhus, 58, 142, 143

Uganda, West Nile virus.
See West Nile virus
United Nations, mapping by, 8
United States: cancer rates in, 262–63; food-related diarrhea outbreaks in, 25; sunlight and cancer rates in, 271–72
United States Geological Service disease maps, 284n2.5
University of British Columbia Department of Geography, 284n2.7A–C
University of Hawaii, dengue fever at, 275–77
urbanity. *See* rurality and urbanity

Vancouver, diarrhea outbreak in, 19–25, 26, 29, 84, 277
Venice, plague in, 49, 51
Vesalius, Andres, 2, 31–36, 41–42, 45, 46, 70, 71, 72, 86, 282n3.1–3.4; *De Humani Corporis Fabrica*, 31, 34, 41, 45, 284n3.1–3.4
Vibrio cholerae, 22, 214, 218
Vinten-Johansen, Peter, 302n9
Virchow, Rudolf, 142
virus, 2
vitamin D, and cancer, 270–72

Wales, registration districts in, 123
Walkerton, Ontario, disease resulting from contaminated water supply, 276–77

war, and disease incidence, 279
Washington, D.C., West Nile virus in, 18
Washington State: BSE in, 8; West Nile virus in, 18
waterborne diseases, 274; cancer theories, 251; cholera theories, 4, 139–41, 147–54, 163, 174, 180, 185–86, 187, 191, 198–203, 210, 241, 301n7; private water suppliers, 158–59, 172, 174, 181–82, 185, 187–88, 214, 301n7; typhoid fever, 154–63, 219
Watson, Ruth, 296n1 (ch. 3)
Webster, Noah, 80, 83
Wells, Spenser, 246
Wertheimer, Nancy, 269
western equine encephalitis, 9
Western Lancet, 139, 141
West Nile virus, 4, 27, 246, 284n2.2; mapping of, 9–10, 16–19, 25; in New York State, 9, 10
West Virginia, typhoid outbreak in, 219–21
Whitehead, Henry, 214–15, 217, 230, 234, 237; Broad Street cholera outbreak investigation, 4–5, 194–98, 205–9; *The Cholera in Berwick Street*, 195, 197
Whiting, John Joseph, 182
whooping-cough outbreaks, 143
WNV. *See* West Nile virus
Woburn, Massachusetts, cancer clusters in, 304n14
Woo-Suk, Hwang, 229
workplace, and disease incidence. *See* occupational differences in disease incidence
World Health Organization, 295n3; mapping by, 8
Wyld, James, 120, 287n7.2, 288n8.2A–B

yellow fever, 146, 277; economic effects of epidemics, 77–78, 77–79; plague compared to, 74
yellow fever (sites): in Barbados, 73, 82; in Boston, 79, 82; in Charleston, South Carolina, 79; in Guadeloupe, 73; in Halifax, Nova Scotia, 79; in Havana, 73; in Mexico, 73; in New

Orleans, 79; in New Slip area, 84–87; in New York City, 79, 80, 82, 83–87, 87–90, 119, 200; Old Slip area of New York, 87, 89; in Philadelphia, 77, 80–81, 82; and slavery, 73–74; in St. Christopher, 73; trade routes, 78–79, 81, 82–83
yellow fever (theories), 3; as airborne disease, 80, 83–84, 200; anthropogenic nature, 76–77; climactic conditions promoting, 80, 82, 89; environmental theories, 74–77; indigenous nature of disease, 74, 82; and mapping, 3–4; miasmatic theory, 85–90, 89; and mortality tables, 87; multifactoral disease, 83; and sanitary conditions, 77, 80–81, 82, 85–87, 89
York, Jeremiah, 208